Microorganisms for Sustainability

Volume 17

Series Editor

Naveen Kumar Arora, Environmental Microbiology, School for Environmental Science, Babasaheb Bhimrao Ambedkar University, Lucknow, Uttar Pradesh, India

More information about this series at http://www.springer.com/series/14379

Pankaj Kumar Arora
Editor

Microbial Technology for the Welfare of Society

Springer

Editor
Pankaj Kumar Arora
Department of Microbiology
Babasaheb Bhimrao Ambedkar University
Lucknow, Uttar Pradesh, India

ISSN 2512-1901 ISSN 2512-1898 (electronic)
Microorganisms for Sustainability
ISBN 978-981-13-8843-9 ISBN 978-981-13-8844-6 (eBook)
https://doi.org/10.1007/978-981-13-8844-6

This Springer imprint is published by the registered company Springer Nature Singapore Pte Ltd.
The registered company address is: 152 Beach Road, #21-01/04 Gateway East, Singapore 189721, Singapore

Preface

The book *Microbial Technology for Welfare of Society* describes the various aspects of modern microbiology, including microbial enzymes, metagenomics, secondary metabolites, next-generation sequencing, microbial-based biopesticides, biodegradation, bioremediation, and wastewater treatment, and how microbes play an important role for the welfare of the living beings and environment. Many bacteria play a significant role in cleaning our environment and in detoxifying the various xenobiotic compounds. Several microbes produce secondary metabolites that are useful to human beings.

This book covers the different aspects of microorganisms-based biotechnology, the recent methodologies, such as advanced molecular techniques, as well developments in classical microbiological techniques. The authors also explain the utilization of the latest and classical techniques in modern-day microbial biotechnology. The application in the industry, including that of microbial enzymes, in medical biotechnology is explained. The chapters contributed in this book are of high quality and written by experts from well-known universities, research laboratories, and institutes from around the globe in the area of modern microbiology. In this book, extensive focus has been relied on the recent advances in microbial technology with welfare of living beings and environment.

I hope that this book will be useful to the researchers, environmentalists and scientists, microbiologists and biotechnologists, ecotoxicologists, remediation practitioners and policymakers, industry persons, and master's and doctoral students in the relevant field. Thus, in this book, the readers will find the updated information as well as the future direction for research in the field of microbial technology.

I would like to acknowledge the Department of Biotechnology, India, for providing me Ramalingaswami Re-entry Fellowship.

Lucknow, Uttar Pradesh, India

Pankaj Kumar Arora

Contents

About the Series Editor

Naveen Kumar Arora, PhD, Microbiology, Professor in the Department of Environmental Science, Babasaheb Bhimrao Ambedkar University (A Central University), Lucknow, Uttar Pradesh, India, is a renowned researcher in the field of Environmental Microbiology and Biotechnology. His specific area of research is rhizosphere biology and PGPRs. He has more than 50 research papers published in premium international journals and several articles published in magazines and dailies. He is an editor of 10 books, published by Springer. He is a member of several national and international societies, Secretary General of Society for Environmental Sustainability, in editorial board of four journals, and reviewer of several international journals. He is also the Editor in Chief of the journal *Environmental Sustainability* published by Springer Nature. He has delivered lectures in conferences and seminars around the globe. He has a long-standing interest in teaching at the PG level and is involved in taking courses in bacteriology, microbial physiology, environmental microbiology, agriculture microbiology, and industrial microbiology. He has been an advisor to 118 postgraduate and 8 doctoral students. Recently, he was awarded for excellence in research by the Honorable Governor of Uttar Pradesh. Although an academician and researcher by profession, he has a huge obsession for the wildlife and its conservation and has authored a book *Splendid Wilds*. He is the President of Society for Conservation of Wildlife and has a dedicated website www.naveenarora.co.in for the cause of wildlife and environment conservation.

About the Editor

Dr. Pankaj Kumar Arora, is currently an Assistant Professor and DBT-Ramalingaswami Re-entry Fellow at the Department of Microbiology, Babasaheb Bhimrao Ambedkar University, Lucknow, India. Dr. Arora is also an Editorial Board Member for the journal *Scientific Reports*, an Associate Editor for the journal *Frontiers in Microbiology*, and an Academic Editor for *PLOS ONE*. He is a recipient of several national awards and fellowships including a Young Botanist Award and Dr. Y. S. Murty Medal from the Indian Botanical Society. His major focus area is environmental microbiology; here, he is currently investigating the biodegradation and bioremediation of various xenobiotic compounds including nitrophenols, chlorinated nitrophenols, and indole. He has a total of 36 publications in reputed journals and has 8 years of teaching and research experience at national and international institutes.

Biotechnological Applications of Lipases in Flavour and Fragrance Ester Production

1

Madan L. Verma

Abstract

Lipases are of widespread occurrence throughout the Earth's flora and fauna. However, microbial flora comprising bacteria, fungi and yeast has drawn more attention in biotechnological applications. Lipases (EC 3.1.1.3) include the hydrolase family of enzymes that possess the unique feature of acting as an interface between the aqueous and non-aqueous phases, and this feature distinguishes them from esterase. Owing to the most versatile biocatalyst, they bring about a range of bioconversion reactions such as hydrolysis, interesterification, esterification, alcoholysis, acidolysis and aminolysis.

Lipase is drawing attention due to its potential applications for flavour and fragrance ester synthesis. Flavour and fragrance ester synthesis by biotechnological processes plays nowadays an increasing role in the food- and pharmaceutical-based industries. This is the result, amongst other things, of scientific advances in biological processes, making use of microorganisms or enzymes as an alternative to chemical synthesis, combined with recent developments in analytical techniques. In this chapter, a comprehensive and illustrious survey is made of the applied aspects of microbial lipases in flavour and fragrance ester synthesis.

Keywords

Microorganism · Lipase · Applications · Fragrance esters · Food industry

M. L. Verma (✉)
Centre for Chemistry and Biotechnology, Deakin University, Geelong, VIC, Australia

Department of Biotechnology, Dr YS Parmar University of Horticulture and Forestry, Hamirpur, HP, India

P. K. Arora (ed.), *Microbial Technology for the Welfare of Society*, Microorganisms for Sustainability 17,
https://doi.org/10.1007/978-981-13-8844-6_1

1.1 Introduction

The advent of enzyme technology represents an important breakthrough in the biotechnology industry, with the worldwide usage of enzymes being nearly US $ 1.5 billion in 2000 (Gupta et al. 2004; Verma et al. 2008a, b; Kanwar and Verma 2010; Verma and Barrow 2015; Devi et al. 2017; Sharma and Sharma 2018). Hydrolytic enzymes, such as lipases, esterases, proteases, amylases and amidases, occupy the major share of the industrial enzyme market. In recent times, lipases have emerged as key enzymes in swiftly growing biotechnology, owing to their properties such as high activity over a wide temperature and pH range, substrate specificity, diverse substrate range and enantioselectivity (de Souza et al. 2017; Padilha et al. 2018). Their importance is increasing day by day in several industries dealing in foods, detergents, chemicals, pharmaceuticals, etc. (Hasan et al. 2006; Verma and Kanwar 2008, 2010, 2012; Verma et al. 2008a, b, 2011, 2012a, b, 2013; Verma 2009; Verma and Barrow 2015; Dias et al. 2018; Sharma and Sharma 2018).

Lipases are the hydrolytic enzymes that act on the carboxyl ester bonds of the triacylglycerols to release fatty acids and glycerol. Due to low solubility of natural substrate triacylglycerols, lipases work at the interface of aqueous and non-aqueous media. Under micro-aqueous conditions, lipases possess the unique ability to carry out the reverse reaction, leading to esterification, alcoholysis and acidolysis (de Souza et al. 2017; Gao et al. 2018). Besides being lipolytic, lipases also possess esterolytic activity and thus have a very diverse substrate range, although they are highly specific as chemo-, regio- and enantioselective biocatalysts (Kobayashi and Adachi 2004; Dias et al. 2018). The active site of lipase belongs to serine hydrolase group; it comprised of a catalytic triad Ser-Asp/Glu-His as well as consensus sequence of Gly-x-Ser-Gly. The three-dimensional (3-D) structures of lipases revealed the characteristic α/β-hydrolase fold (Nardini and Dijkstra 1999).

Lipases offer a wide variety of possibilities for food flavour ester production. Lipase specificity, whether applied via whole-cell or cell-free systems, enables the production of certain chemicals difficult to synthesise (Salvi et al. 2018). Lipase stereoselectivity is an important advantage for the food industry where a specific optical conformation may be associated to flavour properties. Enzymes may also be used directly as food additives, not only to produce or liberate flavour from precursors but also to correct off-flavours caused by specific compounds, naturally occurring or produced during processing (Dias et al. 2018).

A great deal of research has been directed towards the use of lipases mainly for in vitro flavour synthesis (Patel et al. 2018). Moreover, special interest has been given to using organic solvents as reaction media because, in these conditions, these hydrolytic enzymes work preferably in the synthetic way (Kanwar and Verma 2010; Verma et al. 2011; de Souza et al. 2017). Recently, ester synthesis by means of lipase is an interesting alternative considering that there are many well-known flavour esters in the natural aroma of fruits, traditionally obtained by extraction or by chemical synthesis.

The esters synthesised by lipase find applications in numerous fields such as flavour and fragrance synthesis, biodiesel production, resolution of the racemic drugs, fat and lipid modification and synthesis of enantiopure pharmaceuticals and nutraceuticals. It plays a crucial role in the food processing industries since the process is unaffected by the unwanted by-products. Lipase modifications such as the solvent engineering to suit for the non-aqueous ester synthesis have been discussed. This chapter deals with lipase-catalysed flavour and fragrance ester synthesis, esterification strategies, optimum conditions and their scale-up study in food processing industries.

1.2 Sources and Commercialisation of Lipases

Lipases are the prominent biocatalysts that are produced by many organisms including microorganisms, animals and plants. However microbial lipases hold a great prospect in biotechnological applications due to ease of production and scale-up at industrial setting.

A plethora of native or genetically modified microorganisms are used in industries as the sources of lipase enzymes (Jaeger et al. 1999; Palekar et al. 2000). Microorganisms reported for lipase production are *Acinetobacter* sp., *Achromobacter* sp., *Alcaligenes* sp., *Arthrobacter* sp., *Bacillus* sp., *Brochothrix* sp., *Burkholderia* sp., *Chromobacterium* sp., *Corynebacterium* sp., *Cryptococcus* sp., *Enterococcus* sp., *Lactobacillus* sp., *Microthrix* sp., *Moraxella* sp., *Pasteurella* sp., *Propionibacterium* sp., *Psychrobacter* sp., *Proteus* sp., *Staphylococcus* sp., *Serratia* sp., *Streptomyces* sp., *Sulfolobus* sp. and *Pseudomonas* sp. (Pandey et al. 1999; Beisson et al. 2000; Verma et al. 2008a, b). The lipases from *Pseudomonas* species are widely used for a variety of biotechnological applications (Pandey et al. 1999; Beisson et al. 2000). Several products based on microbial lipases have been launched successfully in the market in the past few years (Table 1.1). A number of such products are from *Pseudomonas* spp. such as Lumafast and Lipomax with their major application as detergent enzymes, while Chiro CLEC-PC, Chirazyme L-1 and Amano P, P-30 and PS have shown excellent applications in non-aqueous media for ester synthesis.

Table 1.1 Commercial lipases, sources, applications and their suppliers

Source	Commercial lipase/supplier	Application	Reference(s)
Pseudomonas mendocina	Lumafast/Genencor International, USA	Detergent	Jaeger et al. (1994) and Jaeger and Reetz (1998)
P. alcaligenes	Lipomax/Gist-Brocades, The Netherlands; Genencor International, USA	Detergent	Jaeger et al. (1994) and Jaeger and Reetz (1998)
P. Glumae	n. s./Unilever, The Netherlands	Detergent	Jaeger et al. (1994)
Bacillus pumilus	n. s./Solvay, Belgium	Detergent	
P. Cepacia	Chiro CLEC-PC, Chirazyme L-1/Altus Biologics, Manheim	Organic synthesis	Jaeger and Reetz (1998)

(continued)

Table 1.1 (continued)

Source	Commercial lipase/supplier	Application	Reference(s)
P. Cepacia	Amano P, P-30, PS, LPL-80, LPL-200S/Amano Pharmaceuticals, Japan	Organic synthesis	
P. Cepacia	Lipase AH/Amano Pharmaceuticals, Japan	Organic synthesis	
P. fluorescens	Lipase AK, YS/Amano Pharmaceuticals, Japan	Organic synthesis	
Pseudomonas sp.	Lipase K-10/Amano Pharmaceuticals, Japan	Organic synthesis	Jaeger and Reetz (1998)
Alcaligenes sp.	Lipoprotein lipase, Lipase PL, QL/QLL, PLC/PLG, QLC/QLG/Meito Sangyo Co., Japan	Organic synthesis and research	Jaeger and Reetz (1998) and Godfrey and West (1996)
C. viscosum	*Chromobacterium viscosum* lipase/Asahi Chemical Biocatalysts; Lipase 50P/Biocatalysts, UK	Organic synthesis	Godfrey and West (1996)
P. fluorescens	Lipase 56P/ Biocatalysts, UK	Biotransformations, chemicals	
Achromobacter sp.	Lipase AL, ALC/ALG/Meito Sangyo Co., Japan	Technical grade	
n.s.	Combizyme 23P (proteinase/lipase mix)/Biocatalysts, UK	Waste treatment	
n.s.	Combizyme 209P (amylase/lipase/proteinase mix); Combizyme 61P (proteinase/lipase mix)/ Biocatalysts, UK	Waste treatment and grease disposal	
n.s.	Greasex (lipase)/Novo Nordisk	Leather	

n.s. not specified

1.3 Nature and Features of Lipase Structure

The three-dimensional structures of lipases were first reported in *Mucor miehei* (Brady et al. 1990) and human pancreas (Winkler et al. 1990), respectively, using x-ray crystallography. Since then, structures of lipases from various sources like *Geotrichum candidum* (Schrag and Cygler 1993), *Candida rugosa* (Grochulski et al. 1993), *Chromobacterium viscosum* (Lang et al. 1996) have also been reported. The main features of these structures include α–/β-hydrolase fold (central, hydrophobic β-sheet that is covered by α-helices from both sides), an active site formed by the Ser-His-Asp/Glu triad, oxyanion hole and, in most cases, a 'lid' formed by an α-helix that covers the active site. The active site of lipases is chemically similar but structurally different than that of serine proteases (Winkler et al. 1990).

In lipases studied to date, the catalytic serine has been found to be located in exactly same place on the central α-sheet. The Asp or Glu of the active site, however, is located at different places in different lipases. The His of the catalytic triad is attached to the other two catalytic residues in a similar fashion in all the lipases (Dodson et al. 1992). When compared with serine proteases, the polarity of the main chain, which supports the active Ser residue, is reversed in the lipases. The seryl hydroxyl group in lipases is oriented differently than in serine proteases, resulting in inverted stereochemistry of the catalytic triad, which may have important mechanistic implications (Dodson et al. 1992). *Pseudomonas aeruginosa* lipase, a 29-kDa protein, was postulated to have a lid that stretched between residues 125 and 148 (Cherukuvada et al. 2005). Further, using molecular dynamics simulations, it was proposed that there exists, in addition to the above-mentioned lid, a novel second lid comprised of 210–222 residues in this lipase which regulates the first lid movement. The two hydrophobic residues, Phe214 and Ala217, played important roles in lid movement. This work also elucidated the interplay of hydrophobic interactions in the dynamics and hence the function of an enzyme.

The Lipase Engineering Database (LED; http://www.led.uni-stuttgart.de) provides information on sequence-structure-function relationships of 92 microbial lipases and homologous serine hydrolases assigned into 15 superfamilies and 32 homologous families (Pleiss et al. 2000). This enzyme engineering database covers the unique kinetic properties of lipase produced through protein engineering and design approaches.

1.4 Mechanism of Lipase-Catalysed Reaction

Like other catalysts, enzymes affect rates of reversible reactions in both directions (Kastle and Loevenhart 1990); the direction taken is determined by thermodynamics. The displacement of the equilibrium in reactions catalysed by simple enzymes such as hydrolases is achieved by changing the concentration of the reactants, the main one being water. By diminishing the concentration or activity of water, it is theoretically possible to use any hydrolase to catalyse the corresponding synthetic reaction. Many strategies have been utilised to displace the equilibrium.

The mechanism of lipase-catalysed esterification involves two tetrahedral intermediates (Fig. 1.1). The first tetrahedral intermediate is formed by nucleophilic attack of serine residue of the catalytic triad on the acid with the formation of an acyl-enzyme complex by releasing of a water molecule. An alcohol molecule attacks the complex (nucleophilic attack) to give another tetrahedral intermediate, which, finally, loses an ester molecule to convert the enzyme in its native form. Both of the tetrahedral intermediates possess an oxyanion that is stabilised by hydrogen bonds to protein atoms of the oxyanion hole. There are several studies that showed how the substrate binds to lipases (Grochulski et al. 1993).

The phosphonates resemble the transition state during acylation, whereas the sulphonates resemble the transition state during the deacylation step. It was observed that the alkyl chains of the inhibitors bind in a long, hydrophobic tunnel having an L shape (first projecting towards the middle of the protein and then bending towards its surface), which is characteristic of *Candida rugosa* lipase. The enzyme selectively formed a complex with the SP diastereomer of phophonochloridate, suggesting that in ester hydrolysis reactions, catalytic 'Ser' attacks the *re* face of the ester

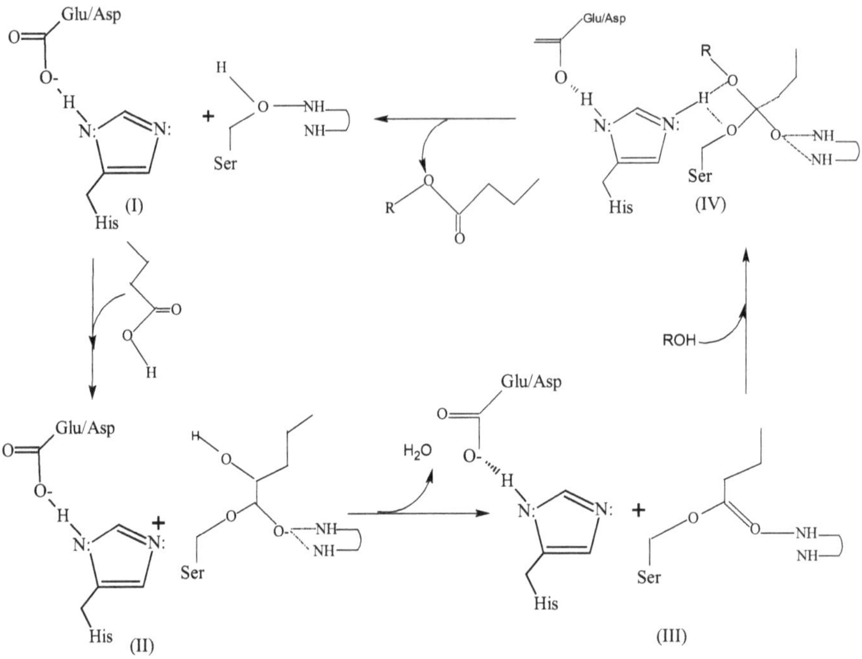

Fig. 1.1 Mechanism of lipase-catalysed esterification/hydrolysis reaction (Verma et al. 2008a). (*I*) Binding of substrate, activation of nucleophilic serine residue by neighbouring histidine and nucleophilic attack of the substrate's carbonyl carbon atom by Ser. (*II*) Transient tetrahedral intermediate that loses a water molecule to give an acyl enzyme complex. (*III*). An alcohol molecule attacks the acyl enzyme complex (nucleophilic attack) to give (*IV*) another tetrahedral intermediate that releases an ester molecule to convert the lipase to its native form

carbonyl. On the other hand, from the structure of the sulphonate complexes, it was concluded that in deacylation of the acyl-enzyme complex, water attacks the *si* face of the complex (Grochulski et al. 1994).

1.5 Microbial Lipases and Fragrance Ester Synthesis

The discovery that enzymes can function in organic solvents broadened the scope of biocatalysis, making enzymes highly useful tools for organic chemists. Non-aqueous enzymology offers many biotechnological applications. Lipases bring about a range of bioconversion reactions (Fig. 1.2) such as hydrolysis, interesterification, esterification, alcoholysis, acidolysis and aminolysis (Bornscheuer et al. 2002; Gupta et al. 2004; Hasan et al. 2006; Verma et al. 2008a, b; Kanwar and Verma 2010). They also possess the unique feature of acting at an interface between the aqueous and non-aqueous (i.e. organic) phase; this feature distinguishes them from esterases. The lipase shows unique interfacial activation behaviour due to nature of substrate aggregation. It is believed that activation involves unmasking

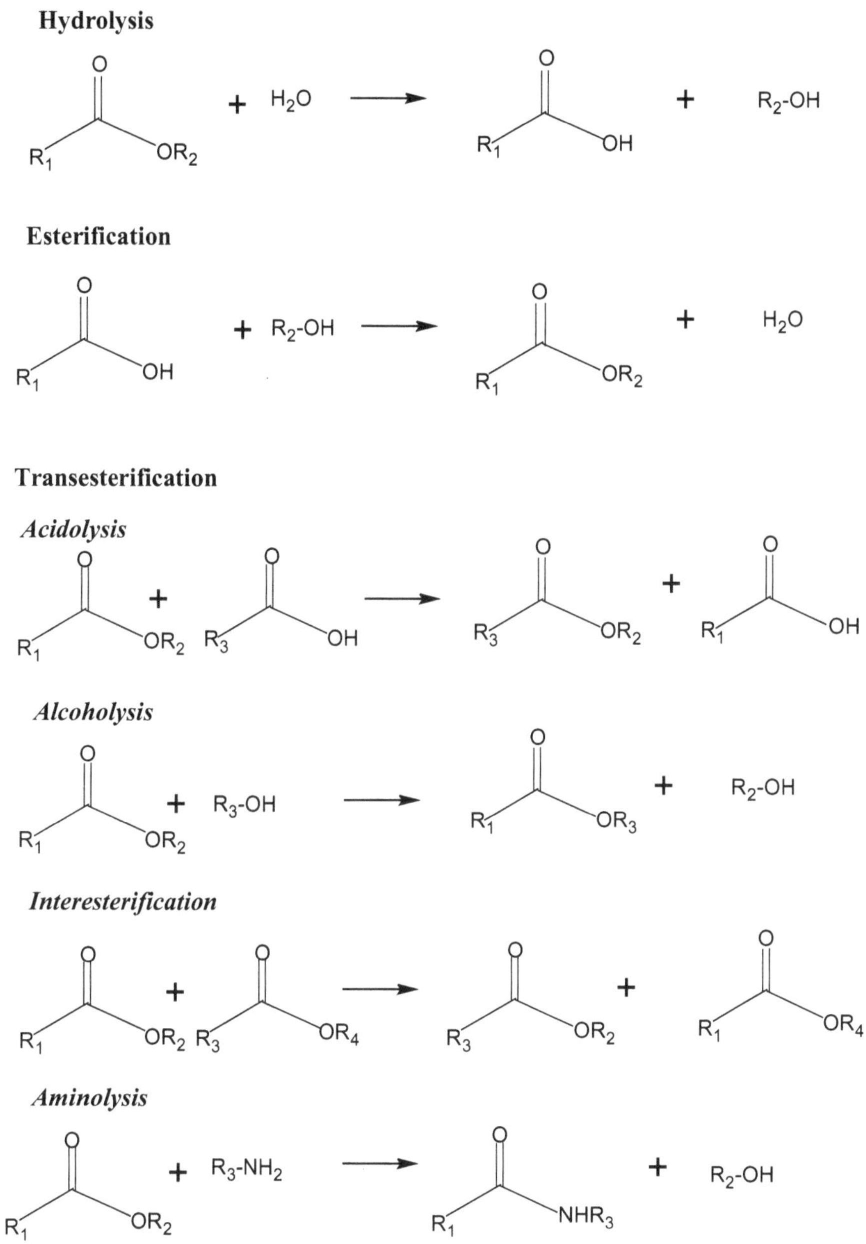

Fig. 1.2 Lipase-catalysed reactions

and structuring of the enzyme's active site through conformational changes that require presence of oil-in-water droplets. Lipase behaviour in non-aqueous media shifts the bioprocess towards the synthesis of esters. Inada et al. (1984) first described ester synthesis catalysed by lipase. Esters based on alcohol moieties such as aliphatic and polyol are commonly used in many chemicals, medicines and foods by selecting unique properties of alcohol and fatty acid. Biosynthesis of esters can be achieved by using the substrate of a fatty acid and an alcohol, a process known as esterification and transesterification.

A variety of flavour and fragrance esters (Table 1.2) are now being produced using immobilised lipase in non-aqueous solvents (Kanwar et al. 2005, 2006, 2007a, b, 2008a, b, 2015; Verma and Kanwar, 2008; Verma et al. 2008a, b, 2011; Kanwar and Verma 2010). The esters produced from long-chain fatty acids (12 to 20 carbon atoms) and short-chain alcohols (three to eight carbon atoms) have been used in food, detergent, cosmetic and pharmaceutical industries (Bauer et al. 1990). Esters prepared by reaction of long-chain acids with long-chain alcohols have important applications as plasticisers and lubricants (Gandhi et al. 1995). Similarly, alcoholic esters of short-chain fatty acids are important flavour and aroma compounds, whereas esters of long-chain fatty acids are being explored for their use as fuel (biodiesel) and as waxes in the oleo-chemical industries (Pandey et al. 1999; Sharma et al. 2001; Jaeger and Eggert 2002; Hasan et al. 2006; Verma et al. 2008b, Kanwar and Verma 2010).

Extracting of esters from natural sources such as sperm whale oil and jojoba oil is an expensive process that provides a low yield of esters. Microbial lipases offer an excellent advantage to overcome these issues of low yields. Thus, non-aqueous enzymology of lipase is the most promising route of high value-added ester production (Sheldon 1996).

The chemical methods are still employed for the bulk synthesis of esters. However, with involvement of high temperature or high pressure, it is difficult in many cases to esterify unstable substances, such as polyunsaturated fatty acids (PUFAs), ascorbic acid and polyols. Further, the regio-specific acylation of polyols requires the protection and deprotection steps (Arcos et al. 1998). These steps cause a rise in manufacturing costs. Lipase-catalysed condensation has the advantages like mild reaction conditions, one-step synthesis without protection and deprotection steps and easy application to food processing. Lipase behaviour in aqueous and non-aqueous media is quite different that is controlled by crucial factors such as substrate, product, water content, temperature and pressure. Although an enzyme-catalysed reaction is usually performed in an aqueous solution, hydrolysis predominately causes the production of desired product to fail when a lipase-catalysed reaction is attempted in an aqueous solution. The controlling the water content in the esterification process catalysed by lipase is a desirable parameter to get a higher yield of esters.

For this reason, lipase-catalysed condensation in a non-aqueous medium, i.e. organic solvents and ionic liquids, has attracted much attention (Sheldon 2001; Moniruzzaman et al. 2010). Fatty acid esters of hydroxy acids like lactic and citric acids and alkyl lactates constitute a very interesting group of surfactants in the food industry (Angello and Vercellotti, 1989; Verma et al. 2008a, b). They are used

Table 1.2 List of lipase-catalysed flavour and fragrance ester production

Name of fragrance esters	Name of flavour	Source of lipases	References
Ethyl butyrate	Fruity flavour	*Candida rugosa* lipase	Devi et al. (2017)
Cinnamyl acetate	`	*Burkholderia ambifaria* lipase	Gao et al. (2018)
Isoamyl acetate	Banana flavour	*Burkholderia cepacia* lipase	Padilha et al. (2018)
Geraniol esters	Fruity flavour	*Candida antarctica* lipase B	Salvi et al. (2018)
Pentyl valerate	Pineapple/ apple flavour	*Pseudomonas* sp. lipase	Patel et al. (2018)
Methyl and ethyl butyrate	Pineapple/ banana flavour	*Candida antarctica* lipase B	de Souza et al. (2017)
Ethyl acetate	Pear flavour	*Bacillus cereus* lipase	Verma et al. (2009)
Isopropyl myristate, geranyl acetate, geranyl butyrate	Fruity flavour	*Bacillus cereus* lipase	Verma (2009)
Isoamyl acetate	Banana flavour	*Candida antarctica*	Langrand et al. (1990)
		Lipolase 100 T, Novozym 435	Kanwar and Goswami (2002) and Kumar et al. (2005)
Geranyl acetate	Floral flavour	*Bacillus cereus*	Verma and Kanwar (2008)
Geranyl butyrate	Citrus flavour	*Pseudomonas aeruginosa*	Verma et al. (2008a, b)
Ethyl acetate	Fruity flavour	*Bacillus cereus*	Verma et al. (2011)
Ethyl laurate	Fruity floral flavour	*Bacillus coagulans*	Kanwar et al. (2005)
(Z)-hexen-1-yl caproate	Green top-flavour	Rapeseed lipase	Liaquat (2011)
2-Phenylethyl esters	Fruity floral flavour	Five commercial fungal lipases (Palatase 20,000 L, lipase AYS 'Amano', lipase A 'Amano' 12, Piccantase A and Piccantase AN)	Tan et al. (2010)
Isoamyl butyrate	Banana flavour	*Rhizomucor miehei*	Mestri and Pai (1994)
Isoamyl propionate	Banana flavour	*Rhizomucor miehei*	Chowdary et al. (2002)
Isoamyl isovalerate	Apple flavour	*Rhizomucor miehei*	Chowdary et al. (2002)
Isobutyl isobutyrate	Pineapple flavour	*Rhizomucor miehei*	Hamsaveni et al. (2001)
Methyl propionate	Fruity flavour	*Rhizomucor miehei*	Perraud and Laboret (1989)

(continued)

Table 1.2 (continued)

Name of fragrance esters	Name of flavour	Source of lipases	References
Ethyl butyrate	Pineapple flavour	*Candida cylindracea*	Yadav and Lathi (2003)
Butyl isobutyrate	Sweet fruity flavour	*Candida cylindracea*, porcine pancreas lipase (PPL) and *Aspergillus niger Rhizomucor miehei*	Welsh et al. (1990)
Short-chain fatty acid esters	Fruity flavour	PPL	Xu et al. (2002)
Long-chain alcoholic esters of lactic acids	Fruity flavour	*Candida antarctica*	From et al. (1997) and Torres and Otero (1999)
Methyl benzoate	Exotic fruity and berry flavour	*Candida rugosa*	Leszczak and Tran-Minh (1998)
Tetrahydrofurfuryl butyrate	Fruity favour	Novozym 435	Yadav and Devi (2004)
Cis-3-hexen-1-yl acetate	Fruity flavour	*Rhizomucor miehei*	Chiang et al. (2003)
Tolyl esters	Honey flavour	*Rhizomucor miehei*, PPL	Suresh Babu et al. (2002) and Manohar and Divakar (2002)
Anthranilic acid esters of C2–C18 alcohols	Flowery flavour of jasmine	*Candida cylindracea* PPL	Kittleson and Pantaleone (1994) and Suresh Babu and Divakar (2001)
4-t-Butylcyclohexyl acetate	Woody and intense flowery flavour	PPL	Manohar and Divakar (2004)
Geranyl methacrylate	Floral fruity flavour	*Rhizomucor miehei*, PPL, *Pseudomonas cepacia*	Athawale et al. (2002)
Citronellyl acetate	Fruity rose flavour	*Candida antarctica* SP435, *Pseudomonas fragi*	Marlot et al. (1985), Mishio et al. (1987), and Claon and Akoh (1994)
Citronellyl propionate			
Citronellyl valerate			
Terpene esters	Fruity flavour	*Candida antarctica* lipase B	Patil et al. (2011)
Farnesol and phytol esters	Fruity flavour	*Candida rugosa*	Sheih et al. (1996)
α-Terpinyl esters	Fruity flavour	*Rhizomucor miehei*	Rao and Divakar (2002)

extensively in the manufacture of cereal products, e.g. yeast-raised baked bread, cakes, doughnuts, noodles and puddings, and dairy products, e.g. ice cream, coffee whiteners and liquid and dry whipped toppings, and also as bactericidal and fungicidal agents. Fatty acid esters of sugars and sugar alcohols find applications as surfactants/emulsifiers in food, detergent, cosmetic and pharmaceutical industries owing to their biodegradability and low toxicity (Saxena et al. 1999; Hasan et al. 2006; Verma et al. 2008a, b; Kanwar and Verma 2010).

Carbohydrate fatty acid esters, synthesised from renewable resources, have a range of applications in the food, cosmetic, oral-care, detergent and pharmaceutical industries (Watanabe et al. 1999). Flavonoids have gained recent interest because of their broad activities (Gao et al. 2001; Havsteen 2002; Tapiero et al. 2002). Lipase-catalysed transesterification, which can be classified into alcoholysis, acidolysis and interesterification, is another way to synthesise esters (Kobayashi and Adachi 2004).

In the lipase-catalysed alcoholysis, methyl, ethyl and vinyl esters have been widely used as substrates. Amongst the esters, vinyl ester has been most extensively used as a substrate because the by-product, vinyl alcohol, is almost irreversibly converted to acetaldehyde to increase the conversion (Hazarika et al. 2003; Yadav and Trivedi 2003). The conversion in the transesterification using other esters as a substrate is also generally high compared to that in the condensation reaction (Compton et al. 2000; Maugard et al. 2000). However, there is a problem in that acetaldehyde forms a Schiff base with a lipase. Furthermore, the condensation reaction has advantages in that the free carboxylic acid can be directly esterified to reduce the cost and that the reaction affords only a harmless by-product, i.e. water, which will simplify the purification processes.

1.6 Factors that Control the Flavour and Fragrance Ester Synthesis

1.6.1 Effect of Solvent on Equilibrium Position

The increasing use of predominantly organic reaction systems has motivated research into solvent selection, a very important factor for the successful application of the biocatalyst. There has been much interest in the development of rules to predict the effects of various solvents on the biocatalyst (Valivety et al. 1991). Lipase-catalysed esterification reaction depends on the log P value which is the function of n-octanol-water partition coefficient system (Laane et al. 1987). When $\log P < 2$, distortion of water structure occurs; if $2 < \log P < 4$, the effect of solvent is unpredictable; and if $\log P > 4$, water structure is intact.

Although the equilibrium position for lipase-catalysed esterification reactions is independent of the enzyme, it is interesting to note that it is not independent of solvent (Kuo and Parkin 1996). A good correlation was found between the ester mole fraction at equilibrium and log P of the solvent (partition coefficient between n-octanol and water). The results have been attributed to the significant difference in polarity of the substrates and products, which often gave rise to dramatic differences in their relative thermodynamic activities in various solvents. During the esterification reaction between glycerol and decanoic acid, high mole fractions of monoacylglycerol and low mole fractions of triacylglycerol were measured in polar solvents, whereas in nonpolar solvents, the measured differences in the mole fractions of mono-, di- and triacylglycerol were much less (Kobayashi et al. 2003).

1.6.2 Effect of Solvent on Kinetics of Lipase Catalysis

Various researchers have studied the effect of organic solvent on lipase kinetics (van Tol et al. 1995), and it was observed that the rates of reaction are greater in hydrophobic solvents than in hydrophilic solvents. Two main reasons have been assigned for the higher activities observed in hydrophobic solvents. First, the solvent affects the thermodynamic activity coefficient of the substrate, or the ground-state free energy of the substrate. Even if the solvent does not affect the enzyme-substrate interaction, the activation energy barrier is bound to be different in various solvents due to the differences in the ground-state free energy (Wangikar et al. 1993). Moreover, the solvent greatly affects the hydration state of the enzyme and, in turn, its ability to interact with the substrate, or its intrinsic activity (Gorman and Dordick 1992). It has been argued that the lower activities in organic solvents are due to the restricted flexibility of protein in these solvents (Almarsson and Klibanov 1996). Further, the rates of enzymatic reactions in anhydrous organic solvents could be increased significantly by the addition of denaturing cosolvents such as methyl sulphoxide or formamide. The reaction medium is generally chosen largely based on experience and/or trial and error.

1.6.3 Effect of Solvent on Substrate Specificity

Many types of solvents have been used in lipase-catalysed condensation, which can be classified into water-miscible and immiscible solvents. Hexane, ether, chloroform and toluene are widely used as water-immiscible solvents. On the other hand, acetone, acetonitrile and tertiary alcohols, such as *t*-butyl alcohol, are used as water-miscible solvents. Water-miscible solvents have an advantage in that they can solubilise hydrophilic substrates such as saccharides to a certain degree without the addition of solubilising reagents thereby facilitating the esterification of the hydrophilic substrates (Degn and Zimmermann 2001; Otero et al. 2001; Castillo et al. 2003). However, a water-miscible solvent may remove water from a lipase molecule, which is essential for its catalytic activity (Krishna et al. 2001), and this removal may deactivate the lipase. For this reason, lipases available for condensation in water-miscible solvents are rather limited at present.

The solvent dramatically influences the substrate specificity of most enzymes (Wangikar et al. 1993; Wescott and Klibanov 1993). The relative variation of the ground-state free energy of the substrate can alter substrate specificity of the enzyme (Wescott and Klibanov 1997). It has been argued that the $kcat/Km$ ratio for two substrates should be proportional to their Raoult's law activity coefficients. This approach has been validated for two different proteases. Lipases, on the other hand, are however known to demonstrate a somewhat complex substrate specificity pattern.

Although an optimum chain length for fatty acid has been observed for a number of lipases, a bimodal pattern for rate parameter versus chain length is observed for lipases from *P. cyclopium*, *A. niger*, *R. delemar*, *C. rugosa* and *M. miehei*. For bimodal pattern, when the rate parameter (e.g. *Vmax/Km*) is plotted against the acyl carbon chain length, two distinct optima are observed, whereas the relative specificities are found to be a function of the solvent (Parida and Dordick 1991). Interestingly, the *kcat/Km* ratios for C8/C6 varied from 3.3 in cyclohexane to 12 in diethyl ether, whereas the *kcat/Km* ratios for C8/C4 varied in an opposite trend (i.e. 2.4 in cyclohexane to 1.0 in diethyl ether). These opposite trends cannot be explained by the simple methodology proposed by Klibanov and coworkers (Ke et al. 1996) based on the ratios of Raoult's law activity coefficients. In general, lipases prefer straight-chain acids to branched-chain ones (Rees et al. 1991). Also, the position of the branching has an effect on the reaction rates. The effect of branching diminishes as the substituent is shifted further away from the carboxylic group. Unsaturated acids have shown lower enzymatic activity compared to the corresponding saturated acids with the α-double bond being more deleterious than a double bond at the β-position (Miller et al. 1988). Similarly, primary alcohols are preferred over secondary ones, whereas tertiary alcohols are not accepted as substrates by the majority of lipases (Claon and Akoh 1993).

1.6.4 Effects of Substrate and Solvent on Lipase Regioselectivity

Although lipase is a useful enzyme for condensation in a water-miscible solvent, yet there are cases where the reaction rate is rather low, e.g. synthesis of ferulic acid esters (Compton et al. 2000). This low reaction rate would be attributed to the electrical property of the carboxyl group in a ferulic acid molecule. Some reports have dealt with the electrical and steric effects of substrates (Bevinakatti and Banerji 1988). However, the studies that numerically correlate these effects with substrate selectivity are limited. Therefore, a kinetic analysis of the lipase-catalysed condensation of *p*-methoxyphenethyl alcohol and several carboxylic acids having various electrical or steric properties was performed that correlated the rate constants with these properties (Kobayashi et al. 2003). The projection area of the non-carboxylic region in a carboxylic acid molecule was considered as a steric property and the electron density of a carboxyl carbon as an electrical property.

Enzymes show remarkable regioselectivity or selectivity based on molecular position of the functional group. Regioselectivity becomes extremely important in the synthesis of sugar esters and their derivatives. Sugars contain multiple hydroxyl groups, all of which are capable of being acylated by acidic moieties. Lipases are known to regioselectively acylate monosaccharides as well as disaccharides in organic solvents (Otto et al. 1998; Degn et al. 1999). As one can expect, the regioselectivity was found to be strongly dependent on the source of the lipase (Nicotra et al. 1989) as well as on the type of saccharide used (Woundenberg-van Oosterom et al. 1996). In addition, the regioselectivity has been shown to be strongly dependent on the solvent used (MacManus and Vulfson 1997), the level of hydration (Kim et al. 1998) and the type of substrate chemistry, i.e. the acylating agents (Rich et al. 1995).

1.6.5 Effect of Water Present in Organic Solvents on Lipase Catalysis

The degree of solvent hydration has a profound effect on lipase catalysis in anhydrous organic solvents. Water not only affects the intrinsic activity of lipases but also acts as a competitive nucleophile (Wangikar et al. 1995). The control of water quantity released during the esterification process regulates the shifting of equilibrium position towards synthetic as well as hydrolytic activity along with rate of the reaction (Svensson et al. 1994; Dudal and Lortie 1995). An important question in non-aqueous enzymology is whether the monolayer of water molecules around the enzyme molecule is essential for activity. For enzymatic reactions in essentially non-aqueous media, activity increases very rapidly with the addition of small amounts of water to the organic media (Zaks and Klibanov 1988).

The enzymatic activity in anhydrous solvents may be correlated to the thermodynamic activity (*aw*) rather than the concentration of water (Wehtje and Adlercreutz 1997). This is readily seen from the effect of water content on catalytic activity of lipase (Halling 1994). The lipase showed a similar optimum at thermodynamic water activity of about 0.55 when used in solvents ranging from hexane to pentanone. The optimum varied widely in terms of total water content. However, for polar organic solvents such as dioxane, acetonitrile, acetone and so forth, it has been observed that water activity failed to predict the critical hydration level for enzyme activity (Bell et al. 1997). Thus, one might conclude that enzymatic activity could be correlated with *aw* only for water-immiscible organic solvents. Also, it should be noted that lipases from different microbial sources showed widely varying optimum *aw* ranging from 0.3 to 1.0 (Valivety et al. 1992). Interestingly, the optimum *aw* also depends on the type of support used for immobilisation for a given lipase in a given solvent. However, a report by Oladepo et al. (1994) contradicted this statement and indicated that the reaction rates for lipase catalysis in organic media showed a similar dependence on water activity irrespective of the support used. Thus, the optimum water requirement might be dictated by the biological source of the enzyme, the organic solvent and possibly the type of support used.

1.6.6 Effect of Water Removal during Lipase Catalysis in Organic Solvents

It is important to carry out lipase catalysis in organic solvents at an optimal hydration level. The control of water content during lipase catalysis in organic solvents is checked by salt hydrates, pervaporation, saturated salt solutions and molecular sieves. Salt hydrates maintain the *aw* of the organic systems used for reverse hydrolysis reactions by undergoing transition between their various states of hydration (Zacharis et al. 1997). Similarly, other salt hydrate pairs might be chosen to maintain the thermodynamic water activity at the desired level. The esterification reaction could be carried out in a membrane reactor with the reaction mixture on one side of the membrane and air, used as an extractive medium for water removal, on the other side (Van der Padt et al. 1993). By using a suitable membrane, selective pervaporation of the reaction water through the membrane could be achieved (Mulder 1991). The air is re-circulated through the reactor after condensing out the permeated water. Water-activity control could be achieved by controlling the temperature of the condenser. This method effectively esterified oleic acid with *n*-butanol in the presence of Lipozyme in isooctane using the cellulose acetate membrane (Kwon et al. 1995). A modification of this technique employed a hollow-fibre membrane reactor with reaction mixture on the lumen side and controlled humidity air on the shell side (Ujang et al. 1997).

Saturated salt solutions possessed a fixed thermodynamic activity at a given temperature. Water activity in an organic solvent can be adjusted by equilibrating with a saturated salt solution via vapour-phase equilibria. Water-producing or water-consuming reactions could be performed in chambers of controlled water activity using these saturated salt solutions (Goderis et al. 1987; Dudal and Lortie 1995). Such systems have limitations on the rate of transfer of water and hence are not suitable for fast reactions. A newer method was developed where water activity could be controlled by circulation of a saturated salt solution through a silicone tubing which was submerged in the reaction medium (Wehtje and Adlercreutz 1997). The silicone membranes are permeable to water vapours and impermeable to ions. Thus, only water and organic substances are transferred through them. Mass-transfer properties for different types of silicone tubing have been characterised for a water-activity-controlling system in organic solvents using saturated salt solutions (Kaur et al. 1997). A modification of this technique involved circulation of a saturated salt solution through microporous hollow-fibre polypropylene membranes while the reaction medium is circulated on the shell side (Rosell et al. 1996).

Water formed by condensation exists in solubilised form and acts as a substrate for the hydrolysis reaction when the reaction was performed in a water-miscible solvent (He et al. 2002; Sonwalkar et al. 2003). Therefore, elimination of water by desiccant/molecular sieves was quite effective for increasing the conversion in lipase-catalysed condensation reaction (Arcos et al. 1998; Maugard et al. 2000; Giacometti et al. 2001; Kuwabara et al. 2003).

However, there was no criterion for the amount of desiccant to be added to the reaction system. The quantitative relationship between the equilibrium conversion and the concentration of a molecular sieve was discussed using the adsorption isotherm of water onto molecular sieves, the apparent reaction constant, the solubility of a hydrophilic substrate in the solvent and the mass balance equation in terms of water into consideration to obtain the criterion. Although the addition of molecular sieves or silica gel usually improves the equilibrium conversion in many cases, negative effects such as the formation of diesters and degradation of unstable substrates have also been reported (Sonwalkar et al. 2003). Thus, nature and quantity of desiccant play a vital role in the enhancing of lipase-catalysed esterification reactions.

1.7 Current Status of Fragrance Ester Production Using Lipases

Biocompatibility, biodegradability and environmental acceptability of biotechnological production of polyesters are desired properties in agricultural and medical applications (Evans and Sikdar 1990). Amongst carbohydrate fatty acid esters, sucrose esters are the most developed carbohydrate esters, which are being produced at about 4000 Tm/year (Hill and Rhode 1999). Amongst fatty acid esters of sugars and sugar alcohols, the fatty acid esters of sorbitol are the second largest class of carboxylic acid esters employed as surfactants (Arcos et al. 1998). Limited studies have been done on large-scale synthesis of esters using lipase as a catalyst (Bloomer et al. 1992; Bourg-Garros et al. 1998; Radzi et al. 2005). These studies would be helpful in potential commercialisation of the enzymatic esterification processes. Scaled-up lipase-catalysed synthesis of (Z)-3-hexen-1-yl acetate, green, fruity and citrus notes, in hexane from laboratory scale (3–5 g) to pilot scale (1–5 kg) has been reported (Bourg-Garros et al. 1998). When the reaction volume was increased (from 20 to 2000 ml), the time required to achieve certain conversion (90%) also increased (from 24 to 48 h). When the reaction volume was 13.5 L, the conversion was low (43%) even after 56 h. Removal of water that was formed in the esterification reaction by azeotropic distillation improved the performance to some extent (50% conversion in 48 h). A similar study for synthesis of ethyl stearate has also been reported (Bloomer et al. 1992). Investigation in large-scale production for oleyl oleate, a liquid wax ester, was performed in batch mode of stirred tank reactor (STR) with one multi-bladed impeller (Radzi et al. 2005). As these reactions were carried out in a rotary evaporator, such studies are not useful in enhancing production level from pilot scale to industrial scale.

There is a need to develop production and downstream-processing systems, which are cost-effective, simple and least time-consuming. The growing demand for lipases has shifted the trend towards prospecting for novel lipases, improving the properties of existing lipases for established technical applications and producing new tailor-made enzymes for entirely new areas of application. Advancement in molecular biology has contributed a large number of recombinant enzymes aligned with industrial needs. Rational protein engineering, by way of mutagenesis and directed evolution, has provided a new and valuable tool for improving or adapting enzyme properties to the desired requirements (Jaeger et al. 1999; Bornscheuer et al. 2002). Recently advancing in the nanoscale supports for enzyme immobilisation has further improved the efficiency of lipase-catalysed bioprocessing in the direction of esterification as well as transesterification studies (Verma et al. 2016; Verma 2017a, b).

1.8 Conclusion and Future Perspectives

Research in the areas of lipase biotechnology has been influenced by the increasing demand of consumers for nutritious food and natural food ingredients. In this chapter, the impact of enzyme technology on the production of flavour and fragrance esters has been discussed. These applications still need further research and to show economic feasibility in order to reach the commercial scale. Genetic and protein engineering may contribute to overcome some of these drawbacks like production capacity and enzyme instability, but it is important to consider the development costs including eventual toxicological studies and market sizes. Advantages of enzyme-catalysed synthesis (mild temperature, pressure and pH conditions, high enantio- or regio-specificity, lack of contaminating by-products) are attractive for the production of flavour and fragrance esters. It may be concluded that the use of enzymes as additives for flavour production and/or modification may become an important part of enzyme technology as shown by some recent developments. Advances in the use of enzymes in non-aqueous media offer a great potential for flavour and fragrance ester syntheses in the near future. Thus, advances in nanotechnology and protein engineering and design domains have developed novel nanobiocatalysts which are the new avenue to be explored for enzymatic catalysis in non-aqueous media for the synthesis of fragrance esters.

Acknowledgements The author wishes to thank the director of the Centre for Chemistry and Biotechnology, Deakin University, Australia, for the kind support to pursue this research work. The financial support from the Department of Biotechnology, Ministry of Science and Technology, Government of India, in the form of predoctoral fellowship to pursue this study is also thankfully acknowledged.

References

Almarsson O, Klibanov AM (1996) Remarkable activation of enzymes in nonaqueous media by denaturing organic cosolvents. Biotechnol Bioeng 49:87–92

Angello AJ, Vercellotti JR (1989) Phospholipids and fatty acid esters of alcohols. In: Charalambous G, Doxastakis G (eds) Food emulsifier chemistry, technology, functional properties and applications. Elsevier, Amsterdam

Arcos JA, Bernabe M, Otero C (1998) Quantitative enzymatic production of 6-O-acylglucose esters. Biotechnol Bioeng 57:505–509

Athawale V, Manjrekar N, Athawale M (2002) Lipase-catalyzed synthesis of geranyl methacrylate by transesterification: study of reaction parameters. Tetrahedron Lett 43:4797–4800

Bauer AR, Garbe D, Surgurg H (1990) Common fragrance and flavour materials, 2nd revised edn. VCH, New York, pp 67–81

Beisson F, Tiss A, Rivière C, Verger R (2000) Methods for lipase detection and assay: a critical review. Eur J Lipid Sci Technol 2:133–153

Bell G, Janssen AEM, Halling PJ (1997) Water activity fails to predict critical hydration level for enzyme activity in polar organic solvents: Interconversion of water concentrations and activities. Enzym Microb Technol 20:471–477

Bevinakatti HS, Banerji AA (1988) Lipase-catalysis: factors governing transesterification. Biotechnol Lett 6:397–398

Bloomer S, Adlercreutz P, Mattiasson B (1992) Kilogram-scale ester synthesis of acyl donor and its use in lipase-catalyzed interesterifications. J Am Oil Chem Soc 69:966–973

Bornscheuer UT, Bessler C, Srinivas R, Krishna SH (2002) Optimizing lipases and related enzymes for efficient application. Trends Biotechnol 20:433–437

Bourg-Garros S, Razafindramboa N, Pavia AA (1998) Large-scale preparation of (*Z*)-3-hexen-1-yl acetate using *Candida antarctica* immobilized lipase in hexane. Biotechnol Bioeng 59:495–500

Brady L, Brzozowski AM, Derewenda ZS, Dodson E, Dodson G, Tolley S, Turkenburg P, Christiansen L, Huge-Jensen B, Norskov L, Thim L, Menge U (1990) A serine protease triad forms the catalytic centre of a triacylglycerol lipase. Nature 343:767–770

Castillo E, Pezzotti F, Navarro A, Lopez-Munguia A (2003) Lipase catalysed synthesis of xylitol monoesters: solvent engineering approach. J Biotechnol 102:251–259

Cherukuvada SL, Seshasayee AS, Raghunathan K, Anishetty S, Pennathur G (2005) Evidence of a double-lid movement in *Pseudomonas aeruginosa* lipase: insights from molecular dynamics simulations. PLoS Comput Biol 1:e28

Chiang WD, Chang SW, Shieh CJ (2003) Studies on the optimized lipase-catalyzed biosynthesis of cis-3-hexen-1-yl acetate in n-hexane. Process Biochem 38:1193–1199

Chowdary GV, Divakar S, Prafulla SG (2002) Modeling on isoamyl isovalerate synthesis from *Rhizomucor miehei* lipase in organic media: optimization studies. World J Microbiol Biotechnol 18:179–185

Claon PA, Akoh CC (1993) Enzymatic synthesis of geraniol and citronellol esters by direct esterification in *n*-hexane. Biotechnol Lett 15:1211–1216

Claon PA, Akoh CC (1994) Effect of reaction parameters on SP435 lipase-catalyzed synthesis of citronellyl acetate in organic solvent. Enzym Microb Technol 16:835–838

Compton DL, Laszlo JA, Berhow MA (2000) Lipase-catalyzed synthesis of ferulate esters. J Am Oil Chem Soc 77:513–519

de Souza MCM, dos Santos KP, Freire RM, Barreto ACH, Fechine PBA, Gonçalves LRB (2017) Production of flavor esters catalyzed by lipase B from *Candida antarctica* immobilized on magnetic nanoparticles. Braz J Chem Eng 34(3):681–690

Degn P, Zimmermann W (2001) Optimization of carbohydrate fatty acid ester synthesis in organic media by a lipase from *Candida antarctica*. Biotechnol Bioeng 74:483–491

Degn P, Pedersen LH, Duus J, Zimmermann W (1999) Lipase-catalyzed synthesis of glucose fatty acid esters in *tert*-butanol. Biotechnol Lett 21:275–280

Devi NA, Radhika GB, Bhargavi RJ (2017) Lipase catalyzed transesterification of ethyl butyrate synthesis in *n*-hexane-a kinetic study. J Food Sci Technol 54(9):2871–2877

Dias ALB, Santos PD, Martínez J (2018) Supercritical CO_2 technology applied to the production of flavor ester compounds through lipase-catalyzed reaction: a review. J CO2 Util 23:159–178

Dodson GG, Lawson DM, Winkler FK (1992) Structural and evolutionary relationships in lipase mechanism and activation. Faraday Discuss 93:95–105

Dudal Y, Lortie R (1995) Influence of water activity on the synthesis of triolein catalyzed by immobilized *Mucor miehei* lipase. Biotechnol Bioeng 45:129–134

Evans JD, Sikdar SK (1990) Biodegradable plastics: an idea whose time has come. ChemTech 20:38–42

From M, Adlercreutz P, Mattiasson B (1997) Lipase catalyzed esterification of lactic acid. Biotechnol Lett 19:315–317

Gandhi NN, Sawant SB, Joshi JB (1995) Specificity of a lipase in ester synthesis: effect of alcohol. Biotechnol Prog 11:282–287

Gao C, Mayon P, MacManus DA, Vulfson EN (2001) Novel enzymatic approach to the synthesis of flavonoid glycosides and their esters. Biotechnol Bioeng 71:235–243

Gao Z, Chu J, Jiang T, Xu T, Wu B, He B (2018) Lipase immobilization on functionalized mesoporous TiO_2: specific adsorption, hyperactivation and application in cinnamyl acetate synthesis. Process Biochem 64:152–159

Giacometti J, Giacometti F, Milin C, Racki DV (2001) Kinetic characterization of enzymatic esterification in a solvent system: adsorptive control of water with molecular sieves. J Mol Catal B Enzym 11:921–928

Goderis H, Ampe G, Feyten M, Fouwe B, Guffens WV, Cauwenbergh S, Tobback P (1987) Lipase-catalyzed ester exchange reactions in organic media with controlled humidity. Biotechnol Bioeng 30:258–266

Godfrey T, West S (1996) The application of enzymes in industry. In: Godfrey T, Reichelt J (eds) Industrial enzymology, 2nd edn. The Nature Press, New York, p 512

Gorman LAS, Dordick J (1992) Organic solvents strip water of enzymes. Biotechnol Bioeng 39:392–397

Grochulski P, Li Y, Schrag JD, Bouthillier F, Smith P, Harrison D, Rubin B, Cygler M (1993) Insights into interfacial activation from an open structure of *Candida rugosa* lipase. J Biol Chem 15:12843–12847

Grochulski P, Li Y, Schrag JD, Cygler M (1994) Two conformational states of *Candida rugosa* lipase. Protein Sci 3:82–91

Gupta R, Gupta N, Rathi P (2004) Bacterial lipases: an overview of production, purification and biochemical properties. Appl Microbiol Biotechnol 64:763–781

Halling PJ (1994) Thermodynamic predictions for biocatalysis in non-conventional media: theory, tests and recommendations for experimental design and analysis. Enzym Microb Technol 16:178–206

Hamsaveni DR, Prafulla SG, Divakar S (2001) Optimization of isobutyl butyrate synthesis using central composite rotatable design. Process Biochem 36:1103–1109

Hasan F, Shah AA, Hameed A (2006) Industrial applications of microbial lipases. Enzym Microb Technol 39:235–251

Havsteen BH (2002) The biochemistry and medical significance of the flavonoids. Pharmacol Ther 96:67–202

Hazarika S, Goswami P, Dutta NN (2003) Lipase catalyzed transesterification of 2-*o*-benzylglycerol with vinyl acetate: solvent effect. Chem Eng J 94:1–10

He XL, Chen BQ, Tan TW (2002) Enzymatic synthesis of 2- ethylhexyl esters of fatty acids by immobilized lipase from *Candida sp.* J Mol Catal B Enzym 18:333–339

Hill K, Rhode O (1999) Sugar-based surfactants for consumer products and technical applications. Fett-Lipid 101:25–33

Inada Y, Nishimura H, Takahashi K, Yoshimoto T, Saha AR, Saito Y (1984) Ester synthesis catalyzed by polyethylene glycol-modified lipase in benzene. Biochem Biophys Res Commun 122:845–890

Jaeger K-E, Eggert T (2002) Lipases for biotechnology. Curr Opin Biotechnol 113:390–397

Jaeger K-E, Reetz MT (1998) Microbial lipases form versatile tools for biotechnology. Trends Biotechnol 16:396–403

Jaeger K-E, Ransac S, Dijkstra BW, Colson C, van HM, Misset O (1994) Bacterial lipases. FEMS Microbiol Rev 15:29–63

Jaeger K-E, Dijkstra BW, Reetz MT (1999) Bacterial biocatalysts: molecular biology, three-dimensional structures and biotechnological applications of lipases. Annu Rev Microbiol 53:315–351

Kanwar L, Goswami P (2002) Isolation of a *Pseudomonas* lipase produced in pure hydrocarbon substrate and its application in the synthesis of isoamyl acetate using membrane-immobilized lipase. Enzym Microb Technol 31:727–735

Kanwar SS, Verma ML (2010) Lipases. In: Encyclopedia of industrial biotechnology. Wiley, New York, pp 1–16

Kanwar SS, Kaushal RK, Verma ML, Kumar Y, Chauhan GS, Gupta R, Chimni SS (2005) Synthesis of ethyl laurate by hydrogel immobilized lipase of *Bacillus coagulans* MTCC-6375. Ind J Microbiol 45:187–193

Kanwar SS, Verma HK, Pathak S, Kaushal RK, Kumar Y, Verma ML, Chimni SS, Chauhan GS (2006) Enhancement of ethyl propionate synthesis by poly (AAc-co-HPMA-cl-MBAm)-immobilized *Pseudomonas aeruginosa* MTCC-4713 exposed to Hg^{2+}, and NH_4^+ ions. Acta Microbiol Immunol Hung 53:195–207

Kanwar SS, Verma ML, Maheshwari C, Chauhan S, Chimni SS, Chauhan GS (2007a) Properties of poly (AAc-co-HPMA-cl-EGDMA) hydrogel-bound lipase of *Pseudomonas aeruginosa* MTCC-4713 and its use in synthesis of methyl acrylate. J Appl Polym Sci 104:183–191

Kanwar SS, Kaushal RK, Verma ML, Kumar Y, Azmi W, Gupta R, Chimni SS, Chauhan GS (2007b) Synthesis of ethyl oleate employing synthetic hydrogel-immobilized lipase of *Bacillus coagulans* MTCC-6375. Indian J Biotechnol 6:68–73

Kanwar SS, Gehlot S, Verma ML, Gupta R, Kumar Y, Chauhan GS (2008a) Synthesis of geranyl butyrate employing *poly* (AAc-*co*-HPMA-*cl*-EGDMA) hydrogel-immobilized lipase of *Pseudomonas aeruginosa* MTCC-4713. J Appl Polym Sci 110:2681–2692

Kanwar SS, Sharma C, Verma ML, Chauhan S, Chimni SS, Chauhan GS (2008b) Short-chain ester synthesis by transesterification employing *poly* (MAc-*co*-DMA-*cl*-MBAm) hydrogel-bound lipase of *Bacillus coagulans* MTCC-6375. J Appl Polym Sci 109:1063–1071

Kanwar SS, Verma ML, Puri S, Chauhan GS (2015) Synthetic hydrogel: characteristics and applications. In: Emerging areas in biotechnology. Nirmal Book Agency Publisher, Kurukshetra, pp 173–212

Kastle JH, Loevenhart AS (1990) Concerning lipase, the fat-splitting enzyme, and the reversibility of its action. Am Chem J 24:491–525

Kaur J, Wehtje E, Adlercreutz P, Chand S, Mattiasson B (1997) Water activity control in enzymatic esterification processes. Enzym Microb Technol 21:502–510

Ke T, Wescott CR, Klibanov AM (1996) Prediction of the solvent dependence of enzymatic prochiral selectivity by means of structure-based thermodynamic calculations. J Am Chem Soc 118:3366–3374

Kim JE, Han JJ, Yoon JH, Rhee JS (1998) Effect of salt hydrate pair on lipase-catalyzed regioselective monoacylation of sucrose. Biotechnol Bioeng 57:121–125

Kittleson JR, Pantaleone DP (1994) Enzymic biphasic process for the synthesis of aromatic esters flavoring agents from corresponding carboxylic acid and alcohol by esterification mediated by a lipase from *Candida cylindracea*. US Patent 5,437,991

Kobayashi T, Adachi S (2004) Reaction equilibrium for lipase catalyzed condensation in organic solvent system. Biotechnol Lett 26:1461–1468

Kobayashi T, Furutani W, Adachi S, Matsuno R (2003) Equilibrium constant for the lipase-catalyzed synthesis of fatty acid butyl ester in various organic solvents. J Mol Catal B Enzym 24-25:61–66

Krishna SH, Divakar S, Prapulla SG, Karanth NG (2001) Enzymatic synthesis of isoamyl acetate using immobilized lipase from *Rhizomucor miehei*. J Biotechnol 87:193–201

Kumar R, Modak J, Madras G (2005) Effect of the chain length of the acid on the enzymatic synthesis of flavors in supercritical carbon dioxide. Biochem Eng J 23:199–202

Kuo S-J, Parkin KL (1996) Solvent polarity influences product selectivity of lipase-mediated micro aqueous media. J Am Oil Chem Soc 73:1427–1433

Kuwabara K, Watanabe Y, Adachi S, Nakanishi K, Mutsuno R (2003) Synthesis of 6-O-unsaturated acyl L-ascorbates by immobilized lipase in acetone in the presence of molecular sieve. Biochem Eng J16:17–23

Kwon SJ, Song KM, Hong WH, Rhee JS (1995) Removal of water produced from lipase catalyzed esterification in organic solvent by pervaporation. Biotechnol Bioeng 46:393–395

Laane C, Boeren S, Vos K, Veeger C (1987) Roles for optimization of biocatalysis in organic-solvents. Biotechnol Bioeng 30:81–87

Lang D, Hoffman B, Haalack L, Hecht H-J, Spener F, Schmid RD, Schomburg D (1996) Crystal structure of a bacterial lipase from *Chromobacterium viscosum* ATCC 6918 refined at 1.6 Å resolution. J Mol Biol 259:704–717

Langrand G, Rondot N, Triantaphylides C, Baratti J (1990) Short-chain flavor esters synthesis by microbial lipases. Biotechnol Lett 12:581–586

Leszczak JP, Tran-Minh C (1998) Optimized enzymatic synthesis of methyl benzoate in organic medium: operating conditions and impact of different factors on kinetics. Biotechnol Bioeng 60:556–561

Liaquat M (2011) Optimized synthesis of (Z)-3-hexen-1-yl caproate using germinated rapeseed lipase in organic solvent. J Mol Catal B Enzym 68:59–65

MacManus DA, Vulfson EN (1997) Reversal of regioselectivity in the enzymatic acylation of secondary hydroxyl groups mediated by organic solvents. Enzym Microb Technol 20:225–228

Manohar B, Divakar S (2002) Application of central composite rotatable design to lipase catalyzed syntheses of m-cresyl acetate. World J Microbiol Biotechnol 18:745–751

Manohar B, Divakar S (2004) Porcine pancreas lipase acetylation of beta–cyclodextrin anchored 4-t-butylcyclohexanol. Indian J Chem Sect B 43B:2661–2665
Marlot C, Langrand G, Triantaphylides C, Baratti J (1985) Ester synthesis in organic solvent catalyzed by lipase immobilized on hydrophilic supports. Biotechnol Lett 7:647–650
Maugard T, Tudella J, Legoy MD (2000) Study of vitamin ester synthesis by lipase-catalyzed transesterification in organic media. Biotechnol Prog 16:358–362
Mestri S, Pai JS (1994) Synthesis of isoamyl butyrate by lipase of *Mucor miehei*. PAFAI J 2:24–26
Miller C, Austin H, Posorske L, Gonzalez J (1988) Characteristics of an immobilized lipase for the commercial synthesis of esters. J Am Chem Soc 65:927–932
Mishio T, Takahashi K, Yoshimoto T, Kodera Y, Saito Y, Inada Y (1987) Terpene alcohol ester synthesis by poylethylene glycol modified lipase in benzene. Biotechnol Lett 9:187–190
Moniruzzaman M, Nakashima K, Kamiya N, Goto M (2010) Recent advances of enzymatic reactions in ionic liquids. Biochem Eng J 48:295–314
Mulder M (1991) Basic principles of membrane technology. Kluwer Academic, Dordrecht
Nardini M, Dijkstra BW (1999) α/β hydrolase fold enzymes: the family keeps growing. Curr Opin Struct Biol 9:732–737
Nicotra F, Riva S, Secundo F, Zuchelli L (1989) An interesting example of complementary regioselective acylation of secondary hydroxyl groups by different lipases. Tetrahedron Lett 30:1703–1704
Oladepo DK, Halling PJ, Larsen VF (1994) Reaction rates in organic media show similar dependence on water activity with lipase catalyst immobilised on different supports. Biocatalysis 8:283–287
Otero C, Arcos JA, Berrendero MA, Torres C (2001) Emulsifiers from solid and liquid polyols: different strategies for obtaining optimum conversions and selectivities. J Mol Catal B Enzym 11:883–892
Otto RT, Bornscheuer UT, Scheib H, Pleiss J, Syldatk C, Schmid RD (1998) Lipase catalyzed esterification of unusual substrates: synthesis of glucuronic acid and ascorbic acid (vitamin C) esters. Biotechnol Lett 20:1091–1094
Padilha GS, Tambourgi EB, Alegre RM (2018) Evaluation of lipase from *Burkholderia cepacia* immobilized in alginate beads and application in the synthesis of banana flavor (isoamyl acetate). Chem Eng Commun 205:23–33
Palekar AA, Vasudevan PT, Yan S (2000) Purification of lipase: a review. Biocatal Biotransformation 18:177–200
Pandey A, Benjamin S, Soccol CR, Nigam P, Krieger N, Soccol UT (1999) The realm of microbial lipases in biotechnology. Biotechnol Appl Biochem 29:119–131
Parida S, Dordick JS (1991) Substrate structure and solvent hydophobicity control lipase catalysis and enantioselectivity in organic media. J Am Chem Soc 113:2253–2259
Patel V, Deshpande MP, Pandey A, Larroche C, Madamwar D (2018) Nanobiocatalysis for the synthesis of Pentyl Valerate in organic solvents: characterization, optimization and reusability studies. Curr Biotechnol 7(2):105–114
Patil D, Das D, Nag A (2011) Enzymatic synthesis and analytical monitoring of terpene ester by 1H NMR spectroscopy. Chem Pap 65:9–15
Perraud R, Laboret F (1989) Optimization of methyl propionate production catalyzed by *Mucor miehei* lipase. Appl Microbiol Biotechnol 44:321–326
Pleiss J, Fischer M, Peiker M, Thiele C, Schmid RD (2000) Lipase engineering database: understanding and exploiting sequence-structure-function relationships. J Mol Catal B Enzym 10:491–508
Radzi SM, Basri M, Salleh AB, Ariff A, Mohammad R, Rahman MBA, Rahman RNZRA (2005) Large-scale production of liquid wax ester by immobilized lipase. J Oleo Sci 54:4203–4209
Rao P, Divakar S (2002) Response surface methodological approach for the *Rhizomucor miehei* lipase-mediated esterification of α-terpineol with propionic acid and acetic anhydride. World J Microbiol Biotechnol 18:341–345
Rees GD, Nascimento MG, Jenta TRJ, Robinson BH (1991) Reverse enzyme synthesis in microemulsion-based organ-gels. Biochim Biophys Acta 1073:493–501

Rich JO, Bedell BA, Dordick JS (1995) Controlling enzyme catalyzed regioselectivity in sugar ester synthesis. Biotechnol Bioeng 45:426–434
Rosell CM, Vaidya AM, Halling PJ (1996) Continuous *in-situ* water activity control for organic phase biocatalysis in a packed bed hollow fiber reactor. Biotechnol Bioeng 49:284–289
Salvi HM, Kamble MP, Yadav GD (2018) Synthesis of geraniol esters in a continuous-flow packed-bed reactor of immobilized lipase: optimization of process parameters and kinetic modeling. Appl Biochem Biotechnol 184(2):630–643
Saxena RK, Gosh PK, Gupta R, Davidson WS, Bradoo S, Gulati R (1999) Microbial lipases: potential biocatalysts for the future industry. Curr Sci 77:101–115
Schrag JD, Cygler M (1993) 1.8-angstrom refined structure of the lipase from *Geotrichum candidum*. J Mol Biol 230:575–591
Sharma R, Sharma N (2018) Microbial lipase mediated by health beneficial modification of cholesterol and flavors in food products: a review. Recent Pat Biotechnol 12(2):81–91
Sharma R, Chisti Y, Banerjee UC (2001) Production, purification, characterization and applications of lipases. Biotechnol Adv 19:627–662
Sheih CJ, Akoh CC, Yee LN (1996) Optimized enzymatic synthesis of geranyl butyrate with lipase AY from *Candida rugosa*. Biotechnol Bioeng 51:371–374
Sheldon RA (1996) Large-scale enzymatic conversion in non-aqueous media. In: Koskinen AMP, Klibanov AM (eds) Enzymatic reactions in organic media. Blackie Academic and Professional, Glasgow, pp 267–307
Sheldon R (2001) Catalytic reactions in ionic liquids. Chem Commun 23:2399–2407
Sonwalkar RD, Chen CC, Ju LK (2003) Roles of silica gel in poly-condensation of lactic acid in organic solvent. Bioresour Technol 87:69–73
Suresh Babu CV, Divakar S (2001) Selection of alcohols through Plakett-Burman design in lipase catalyzed syntheses of anthranilic acid. J Am Oil Chem Soc 78:49–52
Suresh Babu CV, Karanth NG, Divakar S (2002) Lipase catalysed esterification of cresols. Indian J Chem Sect B 41B:1068–1071
Svensson I, Wehtje E, Adlercreutz P, Mattiasson B (1994) Effects of water activity on reaction rates and equilibrium positions in enzymatic esterification. Biotechnol Bioeng 44:549–556
Tan HSG, Yu B, Curran P, Liu SQ (2010) Lipase-catalysed synthesis of natural aroma-active 2-phenylethyl esters in coconut cream. Food Chem 124:80–84
Tapiero H, Tew KD, Nguyen Ba G, Mathe G (2002) Polyphenols: do they play a role in the prevention of human pathologies. Biomed Pharmacother 56:200–207
Torres C, Otero C (1999) Part 1: enzymatic synthesis of lactate and glycolate esters of fatty alcohols. Enzym Microb Technol 25:745–752
Ujang Z, Sharbati NA, Vaidya AM (1997) Organic-phase enzymatic esterification in a hollow fiber membrane reactor with *in situ* gas-phase water activity control. Biotechnol Prog 13:39–42
Valivety RH, Johnston GA, Suckling CJ, Halling PJ (1991) Solvent effects on biocatalysis in organic systems: equilibrium position and rates of lipase catalyzed esterification. Biotechnol Bioeng 38:1137–1143
Valivety RH, Halling PJ, Macrae AR (1992) Reaction rate with suspended lipase catalyst shows similar dependence on water activity in different organic solvents. Biochim Biophys Acta 1118:218–222
Van der Padt A, Sewalt JJW, Van't Riet K (1993) On-line water removal during enzymatic triacylglycerol synthesis by means of pervaporation. J Membr Sci 80:199–208
Van Tol JBA, Stevens RMM, Veldhuizen WJ, Jongejan JA, Duine JA (1995) Do organic solvent affects the catalytic properties of lipase? Intrinsic kinetic parameters of lipase in ester hydrolysis and formation in various organic solvent. Biotechnol Bioeng 47:71–81
Verma ML (2009) Studies on lipase of *Bacillus cereus* MTCC 8372 and its application for synthesis of esters. PhD thesis, HP University Shimla, India
Verma ML (2017a) Nanobiotechnology advances in enzymatic biosensors for the agri-food industry. Environ Chem Lett 15(4):555–560

Verma ML (2017b) Enzymatic nanobiosensors in the agricultural and food industry. In: Ranjan S, Dasgupta N, Lichfouse E (eds) Nanoscience in food and agriculture 4, sustainable agriculture reviews. Springer, Cham, pp 229–245, isbn:978-3-319-53111-324

Verma ML, Barrow CJ (2015) Recent advances in feedstocks and enzyme-immobilised technology for effective transesterification of lipids into biodiesel. In: Kalia V (ed) Microbial factories. Springer, New Delhi, pp 87–103

Verma ML, Kanwar SS (2008) Properties and application of *Poly*(MAc-2011*co*-DMA-*cl*-MBAm) hydrogel immobilized *Bacillus cereus* MTCC 8372 lipase for synthesis of geranyl acetate. J Appl Polym Sci 110:837–846

Verma ML, Kanwar SS (2010) Purification and characterization of a low molecular mass alkaliphilic lipase of *Bacillus cereus* MTCC 8372. Acta Microbiol Immunol Hung 57:187–201

Verma ML, Kanwar SS (2012) Harnessing the potential of thermophiles: the variants of extremophiles. Dynamic Biochem Process Biotechnol Mol Biol 6(1):28–39

Verma ML, Azmi W, Kanwar SS (2008a) Microbial lipases: at the interface of aqueous and non-aqueous media-a review. Acta Microbiol Immunol Hung 55:265–293

Verma ML, Chauhan GS, Kanwar SS (2008b) Enzymatic synthesis of isopropyl myristate using immobilized lipase from *Bacillus cereus* MTCC-8372. Acta Microbiol Immunol Hung 55:327–342

Verma ML, Azmi W, Kanwar SS (2009) Synthesis of ethyl acetate employing celite-immobilized lipase of *Bacillus cereus* MTCC 8372. Acta Microbiol Immunol Hung 56:229–242

Verma ML, Azmi W, Kanwar SS (2011) Enzymatic synthesis of isopropyl acetate by immobilized *Bacillus cereus* lipase in organic medium. Enzyme Res 2011, Article ID 919386, 7 pages

Verma ML, Kumar S, Devi A, Jana AK (2012a) Engineering enzyme for efficient biocatalysis. In: 2nd international conference on biomedical engineering and assistive technologies at NIT, Jalandhar, India, December 6–7, 2012, pp 1–5, isbn:13:978-81-925454-1-7

Verma ML, Kumar S, Devi A, Jana AK (2012b) Metagenomics approaches for the discovery of novel lipases with biotechnological perspectives. In: 2nd international conference on biomedical engineering and assistive technologies at NIT, Jalandhar, India, December 6–7, 2012, pp 1–6, isbn:13:978-81-925454-1-7

Verma ML, Naebe M, Barrow CJ, Puri M (2013) Enzyme immobilisation on amino-functionalised multi-walled carbon nanotubes: structural and biocatalytic characterisation. PLoS One 8(9):e73642

Verma ML, Puri M, Barrow CJ (2016) Recent trends in nanomaterials immobilised enzymes for biofuel production. Crit Rev Biotechnol 36(1):108–119

Wangikar PP, Graycar TP, Estell DA, Clark DS, Dordick JS (1993) Protein and solvent engineering of subtilisin BPN in nearly anhydrous organic media. J Am Chem Soc 115:12231–12237

Wangikar PP, Rich JO, Clark DS, Dordick JS (1995) Probing enzymic transition state hydrophobicities. Biochemistry 34:12302–12310

Watanabe Y, Adachi S, Matsuno R (1999) Condensation of L-ascorbic acid and medium-chain fatty acids by immobilized lipase in acetonitrile with low water content. Food Sci Technol Res 5:188–192

Wehtje E, Adlercreutz P (1997) Water activity and substrate concentration effects on lipase activity. Biotechnol Bioeng 55:798–806

Welsh FW, Williams RE, Dawson KH (1990) Lipase-mediated synthesis of low molecular weight flavor esters. J Food Sci 55:1679–1682

Wescott CR, Klibanov AM (1993) Solvent variation inverts substrate specificity of an enzyme. J Am Chem Soc 115:1629–1631

Wescott CR, Klibanov AM (1997) Thermodynamic analysis of solvent effect on substrate specificity of lyophilized enzymes suspended in organic media. Biotechnol Bioeng 56:340–344

Winkler FK, d'Arey A, Hunziker W (1990) Structure of human pancreatic lipase. Nature 343:771–774

Woundenberg-van Oosterom M, van Rantwijk F, Sheldon RA (1996) Regioselective acylation of disaccharides in *tert*-butyl alcohol catalyzed by *Candida antarctica* lipase. Biotechnol Bioeng 49:328–333

Xu Y, Wang D, Qing Mu X, Ao-Zhao G, Zhang KC (2002) Biosynthesis of ethyl esters of short-chain fatty acids using whole-cell lipase from *Rhizopus chinensis* CCTCC M201021 in non-aqueous phase. J Mol Catal B Enzym 18:29–37
Yadav GD, Devi KM (2004) Immobilized lipase-catalysed esterification and transesterification reactions in non-aqueous media for the synthesis of tetrahydrofurfuryl butyrate: comparison and kinetic modeling. Chem Eng Sci 59:373–383
Yadav GD, Lathi PS (2003) Kinetics and mechanism of synthesis of butyl isobutyrate over immobilized lipases. Biochem Eng J 16:245–252
Yadav GD, Trivedi AH (2003) Kinetic modeling of immobilized lipase catalyzed transesterification of *n*-octanol with vinyl acetate in non-aqueous media. Enzym Microb Technol 32:783–789
Zacharis E, Omar IC, Partridge J, Robb DA, Halling PJ (1997) Selection of salt hydrate pairs for use in water control in enzyme catalysis in organic solvents. Biotechnol Bioeng 55:367–374
Zaks A, Klibanov AM (1988) Enzymatic catalysis in nonaqueous solvents. J Biol Chem 263:3194–3201

Synthesis and Regulation of Fungal Secondary Metabolites

2

Arvind Kumar and Antresh Kumar

Abstract

Fungi are well known for their ability to produce a multitude of secondary metabolites (SMs) which act as a weapon of defense to protect themselves against parasites and predators. A variety of fungal SMs have proved to serve as an important factor for decades. The synthesis of SMs in fungi is a complex, multi-step process and is stage-specific under specialized conditions. SMs are primarily synthesized by non-ribosomal peptide synthetase (NRPS) or polyketide synthase (PKS) enzymes. The genes encoded for SM synthesis are often located in the cluster form at the sub-telomere region. The SM biosynthetic gene cluster comprises of genes encoding for NRPS/PKS, a transcription factor, and other accessory genes essential for assembly and maturation of SM. The regulation of SM synthesis in fungi can be achieved by pathway-specific (in-clustered transcription factor), global regulatory proteins and chromatin remodeling. The regulatory protein-encoding gene present in each gene cluster is considered to be a crucial regulatory circuit of the SM biosynthetic pathway. Moreover, the regulation of fungal SM biosynthesis is also guided by global regulatory proteins responsive to pH, carbon, nitrogen, light/dark, and other environmental cues. Histone modifications by methylation and acetylation often altered the chromatin structure; as a result, these changes repress or express the genes of SM biosynthetic pathway. Taken together, we conclude that fungal SM ability as antibiotics to toxins is useful to mankind.

Keywords

Persistence toxic substances · Bioremediation · Biodegradation · Genetically modified microorganisms

A. Kumar · A. Kumar (✉)
Department of Biotechnology, Central University of South Bihar, Gaya, Bihar, India
e-mail: antreshkumar@cub.ac.in

P. K. Arora (ed.), *Microbial Technology for the Welfare of Society*, Microorganisms for Sustainability 17,
https://doi.org/10.1007/978-981-13-8844-6_2

2.1 Introduction

The enormous variety and diversity of fungi stand a prime place in the natural world. The geographical diversity of India is the most suitable to fungi that contributes more than half of the total fungal species of the world. However, the flora of the fungal kingdom is still undiscovered; thus it needs to be explored for detection of novel therapeutics for the treatment of a plethora of pathogens. Fungi are well known for their ability to produce a multitude of secondary metabolites (SMs) as products. SMs are small organic molecules (<4 kDa) which are unlikely to be involved in cell growth and development but have a pivotal role in self-defense, niche adaptation, and communication to the environment (Kamei and Watanabe 2005; Keller et al. 2005). SMs isolated from fungi have been considered the most resourceful natural product as they have been exploited by humankind as antimicrobial, antitumor, enzyme inhibitor, immunosuppressive, anticholesterol, and antiparasitic agents (Agathos et al. 1987; Demain 1999; Ramawat and Mérillon 2013). Despite their clinical importance, fungi also synthesize some toxic SM substances which are sometimes deleterious to human beings as they can cause systemic infections (Demain and Fang 2000; García-Estrada et al. 2011). A range of SMs with their effectiveness and producing organisms are listed in Table 2.1. Characteristically, synthesis of SMs in fungi is stage-specific as they are produced by a specific organism/group of organisms under certain growth/environmental conditions such as media, pH, temperature, and nitrogenous sources (Jensen et al. 2007; Fox and Howlett 2008; Brakhage 2013; Keller 2015). SM production most often occurs at the late log phase or stationary phase (Bu'Lock 1961). Earlier observation revealed that environmental conditions for both SM production and spore formation were found to be common, but they were more stringent than vegetative growth (Bu'Lock 1961; Sekiguchi and Gaucher 1977). Therefore, it is believed that synthesis of SMs is essential for sporulation. For example, the dark-brown melanin pigments synthesized by oxidative polymerization of phenolic compounds which are deposited in the cell wall during cell sporulation protect the spores from UV light and are also involved in the virulence (Calvo et al. 2002). Aside from melanins, zearalenone produced in *F. graminearum* also induces the sporulation (Wolf and Mirocha 1973). The fungal SM production is also linked to the asexual to sexual development. The switching from primary to secondary metabolism or asexual to sexual development depends on the psi factor or endogenous ratio of oleic acid- and linoleic acid-derived molecules (Champe and El-Zayat 1989; Calvo et al. 2001). It is well known that complex and diverse SMs derived from fungi are synthesized by sequential assembly of primary metabolic intermediate precursors molecule, i.e., acetyl coenzyme A (acetyl-CoA), 1-deoxyxylulose 5-phosphate, and mevalonic and shikimic acid (Dewick 2002). Structurally, fungal SMs are classified as non-ribosomal peptides, polyketides, indole alkaloids, and terpenes (Keller et al. 2005). The synthesis of SMs is a multi-step process which is initiated by the action of various classes of enzymes. However, a variety of tailoring enzymes, viz., transporters, oxygenases, hydroxylases, and regulatory proteins, are also involved in the biosynthesis of

Table 2.1 Fungal secondary metabolites, sources, and their medicinal importance

S. no.	Name of the species	Compound	Properties	References
1	*Acremonium chrysogenum*	Cephalosporin	Antibacterial	Sándor et al. (1998)
2	*Penicillium chrysogenum*	Penicillin	Antibacterial	Raper (1946)
3	*Acremonium* spp.	Spirobenzofuran	Antibacterial	Kleinwaechter et al. (2001)
4	*Zygosporium masonii*	Zygosporin A	Antibacterial	Hayakawa et al. (1968)
5	*Alternaria* spp.,	Altersetin	Antibacterial	Hellmig et al. (2002)
6	*Penicillium griseofulvum*	Griseofulvin	Antifungal	Grisham et al. (1973)
7	*Cryptosporiopsis* cf. *quercina*	Cryptocandin	Antifungal	Strobel et al. (1999)
8	*Mycogone rosea*	Roseoferin	Antifungal	Degenkolb et al. (2000)
9	*Penicillium simplicissimum*	Xanthoepocin	Antifungal	Igarashi et al. (2000)
10	*Emericella rugulosa*	Echinocandin B	Antifungal	Kumar et al. (2018)
11	*Pestalotia heterocornis*	Paclitaxel (Taxol)	Anticancerous	Noh et al. (1999)
	Taxomyces andreanae			Stierle et al. (1993)
12	*Pestalotiopsis microspora*	Torreyanic acid	Anticancerous	Noh et al. (1999)
13	*Streptomyces verticillus*	Bleomycin	Antitumor	Umezawa et al. (1966)
14	*Aspergillus and Trichoderma* spp.	Gliotoxin	Immunosuppressive	Pahl et al. (1996)
15	*Aspergillus parasiticus*	Aflatoxins	Mycotoxin	Dorner et al. (1984)
	Aspergillus flavus			
16	*Penicillium viridicatum*	Ochratoxin A	Mycotoxin	Scott et al. (1970)
17	*Claviceps purpurea*	Ergotamines	Mycotoxin	Amici et al. (1967)
18	*Aspergillus* and *Penicillium* spp.	Patulin	Mycotoxin	Moss (2002)
19	*Pestalotiopsis microspora*	Pestacin and isopestacin	Antioxidant	Harper et al. (2003)
20	*Nodulisporium* spp.	Nodulisporic acids	Insecticidal	Ondeyka et al. (1997)
21	*Chaetomium globosum NK10*	Chaetoglobosin A	Antinematodes	Hu et al. (2013)
22	*Mycena galericulata*	Strobilurin	Antiparasitic	Hosokawa et al. (2000)
23	*Acremonium tubakii*	Cephaibol	Antiparasitic	Vertesy et al. (2003)
24	*Phycomyces blakesleeanus*	Ferritin	Parkinson's disease	Mann et al. (1994)
25	*Ashbya gossypii*	Riboflavin	Nutrient	Wei et al. (2013)
26	*Gibberella fujikuroi*	Gibberellin	Nutrient	Tudzynski (1999)
27	*Penicillium* spp.	Arisugacin	Acetylcholinesterase inhibitor	Omura et al. (1995)

fungal SMs, and their encoded genes are often arranged in a cluster (Fox and Howlett 2008; Keller 2015). The reason why SM biosynthetic genes present together as clusters is still unanswered. It is speculated that all the essential genes for an SM synthesis are transcriptionally regulated by an intrinsic transcription factor (TF) under certain environmental conditions (Fox and Howlett 2008; Keller 2015).

2.2 Different Classes of Fungal Secondary Metabolites

2.2.1 Non-ribosomal Peptides

Non-ribosomal peptides are an enormous class of SMs which are derived from both proteinogenic and non-proteinogenic amino acids. SMs belonging to this class are synthesized by the large multifunctional and multi-modular enzymes called non-ribosomal peptide synthetases (NRPS) and synthesized independently to ribosomes (Stein et al. 1996). NRPS are organized in the modules, which are highly specific to a particular amino acid and also considered as an assembly unit (Fig. 2.1a). The number of modules depends on the length of the peptides. NRPS module characteristically consists of three basic subunits named as adenylation (A), peptidyl carrier protein or thiolation (PCP or T), and condensation (C) domains (Fig. 2.1b). Adenylation domain tends to activate the amino acid to amino-acyl adenylate by reacting with ATP (Fischbach and Walsh 2006). This activated amino acid is then transferred to the PCP domain which forms a thioester bond with the cysteine-thiol group of the PCP domain (Strieker et al. 2010). However, the condensation domain (C) catalyzes the peptide bond formation between amino acid bonded to the thiolation domain and nascent peptide chain on the preceding module. In addition to the minimal core module, NRPS enzymes also contain associatory domains such as cyclization (Cy), epimerization (E), oxidation (Ox), thioesterase (TE), and methyltransferase (MT) domains essential for maturation of SM molecule (Hoffmeister and Keller 2007; Strieker et al. 2010; Evans et al. 2011). The cyclization domain (Cy) is a variant of the classical C domain which introduces thiazoline or oxazoline ring on β-heteroatom-bearing amino acids such as threonine, serine, and cysteine. The heterocyclization of participatory amino acids of SM prevents them from the action of cellular protease (Marshall et al. 2001; Kelly et al. 2005).

Initiation modules usually lacking C domain and termination modules usually possess an essential thioesterase (TE) domain at their C terminal. TE domain catalyzes product hydrolytic cleavage or complex macro-cyclization and releases the peptide molecule (Fig. 2.1b) (Strieker et al. 2010). A range of non-ribosomal peptides are methylated by MT domain which contain C-methyl or N-methyl groups containing amino acids using *S*-adenosyl methionine (SAM) as co-substrate (Walsh et al. 2001). Epimerization (E) domain is meant for the conversion of L-amino acids to the corresponding D-isomer of the peptides (Linne et al. 2001). Penicillin, cephalosporin, cyclosporine, echinocandin B, cryptocandin, and gliotoxin are the common examples of NRPS-mediated SMs (Cacho et al. 2012).

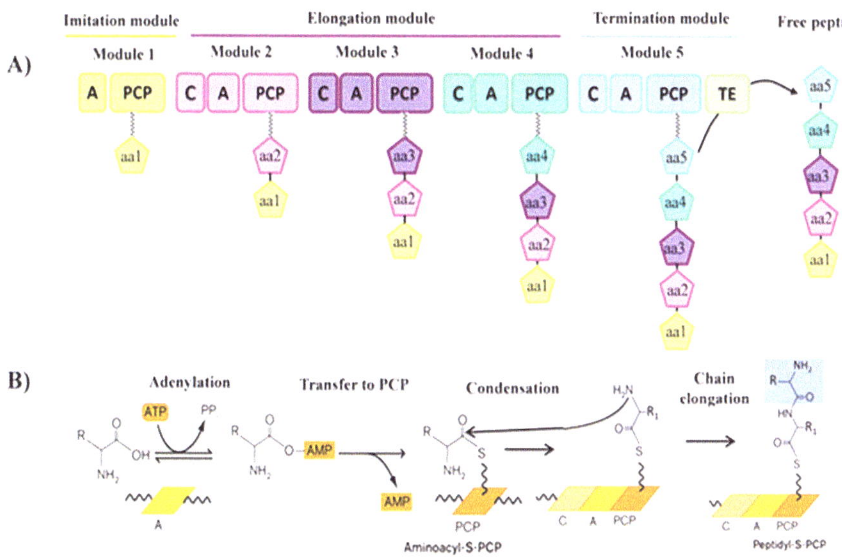

Fig. 2.1 Organization of multi-modular non-ribosomal peptide synthetase (NRPS). (**a**). Arrangement of different enzyme modules of NRPS. Each module contains adenylation (*A*), peptidyl carrier protein or thiolation (*PCP* or *T*), and condensation (*C*) domains. Amino acids are sequentially assembled by NRPS enzyme. Initiation module lacking *C* domain and termination module contains thioesterase domain (*TE*) which is involved in the cleavage. (**b**). Specified domains catalyze the amino acid to the peptides. Adenylation activates amino acid to amino-acyl adenylate by reacting with *ATP*, and it attaches to the *PCP*. Condensation domain catalyzes peptide bond formation between amino acid bonded to the thiolation domain and nascent peptide chain on preceding module

2.2.2 Polyketides

Polyketides are the structurally diverse class of fungal secondary metabolites which are primarily synthesized from the monoketide units (acetyl-CoA, methylmalonyl-CoA, and malonyl-CoA) derived from the primary metabolite pool (Keller et al. 2005; Wakimoto et al. 2011). Fungal polyketides are synthesized or assembled by polyketide synthases (PKS) which are closely related to the fatty acid synthases (FAS) (Hutchinson 2003). Similar to NRPS, PKS are also multi-modular enzymes, having ketoacyl CoA synthase (KS), acyltransferase (AT) and acyl carrier (ACP), and ketone-reducing enoyl reductase (ER), dehydratase (DH), and ketoreductase (KR) domains (Fig. 2.2) (Keller et al. 2005). PKS can be classified into three classes, PKS I, II, and III, which have identical enzymatic functions, but varying chain initiation and termination steps (Gallo et al. 2013). PKS I are complex and the most versatile in nature which can be sub-categorized into iterative and non-iterative type.

Iterative PKS I are multi-enzymes which produce non-reducing (NR), partially reducing (PR), and highly reducing (HR) polyketides. Non-iterative PKS I are organized into modules having multiple domains acting once and catalyze one cycle of polyketide chain elongation. The antibacterials orsellinic acid and erythromycin A and cholesterol-lowering agent

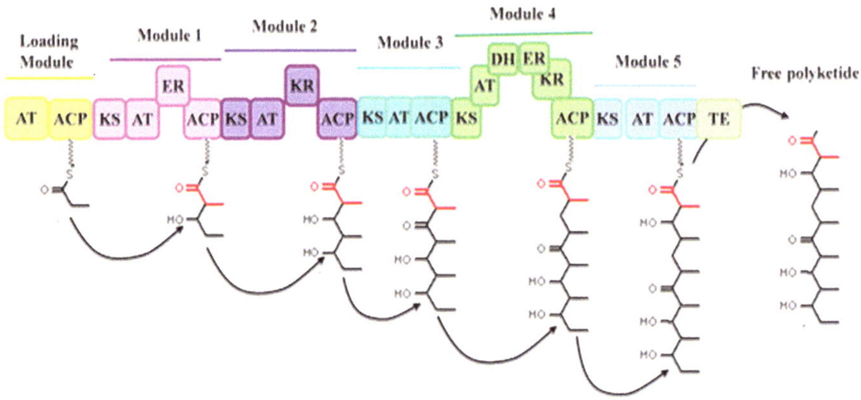

Fig. 2.2 Domain organization of the PKS. The functional domains are represented by box. Loading module lacking KS domain and terminal module has a TE domain. Each module is responsible for the specific polymerization of single monoketide unit, and the specific action of respective domains synthesizes the polyketide molecule. Domains are represented as *AT* acyltransferase, *ACP* acyl carrier protein, *KS* ketosynthase, *KR* ketoreductase, *DH* dehydratase, *ER* enoyl reductase, *TE* thioesterase, and *CoA* coenzyme A

squalestatin S1 are the known examples of iterative PKS I (Weissman 2009). PKS II are the aggregated complex of mono-functional proteins mainly comprising of ketosynthase (KS) α and β and acyl carrier protein (ACP). Each protein has a specific role in the synthesis and acts as iteratively. The iterative action of these proteins led to polyketone synthesis via chain extension to the defined number of cycles. PKS III, also known as chalcone synthases, are the homo-dimeric enzymes, which act iteratively. PKS III have simple, versatile, and broad substrate specificity and are very less abundant in fungi. They mostly utilize CoA thioesters as a substrate for decarboxylative condensation and cyclization reaction (Calne et al. 1989; Staunton and Weissman 2001; Shen 2003; Weissman 2009). The tri- and tetra-ketide pyrones, the penta- and hexa-ketide resorcylic acids, griseofulvin, bikaverin, and the penta-ketide resorcinols are such examples (Hashimoto et al. 2014; Liu et al. 2015).

2.2.3 Terpenoids

Terpenoids or terpenes are the odoriferous SMs which are mostly derived from plants (turpentine and camphor) and less commonly found in fungi (carotenoids, aristolochenes, trichothecenes, gibberellins, and indole-diterpenes). The basic unit of terpene is monomer isoprene (C_5H_8) which undergoes polymerization catalyzed by terpene synthases. The empirical formula of terpenes is $(C_5H_8)_n$, where n represents the number of isoprene units. Structurally, terpenes can be cyclic or linear and unsaturated or saturated. Two abundant building blocks of terpenoids are isopentenyl diphosphate (IPP) and dimethylallyl diphosphate (DMAPP), derived from acetyl-CoA by glyceraldehyde 3-phosphate and pyruvate or mevalonic acid (Fig. 2.3) (Eisenreich et al. 2004).

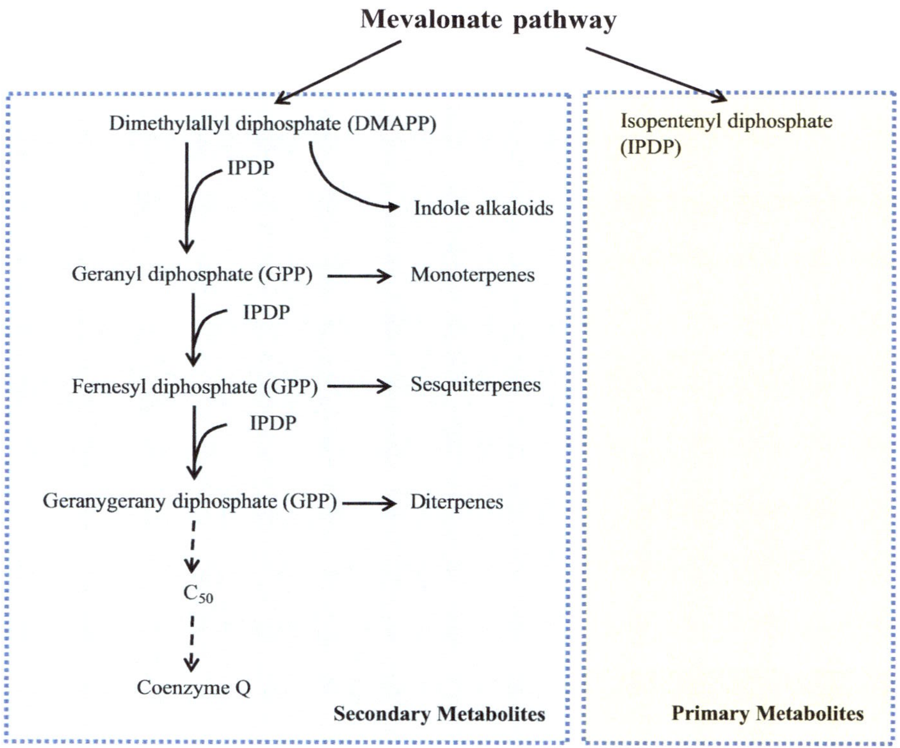

Fig. 2.3 **Terpene biosynthetic pathway.** Isopentenyl diphosphate and dimethylallyl diphosphate, the products of the mevalonate pathway, are the precursors of steroids, carotenoids, and coenzyme Q in many species

Terpenes are classified into three groups: monoterpenes (derived from geranyl pyrophosphate, also known as geranyl diphosphate), sesquiterpenes (derived from farnesyl pyrophosphate, also known as farnesyl diphosphate), and diterpenes/carotenoids (derived from geranyl-geranyl pyrophosphate, also known as geranyl-geranyl diphosphate) (Keller et al. 2005). Most common fungal terpenoids are carotenoids, gibberellins, and trichothecenes which act as a free-radical scavenger, plant hormone, and mycotoxin, respectively (Echavarri-Erasun and Johnson 2002; Alexander et al. 2009; Bömke and Tudzynski 2009).

2.2.4 Alkaloids

Alkaloids set up a diverse group of natural products; around 12,000 alkaloids have been identified that are most commonly produced in plants, animals, bacteria, and fungi. In the case of fungi, *Ascomycota* and *Basidiomycota* have been extensively producing the bioactive alkaloids (Guirimand et al. 2010; Xu et al. 2014). Alkaloids share the basic organic substances that contain at least one nitrogen atom in the heterocyclic ring (Mahmood et al. 2010). Alkaloids are bitter in taste and used in medical science against a diverse range of diseases and toxicants. The first microbial-derived alkaloid is ergot synthesized in *Claviceps purpurea* which causes an ergot of rye disease upon infection to the plant ovaries (Esser and Tudzynski 1978). Alkaloids accurately cannot be classified into families due to their diverse structure and precursor molecules, but based on structural similarities, the precursor molecule, and their source (Waller and Burström 1969; Anaya et al. 2006; Evans 2009; Roberts 2013; Wansi et al. 2013; Aniszewski 2015), alkaloid are separated in different types as shown in Table 2.2.

Indole alkaloids are most commonly produced in fungi which are derived from tryptophan and an isoprenoid precursor (dimethylallyl pyrophosphate). However, intermediate compounds of the shikimic acid pathway such as anthranilate, indole-3-glycerol-phosphate, 4-dimethylallyl tryptophan (4-DMAT), and tryptamine act as a precursor (Fig. 2.4) (Byrne et al. 2002; Xu et al. 2014). The synthesis of indole alkaloids is achieved by NRPS enzymes; adenylation (A) domain of NRPS activates

Table 2.2 Alkaloids classes and associated examples

S. no.	Types of alkaloids	Examples
1	Phenylalkylamine alkaloids	Ephedrine, pseudoephedrine, taxine, and hordenine
2	Pyrrolidine alkaloids	Nicotine and hygrine
3	Piperidine and pyridine alkaloids	Lobeline, coniine, and piperine
4	Tropane alkaloids	Hyoscyamine, atropine, and hyoscine
5	Quinoline alkaloids	Quinine and quinidine
6	Indole alkaloids	Apocynaceae, Rubiaceae, and Loganiaceae
7	Purine alkaloids	Caffeine, theobromine, and methylated xanthine
8	Tropolone alkaloids	Colchicines
9	Diterpenoid alkaloids	Delcosine, lycoctonine type, and ajaconine

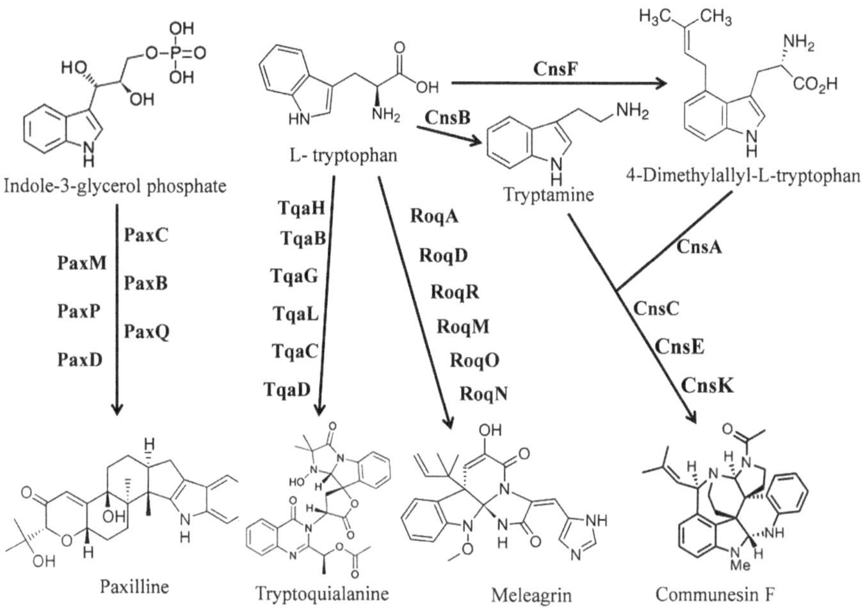

Fig. 2.4 Fungal indole alkaloid biosynthetic pathway. Paxilline biosynthesis from indole-3-glycerol phosphate in *P. paxilli* with the action of paxilline biosynthetic genes (Scott et al. 2013). Tryptoquialanine and meleagrin biosynthesis from L-tryptophan in *P. aethiopicum* and *P. chrysogenum* by respective biosynthetic genes (Gao et al. 2011; García-Estrada et al. 2011). Communesin F biosynthesis from L-tryptophan, tryptamine, and 4-dimethylallyl-L-tryptophan in *P. expansum* with respective biosynthetic genes (Lin et al. 2015)

the individual amino acids to aminoacyl-ADP, transfers to thiolation (T) domain, and forms aminoacyl thioesters. A variety of other enzymes, i.e., geranyl-geranyl pyrophosphate synthases, monooxygenases, methyltransferases, tryptophan decarboxylase, anthranilate synthase, chorismate mutase, strictosidine synthase, and tyrosine decarboxylase, are also involved in indole synthesis (Ziegler and Facchini 2008; Dubouzet et al. 2013; Xu et al. 2014). There are larger NRPS products that contain tryptophan, such as the tetrapeptide apicidin isolated from *Fusarium semitectum* that displays histone deacetylase inhibition activity (Darkin-Rattray et al. 1996; Jin et al. 2010; Xu et al. 2014).

2.3 Genetics of Secondary Metabolite Synthesis

The fungal genome sequencing projects gave insight to the SM research, thus achieving the milestone for biosynthetic gene exploration and regulation (Galagan et al. 2005). Fungal SMs are synthesized by the action of a set of enzymes which are encoded by different genes. It has been reported that the SM gene clusters are randomly distributed in the genome but most often located at the sub-telomeric regions (Fig 2.5a) (Palmer et al. 2010; Cairns and Meyer 2017). The SM biosynthetic genes

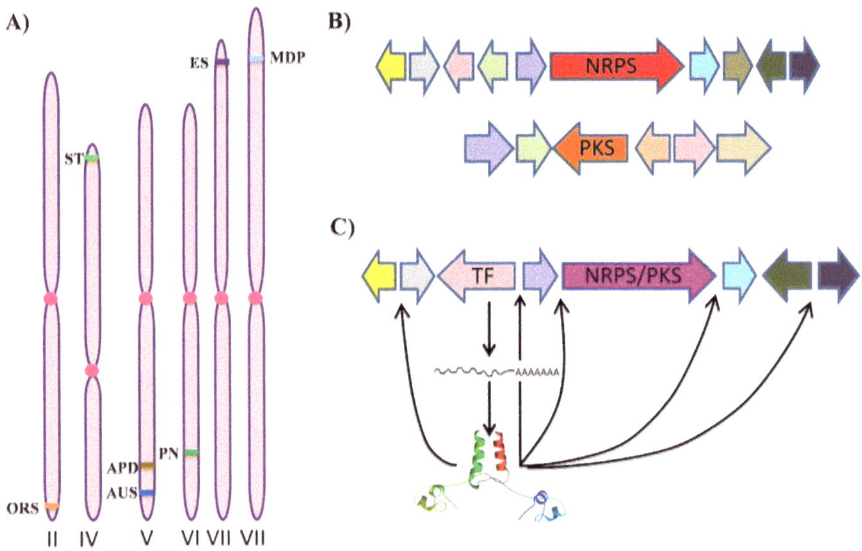

Fig. 2.5 Location, arrangement, and pathway-specific regulation of fungal SM biosynthetic genes. (**a**) Sub-telomeric arrangement of *A. nidulans* SM biosynthetic genes in the respective chromosomes. *ORS* orsellinic acid, *ST* sterigmatocystin, *APD* aspyridone, *AUS* austinol, *PN* penicillin, *ES* emericellamide, *MDP* monodictyphenone. (**b**) Physical arrangement of SM biosynthetic genes in the cluster along with NRPS/PKS. (**c**) Pathway-specific regulation of SM biosynthetic genes by the activation of in-cluster transcription factor

in fungi have special fetchers which are typically clustered in the genome and pooled to a single genetic locus (Fig 2.5b). The gene cluster encoded for SM is comprised of various functional genes. For example, in *A. nidulans*, each cluster contains 6–22 genes with a size ranging from 20 to 60 kb (William et al. 2005). It is expected that genes located in cluster could simplify the horizontal gene transfer and encourage the SM traits in nature (Khaldi and Wolfe 2011). It is also postulated that gene clustering could facilitate co-regulation of cluster's genes, specifically those located in the sub-telomeric regions (Palmer and Keller 2010). The cluster genes encode for NRPS/PKS, regulatory transcription factors, MFS/ABC transporters for translocation, and hydrolases and oxidases in particular for maturation of SM. Two complex proteins either non-ribosomal peptides (NRPS) or polyketides (PKS) are involved in building the general scaffolds of most SMs as described elsewhere (Palmer and Keller 2010; Brakhage 2013).

The architecture and arrangement of SM biosynthetic genes raise a question on their clustered form and sub-telomeric location. For example, there are 24 SM biosynthetic gene clusters present in the *A. fumigatus* genome. Out of 24 gene clusters, 8 were located in the sub-telomeric regions (William et al. 2005). Similarly, *Zymoseptoria tritici* contains 34 gene clusters; among them 12 are located at sub-telomeric regions (Cairns and Meyer 2017). In contrast, the sub-telomeric arrangement of the gene cluster may also influence another level of regulation incurred by telomeric positioning effect (TPE). TPE is a phenomenon in which telomeric or sub-telomeric region genes are transcriptionally repressed due to heterochromatin formation (Gottschling et al. 1990; Castaño et al. 2005; Rosas-Hernández et al. 2008; Smith et al. 2008). TPE covers up to 30-kb region of the telomere in *A. nidulans*. The expression for aflatoxin biosynthetic gene cluster is affected by TPE (Palmer and Keller 2010). The sub-telomeric occurrence of the gene cluster is also a signal for multi-step controlled regulation and production of SMs after stopping the primary metabolism or growth. Heterochromatin starts to turn into euchromatin at stationary phase, and SM gene clusters are more amenable to transcribe and turn on the SM biosynthesis.

2.4 Regulation of Fungal Secondary Metabolites (SMs)

The SM biosynthetic genes are often silent in the initial growth phase; SM genes are transcribed continuously when cells enter in the late exponential stage to the initial decline phase. The fungal SM biosynthetic genes are arranged in a cluster form with a size up to several hundred kilo base pairs (kb) long. The regulation of SM synthesis is either governed by a single or multiple transcriptional factors (TFs). Regulation of SM synthesis in fungi occurs either at pathway-specific level or global regulation level. The in-clustered TF gene located in a pathway-specific gene cluster plays a pivotal role in the SM regulation. AfoAAflR, SirZ, and GliZ are the examples of pathway-specific regulators (Chang et al. 1995; Bok et al. 2006a; Fox et al. 2008). On the other hand, the regulatory factors localized in the genome also respond to different environmental cues for SM synthesis. VeA, PacC, LaeA, and MeaB are the known fungal global regulators that monitor the expression of SM synthesis (Yin and Keller 2011; Brakhage 2013; Knox and Keller 2015).

2.4.1 Pathway-Specific Regulatory Proteins

Pathway-specific TFs are mostly present within the SM biosynthetic gene clusters, and they tend to regulate the expression of their own biosynthetic genes (Fig. 2.5c). Pathway-specific TFs are mostly Zn2Cys6 type which are abundantly present in fungal SM biosynthetic gene clusters (MacPherson et al. 2006; Chung et al. 2013). Nonetheless, TFs belonging to Cys2His2, bZIP, and winged helix are less common (Trapp et al. 1998; Kihara et al. 2008; Chang and Ehrlich 2013; Schmitt and Kück 2000; Hong et al. 2013). *AflR*, belonging to Zn2Cys6 family TF, specifically interacts with palindromic sequence (TCGN5CGR) of the promoters and activates aflatoxin and sterigmatocystin production in *Aspergillus* spp. (Ehrlich et al. 1999). Similarly, other pathway-specific regulators such as *GliZSirZ*, *ApdR*, *CtnA*, *LovE*, and *AfoA* also interact with the promoter sequence for their activation. Gliotoxin produced from *Aspergillus*, *Trichoderma*, and *Penicillium* spp. was shown to be regulated by an in-clustered TF *GliZ* in *Aspergillus fumigatus* (Bok et al. 2006a; Fox and Howlett 2008). Deletion of *GliZ* resulted in the loss of gliotoxin biosynthesis;

however, its induced expression accelerates the gliotoxin level (Bok et al. 2006a). The SM sirodesmin PL produced from *Leptosphaeria maculans* is regulated by in-clustered TF *SirZ* (Fox et al. 2008). Moreover, *ApdR*, *CtnA*, and *LovE* regulate the biosynthesis of cytotoxic metabolite aspyridone in *A. nidulans* (Bergmann et al. 2007), citrinin in *Monascus purpureus* (Shimizu et al. 2007), and lovastatin in *A. terreus*, respectively (Van Den Berg and Hans 2011). An anti-cancerous drug asperfuranone produced from *A. nidulans* was regulated by *AfoA* (Chiang et al. 2009).

*Tri*6 belonging to Cys2His2 zinc finger TF has a role in the regulation of trichothecene biosynthesis in *Fusarium* spp. (Proctor et al. 1995). The TFs *Bmr1*, *Cmr1*, *Cmr1p*, and *Pig1p*, present in *Bipolaris oryzae*, *Cochliobolus heterostrophus*, *Colletotrichum lagenarium*, and *Magnaporthe grisea*, respectively, belong to the Cys2His2 class that regulates the melanin biosynthesis (Tsuji et al. 2000; Kihara et al. 2008). Despite this, bZIP TFs were also reported in pathway-specific regulation in fungi. These TFs are characterized by the presence of basic and leucine zipper regions. The basic region comprises of basic amino acids (arginine and lysine) or hydrophobic residues involved in the sequence-specific DNA binding; this region is unstructured in the absence of DNA but forms an α-helical structure upon binding the DNA target site and is also considered as an essential domain. The leucine zipper (heptad repeats of leucine residues) region mediates dimerization of the protein (Amoutzias et al. 2006). TOXE, a unique TF present in the HC toxin gene cluster, regulates the HC toxin biosynthesis by interacting with TOX2 gene promoters in *Cochliobolus carbonum* (Pedley and Walton 2001).

2.4.2 Global Regulatory Proteins

It was well documented that different regulatory proteins, other than in-clustered TFs, are also involved in the regulation of fungal SM biosynthesis in response to environmental cues. These cues include carbon and nitrogen source, light, pH, and temperature (Fig. 2.6). These factors are triggered in response to SM production (Brakhage 2013; Macheleidt et al. 2016). For example, *CreA*, *AreA*, *PacC*, *LaeA*, and *HapX* are triggered in response to environmental carbon, nitrogen, pH, light, and iron starvation condition, respectively (Dowzer and Kelly 1991; Eisendle et al. 2004; Bayram et al. 2008; Schönig et al. 2008; Schrettl et al. 2010; Brakhage 2013).

2.4.2.1 Regulation by Nitrogen Source

Nitrogen is a crucial nutrient, required for growth and development. Microbes have the ability to metabolize a diverse range of nitrogen sources to colonize at diverse environmental niches and survive under nutrient limitations. Fungi most preferably utilize glutamine and ammonium, as favored nitrogen sources. In absence of these sources, non-preferred or complex nitrogen sources such as nitrate, uric acid, urea, amines, purines, pyrimidines, and amides may also be consumed (Wong et al. 2007). The selective utilization of complex nitrogenous sources is a state called nitrogen metabolite repression which activates the global regulatory circuit leading to transcriptional

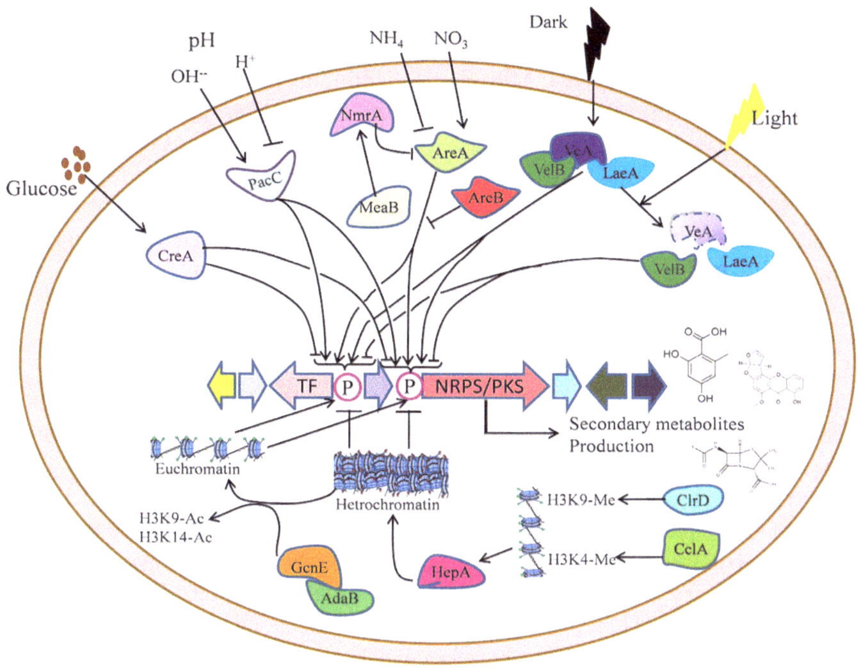

Fig. 2.6 Global regulatory proteins involved in the regulation of secondary metabolism. The NRPS- or PKS-mediated SMs are regulated in the different environmental cues such as light, pH carbohydrates, and carbon and nitrogenous source. The stimulus-responsive regulatory factors (global regulators) may also able to activate or repress the biosynthetic gene cluster of the SMs

activation of genes involved in the uptake and metabolism and ensures degradation (Marzluf 1997; Fraser et al. 2001; Magasanik and Kaiser 2002). The nitrogen metabolite repression is mediated by the GATA family TFs. Both the key regulatory genes named as *AreA* and *nit-2* in *A. nidulans* and *N. crassa*, respectively, are similar in sequence and closely related to each other. They contain Cys_2/Cys_2-zinc finger motifs, which preferentially bind to 5′HGATAR3′ DNA sequence (Kudla et al. 1990; Ravagnani et al. 1997). The activity of *AreA* is regulated by both extracellular and intracellular nitrogen levels (Caddick et al. 2006). At nitrogen-limiting conditions, the level of AreA gets up-regulated, while it was down-regulated at nitrogen-sufficient conditions (Tudzynski 2014). In another level of regulation, AreA was down-regulated when it interacts with co-repressor NmrA/Nmr1, whereas NmrA/Nmr1 is regulated by a bZIP TF MeaB in *A. nidulans*, *N. crassa*, and *F. fujikuroi* (Fig. 2.6) (Wong et al. 2007; Tudzynski 2014). AreB and NreB are also nitrogen-responsive regulators present in *A. nidulans* and *P. chrysogenum*, respectively. AreA/NIT2 is responsible for activation of non-preferred nitrogen-metabolizing genes, whereas AreB/NreB acts antagonistically and represses the same set of genes, most probably via DNA-binding competition (Haas et al. 1997; Wong et al. 2009; Tudzynski 2014).

In *F. fujikuroi*, the production of growth hormone gibberellic acid and red pigment bikaverin induced under nitrogen-limiting conditions is positively regulated by AreA (Bu'Lock et al. 1974; Mihlan et al. 2003). In *F. verticillioides*, fumonisins are regulated at nitrogen-repressing condition mediated by the *AreA* (Kim and Woloshuk 2008; Picot et al. 2010). In *F. graminearum*, mycotoxins deoxynivalenol (DON), zearalenone, and fusarielin H are produced at nitrogen-repressing condition under the control of AreA, whereas they are slightly negatively regulated by NmrA (Giese et al. 2013). Production of aflatoxin is mediated by nitrate repression in *A. parasiticus*, whereas in *A. flavus* it is mediated by ammonium induction with response to AreA (Feng and Leonard 1998; Ehrlich and Cotty 2002).

2.4.2.2 Regulation by Carbon Source

Carbon is an essential nutrient which drastically affects the growth and development of microorganisms. Fungi preferentially utilize simple sugars such as D-glucose, sucrose, D-xylose, and acetate. Moreover, other complex carbon sources such as fructose, cellulose, xylan, pectins, alcohols, amino acids, and glycerol were also utilized under non-availability of the simple sugars (Ruijter and Visser 1997). Cells grown with the simple sugars create a carbon catabolic repression which tends to repress the transcription of a gene, involved in the complex carbon catabolism to ensure preferential utilization of glucose.

CreA/Cre1 is a C2H2 type of key regulatory TF protein present in *A. nidulans* and *A. niger* (Tudzynski et al. 2000) which specifically binds to the 5′-SYGGRG-3′ DNA sequence in the promoters of glucose-repressible genes or complex carbon-catabolizing genes, leading to switching off of the expression in the presence of glucose (Dowzer and Kelly 1989; Flipphi et al. 2003). In *A. nidulans*, expression of xylanase (*xlnR*), cellulase, and arabinose encoding genes is also controlled by CreA (Ruijter and Visser 1997; Tamayo et al. 2008). Moreover, CreA also has a repressive role in SM biosynthesis in fungi (Fig. 2.6). The synthesis of penicillin was found to be repressed under sufficient glucose in the media, and also in response to CreA, it negatively regulates the transcription of *pcbAB*, *pcbC*, and *penDE* in *Penicillium chrysogenum* (Gutiérrez et al. 1999), not in *A. nidulans* (Espeso et al. 1993). The antibiotic pleuromutilin production was repressed in the presence of glucose, whereas it was enhanced by soybean oil in *Pleurotus mutilus* (Hu et al. 2009). Negative regulation of glucose on other SMs such as cephalosporin and lovastatin was also reported in *A. chrysogenum* and *A. terreus*, respectively (Jekosch and Kück 2000; Lai et al. 2007). However, unlike other SMs, AF production was induced by glucose (Wiseman and Buchanan 1987).

2.4.2.3 pH-Mediated Regulation

Extracellular pH is an important factor which influences the cell growth by alteration of the expression level of a variety of genes in fungi. The modulation of gene expression over a wide range of pH is controlled by the genetic regulatory system which is mediated by a global regulatory protein PacC (Caddick et al. 1986). The zinc finger PacC is constituently expressing protein, but full-length protein is inactive and unable to activate the responsive genes. The activation of PacC is mediated by the six pal gene products at alkaline ambient pH (Fig. 2.6). In acidic conditions, the full-length PacC (72-kD) protein predominantly is in a closed conformation in *A. nidulans*, but at alkaline to neutral pH, it goes through two rounds of proteolytic cleavage and yields a functional PacC 27 (27-kDa) TF. This PacC 27 activates the alkaline-expressed genes and represses the acid-expressed genes (Díez et al. 2002). A variety of SM biosynthetic genes are controlled by pH-mediated regulation (Fig. 2.6). In *Trichoderma virens*, 5% of the transcriptomes are pH-dependent; among these around 25% are PacC dependent, whereas secondary metabolism-related genes are predominantly regulated by PacC (Trushina et al. 2013).

SMs are mostly produced under acidic pH conditions, which are negatively regulated by PacC (Fig. 2.6). For example, a mycotoxin, fumonisin, is negatively regulated by Pac1, and its production was repressed under alkaline conditions in *F. verticillioides* (Flaherty et al. 2003). The mycotoxin, ochratoxin A, synthesized by *A. ochraceus* is another example which is synthesized under acidic pH conditions (O'Callaghan et al. 2006). On that note, the mycotoxins, aflatoxin and sterigmatocystin, induced their production by fivefold at pH 4–5 as compared to pH 8 in *Aspergillus* spp. (Keller et al. 1997). The mycotoxin deoxynivalenol (DON) regulates the production at low pH in *F. graminearum* (Gardiner et al. 2009). In contrast, PacC also positively regulates the SM production in few cases, such as penicillin biosynthesis at alkaline pH (Espeso et al. 1993).

2.4.2.4 Light-Mediated Regulation

Light is an information carrier of nature; the cellular/molecular machinery converts its electromagnetic energy (photons) into the chemical language and transmits their dynamic signals to cellular functions. The plants used it as an energy source, whereas other organisms used it as a source of information to execute biological signal for the growth and development. In fungi, light is the essential signal for asexual sporulation and induces the conidiation at red light, whereas it is suppressed at far-red light (Mooney and Yager 1990; Corrochano 2007). The reception and transmission of the light signal are modulated by the complex proteins, consisting of LOV (light-oxygen-voltage) sensing domain (that senses blue light) and PAS (Per-Arnt-Sim) domain (involved in the dimerization of proteins), which form a complex regulatory network (Ballario et al. 1998). In *N. crassa*, two major light response regulatory proteins are White Collar-1 (WC-1) and White Collar-2 (WC-2) that contain photoreceptors which dimerize by the exposure of blue light and act together as a TF complex (white collar complex, WCC). The homolog of WC-1 and WC-2 in *A. nidulans* is named as LreA (light response) and LreB (Ballario et al. 1998; Purschwitz et al. 2008). These photoreceptors activate the clock-controlled genes and regulate the circadian clock (Ballario et al. 1996; Crosthwaite et al. 1997). The LOV domain of WC-1 acts as a photoreceptor, responds to blue light, and binds to the flavin molecule (FAD or FMN). The blue light-induced conformational change eventually transduced the light signal to downstream biochemical signals (Crosson and Moffat 2002; Crosson et al. 2003). The downstream signals eventually activate the targeted proteins such as histone acetyltransferase NGF-1/Gcn5p which acetylates the H3K14 residues of light-inducible gene promoter wrap under histone proteins at heterochromatic mark (Grimaldi et al. 2006; Brenna et al. 2012). In dark conditions, WCC forms "dark" WCC which consists of WC-1/WC-2 heterodimer and binds to light-responsive elements (LREs) in the promoters of light-responsive genes, leading to the transcriptional repression (Talora et al. 1999; Froehlich et al. 2002). After elongated exposure to blue light, another photoreceptor PAS protein VIVID (VVD) accumulates to the threshold level and disrupts the WCC complex proteins, leading to switching off of the light-dependent transcription (Schwerdtfeger and Linden 2001, 2003). VVD is a white collar-dependent blue light photoreceptor which represses the light responses and also deals with the changes in light intensities (Schwerdtfeger and Linden 2001, 2003). The disruption of WCC complex is mediated by the conformational change of WC-1 protein by phosphorylation reaction leading to switching off of the gene (Schwerdtfeger and Linden 2000; He and Liu 2005). Sterigmatocystin (ST) biosynthesis in *A. nidulans* was suppressed on exposure of blue or white light and induced on red light exposure greater than the dark condition (Purschwitz et al. 2008).

Velvet (VeA) Family of Light-Responsive Global Regulators

Proteins under velvet (VeA) family are abundantly distributed in ascomycetes and basidiomycetes and play a vital role in the secondary metabolism (Ni and Yu 2007). VeA is a light-dependent regulatory protein; exposure of light governs its localization within the cell. During dark conditions, VeA accumulates in the nucleus and upon exposure of light mobilizes it to the cytoplasm (Stinnett et al. 2007; Bayram et al. 2008). The truncated VeA mutant protein does not respond to the light signals and is effectively unable to synthesize mycotoxin ST and antibiotic penicillin (Kato et al. 2003). VeA also has a crucial role in conidia formation, which is induced by red light (Mooney and Yager 1990). The heterotrimeric velvet complex VelB/VeA/LaeA is also involved in the light-dependent regulation of cellular developmental and secondary metabolism in *A. nidulans*. Upon light exposure, the cellular VeA level and its translocation to the nucleus both get reduced (Fig. 2.6).

The VelB (another developmental regulator like VeA) forms the dimer with VeA (VelB-VeA) in the cytoplasm and transports to the nucleus. VeA acts as a bridge between VelB and a secondary metabolism nuclear master regulator, LaeA. The deletion of either VeA or VelB results in impaired sexual fruiting body formation and SM production (Fig. 2.6) (Bayram et al. 2008; Purschwitz et al. 2008). The similar trimeric complexes have also been reported in other fungal species such as *F. oxysporum* and *P. chrysogenum* (Kopke et al. 2013; López-Berges et al. 2013). Other than these, VosA protein also mediates asexual differentiation via heterodimer formation with VeA or VelB. The heterodimer (VosA-VelB) formation results in inhibition of asexual differentiation in the dark and induces trehalose biosynthesis in spores to protect from stress (Ni and Yu 2007). In *A. nidulans* a sexual development positive regulator, VelC, has also been recently characterized (Park et al. 2014). FphA is another light-responsive transcription regulatory protein which senses red light. It interacts with the VeA, induces the asexual spore formation, and represses the fruiting body formation (Blumenstein et al. 2005; Purschwitz et al. 2009). LreA and LreB act as activators of the sexual cycle and are negatively regulated by light with the action of FphA (Purschwitz et al. 2008).

2.4.2.5 Chromatin-Mediated Regulation

The eukaryotic gene expression and SM biosynthesis are also controlled by another hierarchical level of regulation known as chromatin-mediated regulation. This primarily depends on the degree of chromatin compaction or chromosomal organization of biosynthetic genes. The fungal SM biosynthetic gene clusters are most often present in the telomeric region and usually not expressed in the normal conditions (Reyes-Dominguez et al. 2010; Strauss and Reyes-Dominguez 2011). Telomeres of the chromosomes are normally heterochromatic in nature, which get relaxed into the euchromatic form by the chemical modifications of nucleosomal histone amino acids. These modifications at the epigenetic level dramatically monitor the expression of SM biosynthetic genes (Bok et al. 2009). The majority of chemical modifications are achieved at the basic amino acids of histone proteins located in the nucleosome.

Histone Modifications

The large eukaryotic genome (10–100,000 Mb) is systematically compacted to several thousand folds in the nucleus (Hodgkin 2001). This compaction is mediated by histone proteins which bind to the DNA molecule in a systematic manner and are organized in the chromatin structure (Kornberg 1974). Chromatin consists of a repetitive unit of nucleosome which comprises of 147 bp of DNA molecule, which is wrapped around an octameric histone protein (H3/H4 and H2A/H2B) and thereby forms an 11-nm fiber structure (Thomas and Kornberg 1975; Luger 2003). The thread of nucleosome is further condensed into a 30-nm fiber by linker histone H1 (Brosch et al. 2008). The various associated proteins further condense to higher hierarchical level and form compact chromatin structure. Certain segments of the entire chromosomes become more condensed than the rest of the sections; the condensed region is termed as heterochromatin and the relaxed region as euchromatin. The heterochromatic state is mostly present at the telomeric and centromeric regions of the chromosome, whereas the euchromatic state is present in the rest of the region. The heterochromatic state prevents the transcription of genes, whereas the euchromatic state facilitates to interact with TFs (Sterner and Berger 2000). These core histones (H2A, H2B, H3, and H4) are conserved throughout the eukaryotes; these consist of a globular domain and versatile N-terminal tail (Luger 2003). N-terminal tail can be easily modified by the action of various post-translation modification enzymes as acetylation, phosphorylation, methylation, ubiquitination, ADP-ribosylation, and SUMOylation (Verdone et al. 2006; Brosch et al. 2008; Martinez-Zamudio and Ha 2012; Rossetto et al. 2012; Wilson and Hochstrasser 2016). However, methylation and acetylation have been widely involved in the chromatin remodeling and secondary metabolite biosynthetic gene regulation (Gacek and Strauss 2012).

Histone Methylation

Histone methylation inferred the transfer of a methyl group from S-adenosyl-L-methionine (SAM) to substrate proteins or specific amino acid residue. Methylation of histone proteins is specifically restricted to lysine and arginine at different positions and represents a much more complex language as compared to acetylation. The methylation of some residues such as H3K9, H3K27, and H3K4 is best studied in SM production (Fig. 2.6). Acetylation and methylation events on lysine do not occur simultaneously on the same residue (Gacek and Strauss 2012). The H3K9 in *A. nidulans* is strongly methylated by methyltransferase ClrD; di- and trimethylation of H3K9 subsequently recruits HepA and forms the condensed heterochromatin (Bannister et al. 2001; Reyes-Dominguez et al. 2010). In contrast, deletion of

ClrD in *A. nidulans* decreased the trimethylation state of H3K9 and subsequently resulted to loss of HepA recruitment in the ST cluster. Furthermore, deletion of HepA decreased the heterochromatin formation and increased the AflR expression (Reyes-Dominguez et al. 2010). In addition, H3K4 methylation is also associated with SM production; it occurred by CclA gene in *A. nidulans*. The deletion of CclA resulted in the reduction of H3K4 methylation along with H3K9 trimethylation in some SM gene promoters which increases the production of SMs F9775B and F9775A and expression of their associated genes (Fig. 2.6) (Bok et al. 2009). However, methyltransferase LaeA directly/indirectly induces the secondary metabolism and virulence in the *A. nidulans* and neutralizes the heterochromatic mark establishment. Though LaeA contains the methyltransferase domain, substrate or methyl-accepting residue has not been identified up to date, whereas it self-methylates methionine 207 in *A. nidulans*. Furthermore, self-methylation of methionine 207 did not show any role in its biological function (Patananan et al. 2013). The LaeA knockout strain of *A. nidulans* down-regulates the expression of ST, penicillin, and some indole alkaloid biosynthesis gene clusters in microarray experiments (Bok et al. 2006b; Yin and Keller 2011). Furthermore, LaeA knockout strain of *A. fumigatus* up-regulates the 22 secondary metabolism gene clusters and down-regulates the 13 SM gene clusters (Perrin et al. 2007).

Histone Acetylation

Acetylation of histone proteins under distinct physiological conditions also plays an important role in the regulation of secondary metabolism gene clusters. Histone acetylation marks are subjected to active transcription of genes at several positions; it occurs by histone acetyltransferase (HAT) enzymes, and these marks were removed by histone deacetylases (HDAC). The hyperacetylation of histone commonly responds to euchromatin formation, while hypoacetylation responds to heterochromatin formation and silences the gene transcription. Few histones involved in acetylation such as H3K9, H3K14, and H4K12 greatly influence the SM production and are acquired by multi-subunit Saga/Ada complexes or other associated histone acetyltransferases (Lan et al. 2016).

The GcnE and AdaB histone acetyltransferases are a part of SAGA-ADA complex and acetylate the H3K9 and H3K14 residue at the promoter region of orsellinic acid biosynthesis genes in fungal and bacterial interaction in *A. nidulans*. However, this acetylation mark is also detected in genes other than secondary metabolism gene clusters (Nützmann et al. 2011). In addition, the regulatory subunits of Saga/Ada, GcnE, and AdaB, respectively, are required for induction of conidiation in *A. nidulans* (Cánovas et al. 2014). Furthermore, the acetylation of H3K9 which enhances the orsellinic acid and sterigmatocystin biosynthesis gene clusters is triggered by *S. rapamycinicus* (Reyes-Dominguez et al. 2008). Targeted mutation of the potentially acetylated lysine residues 9, 14, 18, and 23 of histone H3 led to major changes in the transcriptional and metabolite levels of sterigmatocystin, orsellinic acid, and penicillin in *A. nidulans* (Nützmann et al. 2013). EsaA acetylates the H4K12 and enhances the sterigmatocystin, penicillin, terrequinone, and orsellinic acid production (Soukup et al. 2012). Furthermore the SM production is also enhanced by histone deacetylase (HDAC) inhibitor such as suberoylanilide hydroxamic acid (SAHA), trichostatin A, and 5-azacytidine (Fisch et al. 2009; Lim et al. 2012; Zutz et al. 2013; Albright et al. 2015), whereas SM production is blocked by histone acetyltransferase HAT inhibitor anacardic acid and curcumin (Netzker et al. 2015; Wee 2015).

For the activation of cryptic SM gene clusters, HDAC inhibitors (HDACi) have been widely used in fungi. These HDACi are valproic acid and trichostatin A (TSA), and suberoylanilide hydroxamic acid (SAHA) treatment effectively increases the SM production by chromatin modification in fungi (Henrikson et al. 2009; Zutz et al. 2013). In addition to the HDACi, histone demethylase inhibitor such as 5-azacytidine is also used in the SM modulation (Zutz et al. 2013).

References

Agathos S, Madhosingh C, Marshall J, Lee J (1987) The fungal production of cyclosporine A. Ann N Y Acad Sci 506:657–662

Albright JC et al (2015) Large-scale metabolomics reveals a complex response of *Aspergillus nidulans* to epigenetic perturbation. ACS Chem Biol 10:1535–1541

Alexander NJ, Proctor RH, McCormick SP (2009) Genes, gene clusters, and biosynthesis of trichothecenes and fumonisins in *Fusarium*. Toxin Rev 28:198–215

Amici AM, Minghetti A, Scotti T, Spalla C, Tognoli L (1967) Ergotamine production in submerged culture and physiology of *Claviceps purpurea*. Appl Microbiol 15:597–602

Amoutzias G et al (2006) One billion years of bZIP transcription factor evolution: conservation and change in dimerization and DNA-binding site specificity. Mol Biol Evol 24:827–835

Anaya AL, Cruz-Ortega R, Waller GR (2006) Metabolism and ecology of purine alkaloids. Front Biosci 11:2354

Aniszewski T (2015) Alkaloids: chemistry, biology, ecology, and applications. Elsevier, Amsterdam

Ballario P, Vittorioso P, Magrelli A, Talora C, Cabibbo A, Macino G (1996) White collar-1, a central regulator of blue light responses in *Neurospora*, is a zinc finger protein. EMBO J 15:1650–1657

Ballario P, Talora C, Galli D, Linden H, Macino G (1998) Roles in dimerization and blue light photoresponse of the PAS and LOV domains of *Neurospora crassa* white collar proteins. Mol Microbiol 29:719–729. https://doi.org/10.1046/j.1365-2958.1998.00955.x

Bannister AJ, Zegerman P, Partridge JF, Miska EA (2001) Selective recognition of methylated lysine 9 on histone H3 by the HP1 chromo domain. Nature 410:120

Bayram Ö et al (2008) VelB/VeA/LaeA complex coordinates light signal with fungal development and secondary metabolism. Science 320:1504–1506

Bergmann S, Schümann J, Scherlach K, Lange C, Brakhage AA, Hertweck C (2007) Genomics-driven discovery of PKS-NRPS hybrid metabolites from *Aspergillus nidulans*. Nat Chem Biol 3:nchembio869

Blumenstein A et al (2005) The *Aspergillus nidulans* phytochrome FphA represses sexual development in red light. Curr Biol 15:1833–1838

Bok JW et al (2006a) GliZ, a transcriptional regulator of gliotoxin biosynthesis, contributes to *Aspergillus fumigatus* virulence. Infect Immun 74:6761–6768

Bok JW, Hoffmeister D, Maggio-Hall LA, Murillo R, Glasner JD, Keller NP (2006b) Genomic mining for Aspergillus natural products. Chem Biol 13:31–37

Bok JW et al (2009) Chromatin-level regulation of biosynthetic gene clusters. Nat Chem Biol 5:462–464

Bömke C, Tudzynski B (2009) Diversity, regulation, and evolution of the gibberellin biosynthetic pathway in fungi compared to plants and bacteria. Phytochemistry 70:1876–1893

Brakhage AA (2013) Regulation of fungal secondary metabolism. Nat Rev Microbiol 11:21–32
Brenna A, Grimaldi B, Filetici P, Ballario P (2012) Physical association of the WC-1 photoreceptor and the histone acetyltransferase NGF-1 is required for blue light signal transduction in *Neurospora crassa*. Mol Biol Cell 23:3863–3872
Brosch G, Loidl P, Graessle S (2008) Histone modifications and chromatin dynamics: a focus on filamentous fungi. FEMS Microbiol Rev 32:409–439
Bu'Lock J (1961) Intermediary metabolism and antibiotic synthesis. Adv Appl Microbiol 3:293–342
Bu'Lock J, Detroy R, Hošťálek Z, Munim-Al-Shakarchi A (1974) Regulation of secondary biosynthesis in *Gibberella fujikuroi*. Trans Br Mycol Soc 62:377–389
Byrne KM, Smith SK, Ondeyka JG (2002) Biosynthesis of nodulisporic acid A: precursor studies. J Am Chem Soc 124:7055–7060
Cacho RA, Jiang W, Chooi Y-H, Walsh CT, Tang Y (2012) Identification and characterization of the echinocandin B biosynthetic gene cluster from *Emericella rugulosa* NRRL 11440. J Am Chem Soc 134:16781–16790
Caddick MX, Brownlee AG, Arst HN (1986) Regulation of gene expression by pH of the growth medium in *Aspergillus nidulans*. Mol Gen Genet MGG 203:346–353
Caddick MX et al (2006) Opposing signals differentially regulate transcript stability in *Aspergillus nidulans*. Mol Microbiol 62:509–519
Cairns T, Meyer V (2017) In silico prediction and characterization of secondary metabolite biosynthetic gene clusters in the wheat pathogen *Zymoseptoria tritici*. BMC Genomics 18:631. https://doi.org/10.1186/s12864-017-3969-y
Calne R et al (1989) Rapamycin for immunosuppression in organ allografting. Lancet 334:227
Calvo AM, Gardner HW, Keller NP (2001) Genetic connection between fatty acid metabolism and sporulation in *Aspergillus nidulans*. J Biol Chem 276:25766–25774
Calvo AM, Wilson RA, Bok JW, Keller NP (2002) Relationship between secondary metabolism and fungal development. Microbiol Mol Biol Rev 66:447–459
Cánovas D et al (2014) The histone acetyltransferase GcnE (GCN5) plays a central role in the regulation of Aspergillus asexual development. Genetics 197:1175–1189
Castaño I, Pan SJ, Zupancic M, Hennequin C, Dujon B, Cormack BP (2005) Telomere length control and transcriptional regulation of subtelomeric adhesins in Candida glabrata. Mol Microbiol 55:1246–1258
Champe SP, El-Zayat A (1989) Isolation of a sexual sporulation hormone from *Aspergillus nidulans*. J Bacteriol 171:3982–3988
Chang P-K, Ehrlich KC (2013) Genome-wide analysis of the Zn(II)2Cys6 zinc cluster-encoding gene family in *Aspergillus flavus*. Appl Microbiol Biotechnol 97:4289–4300
Chang P-K, Ehrlich KC, Yu J, Bhatnagar D, Cleveland TE (1995) Increased expression of *Aspergillus parasiticus* aflR, encoding a sequence-specific DNA-binding protein, relieves nitrate inhibition of aflatoxin biosynthesis. Appl Environ Microbiol 61:2372–2377
Chiang Y-M, Szewczyk E, Davidson AD, Keller N, Oakley BR, Wang CCC (2009) A gene cluster containing two fungal polyketide synthases encodes the biosynthetic pathway for a polyketide, Asperfuranone, in *Aspergillus nidulans*. J Am Chem Soc 131:2965–2970. https://doi.org/10.1021/ja8088185
Chung H, Choi J, Park S-Y, Jeon J, Lee Y-H (2013) Two conidiation-related Zn(II)2Cys6 transcription factor genes in the rice blast fungus. Fungal Genet Biol 61:133–141
Corrochano LM (2007) Fungal photoreceptors: sensory molecules for fungal development and behaviour. Photochem Photobiol Sci 6:725–736
Crosson S, Moffat K (2002) Photoexcited structure of a plant photoreceptor domain reveals a light-driven molecular switch. Plant Cell 14:1067–1075
Crosson S, Rajagopal S, Moffat K (2003) The LOV domain family: photoresponsive signaling modules coupled to diverse output domains. Biochem(Mosc) 42:2–10
Crosthwaite SK, Dunlap JC, Loros JJ (1997) *Neurospora* wc-1 and wc-2: transcription, photoresponses, and the origins of circadian rhythmicity. Sci 276:763–769

Darkin-Rattray SJ et al (1996) Apicidin: a novel antiprotozoal agent that inhibits parasite histone deacetylase. Proc Natl Acad Sci 93:13143–13147

Degenkolb T et al (2000) Roseoferin, a new aminolipopeptide antibiotic complex from Mycogone rosea DSM 12973, structures and biological activities. J Antibiot 53:184–190

Demain A (1999) Pharmaceutically active secondary metabolites of microorganisms. Appl Microbiol Biotechnol 52:455–463

Demain AL, Fang A (2000) The natural functions of secondary metabolites. In: History of modern biotechnology I. Springer, New York, pp 1–39

Dewick PM (2002) Medicinal natural products: a biosynthetic approach. Wiley, Chichester

Díez E et al (2002) Activation of the *Aspergillus* PacC zinc finger transcription factor requires two proteolytic steps. EMBO J 21:1350–1359

Dorner JW, Cole RJ, Diener UL (1984) The relationship of *Aspergillus flavus* and *Aspergillus parasiticus* with reference to production of aflatoxins and cyclopiazonic acid. Mycopathologia 87:13–15

Dowzer CE, Kelly JM (1989) Cloning of the creA gene from *Aspergillus nidulans*: a gene involved in carbon catabolite repression. Curr Genet 15:457–459

Dowzer C, Kelly JM (1991) Analysis of the *creA* gene, a regulator of carbon catabolite repression in *Aspergillus nidulans*. Mol Cell Biol 11:5701–5709

Dubouzet JG, Matsuda F, Ishihara A, Miyagawa H, Wakasa K (2013) Production of indole alkaloids by metabolic engineering of the tryptophan pathway in rice. Plant Biotechnol J 11:1103–1111

Echavarri-Erasun C, Johnson EA (2002) Fungal carotenoids. In: Applied mycology and biotechnology. Elsevier, Amsterdam, pp 45–85

Ehrlich K, Cotty P (2002) Variability in nitrogen regulation of aflatoxin production by *Aspergillus flavus* strains. Appl Microbiol Biotechnol 60:174–178

Ehrlich K, Montalbano B, Cary J (1999) Binding of the C6-zinc cluster protein, AFLR, to the promoters of aflatoxin pathway biosynthesis genes in *Aspergillus parasiticus*. Gene 230:249–257

Eisendle M, Oberegger H, Buttinger R, Illmer P, Haas H (2004) Biosynthesis and uptake of siderophores is controlled by the PacC-mediated ambient-pH regulatory system in *Aspergillus nidulans*. Eukaryot Cell 3:561–563

Eisenreich W, Bacher A, Arigoni D, Rohdich F (2004) Biosynthesis of isoprenoids via the non-mevalonate pathway. Cell Mol Life Sci 61:1401–1426

Espeso EA, Tilburn J, Arst H Jr, Penalva M (1993) pH regulation is a major determinant in expression of a fungal penicillin biosynthetic gene. EMBO J 12:3947–3956

Esser K, Tudzynski P (1978) Genetics of the ergot fungus *Claviceps purpurea*. Theor Appl Genet 53:145–149

Evans WC (2009) Trease and Evans' pharmacognosy e-book. Elsevier Health Sciences, New York

Evans BS, Robinson SJ, Kelleher NL (2011) Surveys of non-ribosomal peptide and polyketide assembly lines in fungi and prospects for their analysis in vitro and in vivo. Fungal Genet Biol 48:49–61

Feng GH, Leonard TJ (1998) Culture conditions control expression of the genes for aflatoxin and sterigmatocystin biosynthesis in *Aspergillus parasiticus* and *A. nidulans*. Appl Environ Microbiol 64:2275–2277

Fisch K et al (2009) Chemical induction of silent biosynthetic pathway transcription in *Aspergillus niger*. J Ind Microbiol Biotechnol 36:1199–1213

Fischbach MA, Walsh CT (2006) Assembly-line enzymology for polyketide and nonribosomal peptide antibiotics: logic, machinery, and mechanisms. Chem Rev 106:3468–3496

Flaherty JE, Pirttilä AM, Bluhm BH, Woloshuk CP (2003) PAC1, a pH-regulatory gene from *Fusarium verticillioides*. Appl Environ Microbiol 69:5222–5227. https://doi.org/10.1128/aem.69.9.5222-5227.2003

Flipphi M, Kocialkowska J, Felenbok B (2003) Relationships between the ethanol utilization (alc) pathway and unrelated catabolic pathways in *Aspergillus nidulans*. Eur J Biochem 270:3555–3564

Fox EM, Howlett BJ (2008) Biosynthetic gene clusters for epipolythiodioxopiperazines in filamentous fungi. Mycol Res 112:162–169

Fox EM, Gardiner DM, Keller NP, Howlett BJ (2008) A Zn(II)2Cys6 DNA binding protein regulates the sirodesmin PL biosynthetic gene cluster in *Leptosphaeria maculans*. Fungal Genet Biol 45:671–682. https://doi.org/10.1016/j.fgb.2007.10.005

Fraser JA, Davis MA, Hynes MJ (2001) The formamidase gene of *Aspergillus nidulans*: regulation by nitrogen metabolite repression and transcriptional interference by an overlapping upstream gene. Genetics 157:119–131

Froehlich AC, Liu Y, Loros JJ, Dunlap JC (2002) White Collar-1, a circadian blue light photoreceptor, binding to the frequency promoter. Science 297:815–819

Gacek A, Strauss J (2012) The chromatin code of fungal secondary metabolite gene clusters. Appl Microbiol Biotechnol 95:1389–1404

Galagan JE, Henn MR, Ma L-J, Cuomo CA, Birren B (2005) Genomics of the fungal kingdom: insights into eukaryotic biology. Genome Res 15:1620–1631. https://doi.org/10.1101/gr.3767105

Gallo A, Ferrara M, Perrone G (2013) Phylogenetic study of polyketide synthases and nonribosomal peptide synthetases involved in the biosynthesis of mycotoxins. Toxins 5:717–742

Gao X, Chooi Y-H, Ames BD, Wang P, Walsh CT, Tang Y (2011) Fungal indole alkaloid biosynthesis: genetic and biochemical investigation of the tryptoquialanine pathway in *Penicillium aethiopicum*. J Am Chem Soc 133:2729–2741

García-Estrada C et al (2011) A single cluster of coregulated genes encodes the biosynthesis of the mycotoxins roquefortine C and meleagrin in *Penicillium chrysogenum*. Chem Biol 18:1499–1512. https://doi.org/10.1016/j.chembiol.2011.08.012

Gardiner DM, Osborne S, Kazan K, Manners JM (2009) Low pH regulates the production of deoxynivalenol by *Fusarium graminearum*. Microbiology 155:3149–3156. https://doi.org/10.1099/mic.0.029546-0

Giese H, Sondergaard TE, Sørensen JL (2013) The AreA transcription factor in *Fusarium graminearum* regulates the use of some nonpreferred nitrogen sources and secondary metabolite production. Fungal Biol 117:814–821

Gottschling DE, Aparicio OM, Billington BL, Zakian VA (1990) Position effect at *S. cerevisiae* telomeres: reversible repression of Pol II transcription. Cell 63:751–762

Grimaldi B et al (2006) The *Neurospora crassa* White Collar-1 dependent blue light response requires acetylation of histone H3 lysine 14 by NGF-1. Mol Biol Cell 17:4576–4583

Grisham LM, Wilson L, BENSCH KG (1973) Antimitotic action of griseofulvin does not involve disruption of microtubules. Nature 244:294–296

Guirimand G, Courdavault V, St-Pierre B, Burlat V (2010) Biosynthesis and regulation of alkaloids. In: Plant developmental biology-biotechnological perspectives. Springer, Berlin, pp 139–160

Gutiérrez S et al (1999) Transcription of the pcbAB, pcbC and penDE genes of *Penicillium chrysogenum* AS-P-78 is repressed by glucose and the repression is not reversed by alkaline pHs. Microbiology 145:317–324

Haas H, Angermayr K, Zadra I, Stöffler G (1997) Overexpression of nreB, a new GATA factor-encoding gene of *Penicillium chrysogenum*, leads to repression of the nitrate assimilatory gene cluster. J Biol Chem 272:22576–22582

Harper JK et al (2003) Pestacin: a 1, 3-dihydro isobenzofuran from *Pestalotiopsis microspora* possessing antioxidant and antimycotic activities. Tetrahedron 59:2471–2476

Hashimoto M, Nonaka T, Fujii I (2014) Fungal type III polyketide synthases. Nat Prod Rep 31:1306–1317

Hayakawa S, Matsushima T, Kimura T, Minato H, Katagiri K (1968) Zygosporin A, a new antibiotic from *Zygosporium masonii*. J Antibiot 21:523–524

He Q, Liu Y (2005) Molecular mechanism of light responses in *Neurospora*: from light-induced transcription to photoadaptation. Genes Dev 19:2888–2899

Hellmig V, Grothe T, Mayer-Bartschmid A (2002) Altersetin, a new antibiotic from culture of endophytic Alternarla sp. taxonomy, fermentation, isolation, structure elucidation and biological activities. J Antibiot (Tokyo) 55:881–892

Henrikson JC, Hoover AR, Joyner PM, Cichewicz RH (2009) A chemical epigenetics approach for engineering the in situ biosynthesis of a cryptic natural product from *Aspergillus niger*. Org Biomol Chem 7:435–438

Hodgkin J (2001) Genome size. In: Brenner S, Miller JH (eds) Encyclopedia of genetics. Academic, New York, p 865

Hoffmeister D, Keller NP (2007) Natural products of filamentous fungi: enzymes, genes, and their regulation. Nat Prod Rep 24:393–416. https://doi.org/10.1039/b603084j

Hong S-Y, Roze LV, Linz JE (2013) Oxidative stress-related transcription factors in the regulation of secondary metabolism. Toxins 5:683–702

Hosokawa N et al (2000) New strobilurins O and P from a mushroom. J Antibiot 53:297–300

Hu C, Zou Y, Zhao W (2009) Effect of soybean oil on the production of mycelial biomass and pleuromutilin in the shake-flask culture of *Pleurotus mutilus*. World J Microbiol Biotechnol 25:1705–1711

Hu Y, Zhang W, Zhang P, Ruan W, Zhu X (2013) Nematicidal activity of Chaetoglobosin A poduced by *Chaetomium globosum* NK102 against *Meloidogyne incognita*. J Agric Food Chem 61:41–46. https://doi.org/10.1021/jf304314g

Hutchinson CR (2003) Polyketide and non-ribosomal peptide synthases: falling together by coming apart. Proc Natl Acad Sci 100:3010–3012

Igarashi Y et al (2000) Xanthoepocin, a new antibiotic from *Penicillium simplicissimum* IFO5762. J Antibiot 53:928–933

Jekosch K, Kück U (2000) Glucose dependent transcriptional expression of the cre1 gene in *Acremonium chrysogenum* strains showing different levels of cephalosporin C production. Curr Genet 37:388–395

Jensen PR, Williams PG, Oh D-C, Zeigler L, Fenical W (2007) Species-specific secondary metabolite production in marine actinomycetes of the genus *Salinispora*. Appl Environ Microbiol 73:1146–1152

Jin JM et al (2010) Functional characterization and manipulation of the apicidin biosynthetic pathway in Fusarium semitectum. Mol Microbiol 76:456–466

Kamei K, Watanabe A (2005) Aspergillus mycotoxins and their effect on the host. Med Mycol 43:95–99

Kato N, Brooks W, Calvo AM (2003) The expression of sterigmatocystin and penicillin genes in *Aspergillus nidulans* is controlled by veA, a gene required for sexual development. Eukaryot Cell 2:1178–1186

Keller NP (2015) Translating biosynthetic gene clusters into fungal armor and weaponry. Nat Chem Biol 11:671–677

Keller NP, Nesbitt C, Sarr B, Phillips TD, Burow GB (1997) pH regulation of sterigmatocystin and aflatoxin biosynthesis in *Aspergillus* spp. Phytopathology 87:643–648. https://doi.org/10.1094/phyto.1997.87.6.643

Keller NP, Turner G, Bennett JW (2005) Fungal secondary metabolism—from biochemistry to genomics. Nat Rev Microbiol 3:937–947

Kelly WL, Hillson NJ, Walsh CT (2005) Excision of the epothilone synthetase B cyclization domain and demonstration of in trans condensation/cyclodehydration activity. Biochem (Mosc) 44:13385–13393

Khaldi N, Wolfe KH (2011) Evolutionary origins of the fumonisin secondary metabolite gene cluster in *Fusarium verticillioides* and *Aspergillus niger*. Int J Evol Biol 2011 pp 1-7. doi:10.4061/2011/423821

Kihara J, Moriwaki A, Tanaka N, Tanaka C, Ueno M, Arase S (2008) Characterization of the BMR1 gene encoding a transcription factor for melanin biosynthesis genes in the phytopathogenic fungus *Bipolaris oryzae*. FEMS Microbiol Lett 281:221–227

Kim H, Woloshuk C (2008) Role of AREA, a regulator of nitrogen metabolism, during colonization of maize kernels and fumonisin biosynthesis in *Fusarium verticillioides*. Fungal Genet Biol 45:947–953

Kleinwaechter P, Schlegel B, Doerfelt H, Graefe U (2001) Spirobenzofuran, a new bioactive metabolite from Acremonium sp. HKI 0230. J Antibiot 54:526–527

Knox BP, Keller NP (2015) Key players in the regulation of fungal secondary metabolism. In: Zeilinger S, Martín J-F, García-Estrada C (eds) Biosynthesis and molecular genetics of fungal secondary metabolites, vol 2. Springer, New York, pp 13–28

Kopke K, Hoff B, Bloemendal S, Katschorowski A, Kamerewerd J, Kück U (2013) Members of the *Penicillium chrysogenum* velvet complex play functionally opposing roles in the regulation of penicillin biosynthesis and conidiation. Eukaryot Cell 12:299–310

Kornberg RD (1974) Chromatin structure: a repeating unit of histones and DNA. Science 184:868–871

Kudla B et al (1990) The regulatory gene *areA* mediating nitrogen metabolite repression in *Aspergillus nidulans*. Mutations affecting specificity of gene activation alter a loop residue of a putative zinc finger. EMBO J 9:1355–1364

Kumar A, Jaiswal V, Kumar V, Dey A, Kumar A (2018) Functional redundancy in Echinocandin B in-cluster transcription factor ecdB of *Emericella rugulosa* NRRL 11440. Biotechnol Rep 19:e00264. https://doi.org/10.1016/j.btre.2018.e00264

Lai L-ST, Hung C-S, Lo C-C (2007) Effects of lactose and glucose on production of itaconic acid and lovastatin by *Aspergillus terreus* ATCC 20542. J Biosci Bioeng 104:9–13. https://doi.org/10.1263/jbb.104.9

Lan H et al (2016) The *Aspergillus flavus* histone acetyltransferase AflGcnE regulates morphogenesis, aflatoxin biosynthesis, and pathogenicity. Front Microbiol 7:1324

Lim FY, Sanchez JF, Wang CC, Keller NP (2012) Toward awakening cryptic secondary metabolite gene clusters in filamentous fungi. Methods Enzymol 517:303

Lin H-C et al (2015) Elucidation of the concise biosynthetic pathway of the communesin indole alkaloids. Angew Chem Int Ed 54:3004–3007. https://doi.org/10.1002/anie.201411297

Linne U, Doekel S, Marahiel MA (2001) Portability of epimerization domain and role of peptidyl carrier protein on epimerization activity in nonribosomal peptide synthetases. Biochemistry (Mosc) 40:15824–15834. https://doi.org/10.1021/bi011595t

Liu L, Zhang Z, Shao C-L, Wang J-L, Bai H, Wang C-Y (2015) Bioinformatical analysis of the sequences, structures and functions of fungal polyketide synthase product template domains. Sci Rep 5:10463. https://doi.org/10.1038/srep10463

López-Berges MS et al (2013) The velvet complex governs mycotoxin production and virulence of *Fusarium oxysporum* on plant and mammalian hosts. Mol Microbiol 87:49–65

Luger K (2003) Structure and dynamic behavior of nucleosomes. Curr Opin Genet Dev 13:127–135. https://doi.org/10.1016/S0959-437X(03)00026-1

Macheleidt J et al (2016) Regulation and role of fungal secondary metabolites. Annu Rev Genet 50:371–392

MacPherson S, Larochelle M, Turcotte B (2006) A fungal family of transcriptional regulators: the zinc cluster proteins. Microbiol Mol Biol Rev 70:583–604

Magasanik B, Kaiser CA (2002) Nitrogen regulation in *Saccharomyces cerevisiae*. Gene 290:1–18

Mahmood ZA, Ahmed SW, Azhar I, Sualeh M, Baig MT, Zoha S (2010) Bioactive alkaloids produced by fungi I. Updates on alkaloids from the species of the genera boletus, Fusarium and psilocybe. Pak J Pharm Sci 23:349–357

Mann V et al (1994) Complex I, iron, and ferritin in Parkinson's disease substantia nigra. Ann Neurol 36:876–881

Marshall CG, Burkart MD, Keating TA, Walsh CT (2001) Heterocycle formation in vibriobactin biosynthesis: alternative substrate utilization and identification of a condensed intermediate. Biochemistry (Mosc) 40:10655–10663

Martinez-Zamudio R, Ha HC (2012) Histone ADP-ribosylation facilitates gene transcription by directly remodeling nucleosomes. Mol Cell Biol. https://doi.org/10.1128/MCB.06667-11

Marzluf GA (1997) Genetic regulation of nitrogen metabolism in the fungi. Microbiol Mol Biol Rev 61:17–32

Mihlan M, Homann V, Liu TWD, Tudzynski B (2003) AREA directly mediates nitrogen regulation of gibberellin biosynthesis in *Gibberella fujikuroi*, but its activity is not affected by NMR. Mol Microbiol 47:975–991

Mooney JL, Yager LN (1990) Light is required for conidiation in *Aspergillus nidulans*. Genes Dev 4:1473–1482

Moss MO (2002) Mycotoxin review-1. Aspergillus and penicillium. Mycologist 16:116–119

Netzker T et al (2015) Microbial communication leading to the activation of silent fungal secondary metabolite gene clusters. Front Microbiol 6:299

Ni M, Yu J-H (2007) A novel regulator couples sporogenesis and trehalose biogenesis in *Aspergillus nidulans*. PLoS One 2:e970

Noh MJ et al (1999) Isolation of a novel microorganism, *Pestalotia heterocornis*, producing paclitaxel. Biotechnol Bioeng 64:620–623

Nützmann H-W et al (2011) Bacteria-induced natural product formation in the fungus Aspergillus nidulans requires Saga/Ada-mediated histone acetylation. Proc Natl Acad Sci 108:14282–14287

Nützmann H-W, Fischer J, Scherlach K, Hertweck C, Brakhage AA (2013) Distinct amino acids of histone H3 control secondary metabolism in *Aspergillus nidulans*. Appl Environ Microbiol 79:6102–6109

O'Callaghan J, Stapleton PC, Dobson ADW (2006) Ochratoxin A biosynthetic genes in *Aspergillus ochraceus* are differentially regulated by pH and nutritional stimuli. Fungal Genet Biol 43:213–221. https://doi.org/10.1016/j.fgb.2005.11.005

Omura S et al (1995) Arisugacin, a novel and selective inhibitor of acetylcholinesterase from Penicillium sp. FO-4259. J Antibiot 48:745–746

Ondeyka JG et al (1997) Nodulisporic acid A, a novel and potent insecticide from a Nodulisporium sp. isolation, structure determination, and chemical transformations. J Am Chem Soc 119:8809–8816

Pahl HL et al (1996) The immunosuppressive fungal metabolite gliotoxin specifically inhibits transcription factor NF-kappaB. J Exp Med 183:1829–1840

Palmer JM, Keller NP (2010) Secondary metabolism in fungi: does chromosomal location matter? Curr Opin Microbiol 13:431–436

Palmer JM et al (2010) Telomere position effect is regulated by heterochromatin-associated proteins and NkuA in *Aspergillus nidulans*. Microbiology 156:3522–3531

Park H-S, Nam T-Y, Han K-H, Kim SC, Yu J-H (2014) VelC positively controls sexual development in *Aspergillus nidulans*. PLoS One 9:e89883

Patananan AN, Palmer JM, Garvey GS, Keller NP, Clarke SG (2013) A novel automethylation reaction in the *Aspergillus nidulans* LaeA protein generates S-methylmethionine. J Biol Chem 288:14032–14045. https://doi.org/10.1074/jbc.M113.465765

Pedley KF, Walton JD (2001) Regulation of cyclic peptide biosynthesis in a plant pathogenic fungus by a novel transcription factor. Proc Natl Acad Sci 98:14174–14179

Perrin RM et al (2007) Transcriptional regulation of chemical diversity in *Aspergillus fumigatus* by LaeA. PLoS Pathog 3:e50

Picot A, Barreau C, Pinson-Gadais L, Caron D, Lannou C, Richard-Forget F (2010) Factors of the *Fusarium verticillioides*-maize environment modulating fumonisin production. Crit Rev Microbiol 36:221–231

Proctor RH, Hohn TM, McCormick SP, Desjardins AE (1995) Tri6 encodes an unusual zinc finger protein involved in regulation of trichothecene biosynthesis in Fusarium sporotrichioides. Appl Environ Microbiol 61:1923–1930

Purschwitz J et al (2008) Functional and physical interaction of blue-and red-light sensors in *Aspergillus nidulans*. Curr Biol 18:255–259

Purschwitz J, Müller S, Fischer R (2009) Mapping the interaction sites of *Aspergillus nidulans* phytochrome FphA with the global regulator VeA and the White Collar protein LreB. Mol Gen Genomics 281:35–42

Ramawat KG, Mérillon J-M (2013) Natural products: phytochemistry, botany and metabolism of alkaloids, phenolics and terpenes. Springer, Berlin/Heidelberg

Raper KB (1946) The development of improved penicillin-producing molds. Ann N Y Acad Sci 48:41–56

Ravagnani A et al (1997) Subtle hydrophobic interactions between the seventh residue of the zinc finger loop and the first base of an HGATAR sequence determine promoter-specific recognition by the *Aspergillus nidulans* GATA factor AreA. EMBO J 16:3974–3986

Reyes-Dominguez Y et al (2008) Nucleosome positioning and histone H3 acetylation are independent processes in the *Aspergillus nidulans* prnD-prnB bidirectional promoter. Eukaryot Cell 7:656–663

Reyes-Dominguez Y et al (2010) Heterochromatic marks are associated with the repression of secondary metabolism clusters in *Aspergillus nidulans*. Mol Microbiol 76:1376–1386

Roberts MF (2013) Alkaloids: biochemistry, ecology, and medicinal applications. Springer, New York

Rosas-Hernández LL et al (2008) yKu70/yKu80 and Rif1 regulate silencing differentially at telomeres in Candida glabrata. Eukaryot Cell 7:2168–2178

Rossetto D, Avvakumov N, Côté J (2012) Histone phosphorylation: a chromatin modification involved in diverse nuclear events. Epigenetics 7:1098–1108

Ruijter GJ, Visser J (1997) Carbon repression in *Aspergilli*. FEMS Microbiol Lett 151:103–114

Sándor E et al (1998) Allosamidin inhibits the fragmentation of *Acremonium chrysogenum* but does not influence the cephalosporin-C production of the fungus. FEMS Microbiol Lett 164:231–236

Schmitt EK, Kück U (2000) The fungal CPCR1 protein, which binds specifically to β-lactam biosynthesis genes, is related to human regulatory factor X transcription factors. J Biol Chem 275:9348–9357

Schönig B, Brown DW, Oeser B, Tudzynski B (2008) Cross-species hybridization with Fusarium verticillioides microarrays reveals new insights into *Fusarium fujikuroi* nitrogen regulation and the role of AreA and NMR. Eukaryot Cell 7:1831–1846

Schrettl M et al (2010) HapX-mediated adaption to iron starvation is crucial for virulence of *Aspergillus fumigatus*. PLoS Pathog 6:e1001124

Schwerdtfeger C, Linden H (2000) Localization and light-dependent phosphorylation of white collar 1 and 2, the two central components of blue light signaling in *Neurospora crassa*. Eur J Biochem 267:414–422

Schwerdtfeger C, Linden H (2001) Blue light adaptation and desensitization of light signal transduction in *Neurospora crassa*. Mol Microbiol 39:1080–1087

Schwerdtfeger C, Linden H (2003) VIVID is a flavoprotein and serves as a fungal blue light photoreceptor for photoadaptation. EMBO J 22:4846–4855

Scott P, Wv W, Harwig J, Fennell D (1970) Occurrence of a mycotoxin, ochratoxin A, in wheat and isolation of ochratoxin A and citrinin producing strains of *Penicillium viridicatum*. Can J Plant Sci 50:583–585

Scott B et al (2013) Deletion and gene expression analyses define the paxilline biosynthetic gene cluster in *Penicillium paxilli*. Toxins 5:1422–1446

Sekiguchi J, Gaucher GM (1977) Conidiogenesis and secondary metabolism in *Penicillium urticae*. Appl Environ Microbiol 33:147–158

Shen B (2003) Polyketide biosynthesis beyond the type I, II and III polyketide synthase paradigms. Curr Opin Chem Biol 7:285–295

Shimizu T, Kinoshita H, Nihira T (2007) Identification and in vivo functional analysis by gene disruption of ctnA, an activator gene involved in citrinin biosynthesis in *Monascus purpureus*. Appl Environ Microbiol 73:5097–5103

Smith KM et al (2008) The fungus *Neurospora crassa* displays telomeric silencing mediated by multiple sirtuins and by methylation of histone H3 lysine 9. Epigenetics Chromatin 1:1–20. https://doi.org/10.1186/1756-8935-1-5

Soukup AA et al (2012) Overexpression of the *Aspergillus nidulans* histone 4 acetyltransferase EsaA increases activation of secondary metabolite production. Mol Microbiol 86:314–330

Staunton J, Weissman KJ (2001) Polyketide biosynthesis: a millennium review. Nat Prod Rep 18:380–416

Stein T, Vater J, Kruft V, Otto A, Wittmann-Liebold B, Franke P, Panico M, McDowell R, Morris HR (1996) The multiple carrier model of nonribosomal peptide biosynthesis at modular multienzymatic templates. J Biol Chem 271:15428–15435

Sterner DE, Berger SL (2000) Acetylation of histones and transcription-related factors. Microbiol Mol Biol Rev 64:435–459

Stierle A, Strobel G, Stierle D (1993) Taxol and taxane production by Taxomyces andreanae, an endophytic fungus of Pacific yew. Science 260:214–216

Stinnett SM, Espeso EA, Cobeño L, Araújo-Bazán L, Calvo AM (2007) *Aspergillus nidulans* VeA subcellular localization is dependent on the importin α carrier and on light. Mol Microbiol 63:242–255

Strauss J, Reyes-Dominguez Y (2011) Regulation of secondary metabolism by chromatin structure and epigenetic codes. Fungal Genet Biol 48:62–69. https://doi.org/10.1016/j.fgb.2010.07.009

Strieker M, Tanović A, Marahiel MA (2010) Nonribosomal peptide synthetases: structures and dynamics. Curr Opin Struct Biol 20:234–240. https://doi.org/10.1016/j.sbi.2010.01.009

Strobel GA, Miller RV, Martinez-Miller C, Condron MM, Teplow DB, Hess WM (1999) Cryptocandin, a potent antimycotic from the endophytic fungus *Cryptosporiopsis* cf. *quercina*. Microbiology 145:1919–1926. https://doi.org/10.1099/13500872-145-8-1919

Talora C, Franchi L, Linden H, Ballario P, Macino G (1999) Role of a white collar-1-white collar-2 complex in blue-light signal transduction. EMBO J 18:4961–4968

Tamayo EN, Villanueva A, Hasper AA, de Graaff LH, Ramón D, Orejas M (2008) CreA mediates repression of the regulatory gene xlnR which controls the production of xylanolytic enzymes in *Aspergillus nidulans*. Fungal Genet Biol 45:984–993

Thomas JO, Kornberg RD (1975) An octamer of histones in chromatin and free in solution. Proc Natl Acad Sci 72:2626–2630

Trapp S, Hohn T, McCormick S, Jarvis B (1998) Characterization of the gene cluster for biosynthesis of macrocyclic trichothecenes in *Myrothecium roridum*. Mol Gen Genet MGG 257:421–432

Trushina N, Levin M, Mukherjee PK, Horwitz BA (2013) PacC and pH–dependent transcriptome of the mycotrophic fungus *Trichoderma virens*. BMC Genomics 14:138

Tsuji G et al (2000) Novel fungal transcriptional activators, Cmr1p of *Colletotrichum lagenarium* and Pig1p of *Magnaporthe grisea*, contain Cys2His2 zinc finger and Zn (II) 2Cys6 binuclear cluster DNA-binding motifs and regulate transcription of melanin biosynthesis genes in a developmentally specific manner. Mol Microbiol 38:940–954

Tudzynski B (1999) Biosynthesis of gibberellins in *Gibberella fujikuroi*: biomolecular aspects. Appl Microbiol Biotechnol 52:298–310

Tudzynski B (2014) Nitrogen regulation of fungal secondary metabolism in fungi. Front Microbiol 5:656

Tudzynski B, Liu S, Kelly JM (2000) Carbon catabolite repression in plant pathogenic fungi: isolation and characterization of the *Gibberella fujikuroi* and *Botrytis cinerea* creA genes. FEMS Microbiol Lett 184:9–15

Umezawa H, Maeda K, Takeuchi T, Okami Y (1966) New antibiotics, bleomycin A and B. J Antibiot 19:200

Van Den Berg MA, Hans M (2011) Improved statin production. In DSM IP Assets BV, United States

Verdone L, Agricola E, Caserta M, Di Mauro E (2006) Histone acetylation in gene regulation. Brief Funct Genomics 5:209–221

Vertesy L, Kurz M, Schiell M, Hofmann J (2003) Cephaibols: novel antiparasitics from *Acremonium tubakii* process for their production, and use thereof. In Google Patents

Wakimoto T, Mori T, Morita H, Abe I (2011) Cytotoxic tetramic acid derivative produced by a plant type-III polyketide synthase. J Am Chem Soc 133:4746–4749

Waller GR, Burström H (1969) Diterpenoid alkaloids as plant growth inhibitors. Nature 222:576. https://doi.org/10.1038/222576a0

Walsh CT et al (2001) Tailoring enzymes that modify nonribosomal peptides during and after chain elongation on NRPS assembly lines. Curr Opin Chem Biol 5:525–534

Wansi JD, Devkota KP, Tshikalange E, Kuete V (2013) 14 – Alkaloids from the medicinal plants of Africa. In: Kuete V (ed) Medicinal plant research in Africa. Elsevier, Oxford, pp 557–605

Wee J (2015) Regulation and subcellular localization of aflatoxin biosynthesis in *Aspergillus parasiticus*. Michigan State University, Michigan

Wei S, Zhao F, Jiang Z, Zhou D (2013) Microbial conversion of waste cooking oil into riboflavin by *Ashbya gossypii* [Conversão microbiana de óleo de cozinha recolhido em riboflavina por Ashbya gossypii]. Biosci J 29:1000–1006

Weissman KJ (2009) Introduction to polyketide biosynthesis. Methods Enzymol 459:3–16

William CN et al (2005) Secondary metabolite biosynthetic gene clusters in filamentous fungi. In: Fungal genetics conference 23rd. Fungal Genetics Newsletter, California, p 190

Wilson NR, Hochstrasser M (2016) The regulation of chromatin by dynamic SUMO modifications. In: SUMO. Springer, New York, pp 23–38

Wiseman DW, Buchanan RL (1987) Determination of glucose level needed to induce aflatoxin production in *Aspergillus parasiticus*. Can J Microbiol 33:828–830

Wolf J, Mirocha C (1973) Regulation of sexual reproduction in *Gibberella zeae* (*Fusarium roseum 'Graminearum'*) by F-2 (zearalenone). Can J Microbiol 19:725–734

Wong KH, Hynes MJ, Todd RB, Davis MA (2007) Transcriptional control of nmrA by the bZIP transcription factor MeaB reveals a new level of nitrogen regulation in *Aspergillus nidulans*. Mol Microbiol 66:534–551

Wong KH, Hynes MJ, Todd RB, Davis MA (2009) Deletion and overexpression of the *Aspergillus nidulans* GATA factor AreB reveals unexpected pleiotropy. Microbiology 155:3868–3880

Xu W, Gavia DJ, Tang Y (2014) Biosynthesis of fungal indole alkaloids. Nat Prod Rep 31:1474–1487

Yin W, Keller NP (2011) Transcriptional regulatory elements in fungal secondary metabolism. J Microbiol 49:329–339

Ziegler J, Facchini PJ (2008) Alkaloid biosynthesis: metabolism and trafficking. Annu Rev Plant Biol 59:735–769

Zutz C, Gacek A, Sulyok M, Wagner M, Strauss J, Rychli K (2013) Small chemical chromatin effectors alter secondary metabolite production in *Aspergillus clavatus*. Toxins 5:1723–1741

Microbial-Based Cancer Therapy: Diagnostic Tools and Therapeutic Strategies

3

Saima Shakil Malik, Nosheen Masood, Iffat Fatima, and Zehra Kazmi

Abstract

It has been more than a century that researchers are focusing on microbial-based cancer treatment and trying to explore novel approaches that can overcome the devastating effects associated with conventional cancer treatments. Great effort has been put to develop live, attenuated, or engineered bacterial strains that can treat cancer by inhibiting or reducing tumor growth. However, uses of bacterial strains have been associated with infections, toxicity, and other harmful effects. Therefore, scientists move toward the idea of using specifically engineered microbial strains and their protein products and compounds that have specific antitumor activity with no or minimal side effects and can be used as antitumor agents. In this way attenuated bacterial strains with embedded foreign genes have been used to convert nontoxic pro-drugs to cytotoxic drugs. Other novel microbial-based cancer treatment modalities include the use of bacterial products like enzymes, proteins, secondary metabolites, and immunotoxins which are capable of targeting cancer cells and reducing/inhibiting their growth either by cell cycle arrest or inducing apoptosis. Current chapter focuses on the available knowledge and future directions regarding microbial-based cancer treatment.

Keywords

Microbiome · Chemotherapy · Attenuated bacteria · Azurin

S. S. Malik (✉)
Fatima Jinnah Women University, Rawalpindi, Pakistan

Armed Forces Institute of Pathology, Rawalpindi, Pakistan

N. Masood · Z. Kazmi
Fatima Jinnah Women University, Rawalpindi, Pakistan

I. Fatima
Quaid-i-Azam University, Islamabad, Pakistan

P. K. Arora (ed.), *Microbial Technology for the Welfare of Society*,
Microorganisms for Sustainability 17,
https://doi.org/10.1007/978-981-13-8844-6_3

3.1 Introduction

Cancer is a crucial public health issue and the second leading cause of cancer-related morbidity and mortality throughout the world. According to an estimate, 1,735,350 cancer cases will be diagnosed only in the United States by this year which means diagnosis of approximately forty-seven hundred new cases per day (Siegel et al. 2018). One of the main reasons behind continuously rising burden is promptly aging population and demands a comprehensible and synchronized response from researchers, legislators, oncologists, and public health professionals (Bernardes et al. 2010). Tumor kinetics or growth depends on various closely linked factors including cell cycle time, i.e., average time of mitotically dividing cell, growth fraction, and total cancer burden (White and Pharoah 2014). Different tumors have variable growth rate due to variations in these factors. Tumors with small volumes tend to grow rapidly and cell proliferation becomes slow with increased tumor burden due to complex process dependent on oxygen and blood supply along with cell loss (McCance and Huether 2018). Conventional cancer treatments like chemotherapy, surgery, and radiotherapy frequently fail to accomplish complete cancer remission and have been reported to have numerous side effects (Miller et al. 2016). This has triggered researchers to develop many other approaches for cancer treatment.

Microbes have inestimable metabolic potential and can adapt to an immense range of different environments therefore; complex relationship exists between microbes and cancer. Cancer is considered as a disease of environmental and host genetics, but it was found that microorganisms are involved in more than twenty percent of human malignancies (De Martel et al. 2012). Microorganisms residing at mucosal sites have potential to cause aerodigestive tract cancers by interacting with tumor microenvironment, and intratumoral microorganisms can affect growth of cancerous cells and metastasize into different body parts (Schwabe and Jobin 2013; Sears and Garrett 2014; Irrazábal et al. 2014). Gut microbiota is also involved in reducing inflammation, detoxifying dietary components, and maintaining a proper balance between growth and proliferation of host cell (Fig. 3.1) (Garrett 2015; Marchesi et al. 2015). Microbial-based cancer treatment has emerged from William Coley's discovery that advanced-stage metastatic cancer can be treated with bacterial extracts comprising of *Streptococcus pyogenes* and *Serratia marcescens* (Cann et al. 2003). Since then a large number of studies have reported correlation between tumor regression and live/attenuated bacteria (Garrett 2015). Successful cancer treatment can be achieved if bacteria are nontoxic, are able to replicate in cancerous cells, are nonimmunogenic, and can efficiently lyse tumor cells (Luginbuehl et al. 2018). Scientists have made significant progress in microbial-based cancer treatment and enter the era of designer microbes using synthetic biology (Hosseinidoust et al. 2016). Genetic predispositions also lead toward the increased risk of cancer development (Lukasiewicz and Fol 2018). Therefore, a holistic approach is required for the investigation of microbiota in cancer treatment.

Cancer research focuses on prevention, early diagnosis, and development of new therapies leading to personalized medicine targeting patient's immune system and

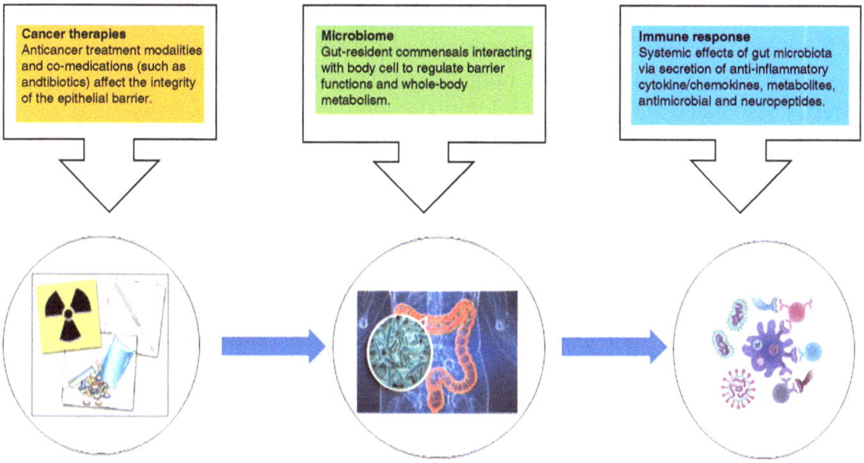

Fig 3.1 Microbiome at the crossroads between cancer pathophysiology

molecular biology of respective tumor (Friedman et al. 2015; Sharma and Allison 2015; Kensler et al. 2016). Microbial-based cancer treatment is one of the most popular and hazard-free anticancer therapies. Therefore, this chapter discusses the association of microbiome with cancer; conventional cancer therapies; adjuvant therapy with bacteria; role of live, attenuated, and engineered bacterial strains in cancer treatment; bacteria-derived anticancer agents; and complications associated with microbial contaminations during hematopoietic transplants. Focus of the chapter is to explore therapeutic potential of microorganisms and their use in clinical trials.

3.2 Microbiome and Cancer

When microbes interact with host immune system due to breaching of mucosal barriers, interaction of microbes can result in immunosuppressive or pro-inflammatory response. Various inflammatory factors are responsible for inducing tumor growth and regulating different receptors which play a role in activation of NF-κB. Various microorganisms are also reported to induce NF-κB signaling which is the main regulatory body of different cancers caused by inflammation (Garrett 2015). Microbial interaction in a tumor environment can behave in both ways either by inducing (Schwabe and Jobin 2013; Vogtmann and Goedert 2016) or preventing carcinogenesis (Bultman 2013, 2016; Ambalam et al. 2016). Different microbiomes which have increased activity associated with a particular type of cancer can also act as biomarkers for the early detection of specific carcinoma (Ambalam et al. 2016).

Augmented microbial infections are found to be linked with promoting different risk factors for lung carcinoma (Schwabe and Jobin 2013). When a healthy lung was examined through different techniques, *Bacteroides*, *Firmicutes*, and *Proteobacteria* were found to be high in numbers. It has been reported that samples of patients with chronic obstructive pulmonary disease (COPD) have enhanced numbers of 16S rRNA which was attributed to the presence of microbial infections. COPD has been a well-established risk factor for lung cancer (Dickson et al. 2013). Lung cancer is directly associated with different microbial infections *like M. tuberculosis*, *M. pneumoniae*, and *Firmicutes* and less prevalence of *Spirochaeta* and *Synergistetes* (Vogtmann and Goedert 2016). Different mechanisms by which microbiomes induce lung carcinoma include dysbiosis of microbiome with the host and increased multiplication of pro-inflammatory infections, enhanced production of ROS which are responsible for DNA damage and ultimately carcinoma, altering metabolic pathway and inducing production of different carcinogens, triggering inflammation signaling, and modifying immune responses (Mao et al. 2018). Colorectal cancer samples indicated high levels of *Fusobacterium* and *Porphyromonas* and low levels of *Ruminococcus* (Vogtmann and Goedert 2016). *Coriobacteria* was found to be associated with CRC forming physiological adaptations in tumor environment, while *Enterobacteria* prevalence was low in tumors (Marchesi et al. 2011). High occurrence rates of *Neisseria elongata*, *Streptococcus mitis*, and *P. gingivalis* are found to be associated with pancreatic cancer (Vogtmann and Goedert 2016). Diverse microbiome was found in pancreatic carcinoma in mice and human models which is found to be associated with suppression of immune response both by innate and adaptive mechanisms (Pushalkar et al. 2018).

Different cancers have been characterized by particular microbiome incidence (Zitvogel et al. 2017). Recent advances in probiotics and prebiotics have revealed their role in prevention of different cancers. Similarly attenuated microbiome has great potential in treatment of cancer by regulating many immunomodulatory responses (Ambalam et al. 2016; Bultman 2013; Kaimala et al. 2018). Many microorganisms secrete metabolites that act as toxins for cancerous cells due to anti-inflammatory or apoptotic property; this feature is adaptive in some microorganisms for competing against other organisms in the environment (Zitvogel et al. 2017). In some cases microbiota works to enhance the efficacy of therapy against carcinomas (Zitvogel et al. 2016; Routy et al. 2018b). Attenuated *Salmonella* species has been reported to have antitumor effects in lung cancer by enhanced production of CD4+ and CD8+ cells. Due to treatment with *Salmonella*, at tumor site it caused enhanced activity of IFNγ which is the mediator of producing CD4+ and CD8+ cells. Vaccine containing attenuated *Salmonella* species also revealed reduced tumor growth in mice (Kaimala et al. 2018). *Lactobacillus* and *Streptococcus* are more prevalent in healthy breast tissue as compared to cancerous because they have reported role as natural killer and of inducing tumor suppression. Similarly *Streptococcus* species reduce DNA damage (Malik et al. 2018). Probiotics (POB) and prebiotics (PEB) can act as cancer-preventive therapeutics in colorectal cancer by different mechanisms. For example *Lactobacillus rhamnosus* can inactivate carcinogens by binding with them or by biotransformation of cancer-causing agents through metabolic pathways. PEB in gut produce acidic compounds like butyrate regulating apoptosis, modulating cell growth, and transforming reactive oxygen species. *B. breve*, *B. longum*, *L. casei*, and *L. rhamnosus* have reported immunomodulatory effects by enhancing the competence of tumor-controlling antibodies, increasing NF-κB functionality, and impacting T-cell progression, respectively. POB alter signaling pathways of carcinogens and cause apoptosis (Ambalam et al. 2016). Genetically modified *L. monocytogenes* strain successfully treated (90%) mice with colorectal cancer. *L. monocytogenes* was designed to produce CD8+ cells due to expression of antigen associated with tumor suppression. Aforementioned species also protects against the recurrence of tumor growth by improving the memory T cells in mice (Kaimala et al. 2018). *T. gondii* was reported to have a potential role in treatment of ovarian cancer by inducing increased production of CD4+ and CD8+ T cells. *L. monocytogenes* strain induced increased entry of macrophages at tumor sites and enhanced antitumor activity of macrophages in mice with ovarian cancers. *Listeria monocytogenes* mediated relatively less critical antitumor response in mice with cervical cancer as its tumor-suppressive response was conditional with activity of αβ-T cells (Kaimala et al. 2018).

3.3 Conventional Cancer Therapies

3.3.1 Chemotherapy

Chemotherapy has now been considered a conventional treatment to cure cancer because it has been practiced since the 1940s. Research in cancer medicine has allowed the treatment of previously fatal cancers. Discoveries of novel therapeutic agents along with increasing technological advancements may lead to achieving the goal of precise and targeted cancer therapy (Wishart 2016). A variety of pharmacological compounds have been used for chemotherapy and are categorized into different groups based on their chemical structure and modes of action. These include alkylating agents, antimetabolites, microtubule target agents, hormones, antibiotics, etc. (Bharti et al. 2018). Alkylating drugs induce mutations through interference with central dogma of cells. This property is utilized to promote DNA damage in cancer cells leading to apoptosis. Alkyl radicals are produced which replace the hydrogen atoms from one molecule to the other. These alkylating agents are further subdivided into nitrogen mustards and platinum coordination complexes (Paul and Bhattacharya 2012; Madkour 2017).

Nitrogen mustards were among the earliest ones to be clinically used in chemotherapy (Gilman and Philips 1946). In addition, derivatives of these compounds such as cyclophosphamide are still widely used as effective antitumor drugs (Shaked et al. 2005; Rau et al. 2015). Cyclophosphamide is comprised of a unique chemical structure which enables it to specifically target cancer cells (Fulda et al. 2010). Development of this drug was based on the fact that higher levels of phosphamidase are expressed in tumor cells which catalyze cleavage of phosphorus-nitrogen bond resulting in release of nitrogen mustard within cancerous cells (Calinski 2013). Several clinical trials have proven the efficacy of cyclophosphamide as an anticancer compound (van Laar et al. 2014; Pasquini et al. 2015; Sikov et al. 2015; Laport et al. 2016). Currently it is used to treat different cancer types including lymphoma, leukemia, and ovarian and breast cancer (Clowse et al. 2009; López-Tarruella and

Martín 2009; Hillmen 2011; Mey et al. 2012). Ifosfamide is another derivative, which has been clinically investigated in comparison to cyclophosphamide. It was reported that ifosfamide is less toxic as compared to cyclophosphamide (Fresneau et al. 2017).

Nitrosoureas, another group of alkylating compounds, react with both nucleic acids and proteins through alkylation and carbamylation (Stankiewicz-Kranc et al. 2010). These are initially decomposed to form highly reactive 2-chloroethyl carbonium ion which causes alkylation of nitrogenous bases of DNA molecule (dos Santos Guimarães et al. 2013; Puyo et al. 2014). Nitrosoureas are further subdivided into different categories of which lipid-soluble agents are commonly used in chemotherapy (Tripathi 2013).

Triazenes are cytotoxic and mutagenic drugs containing DNA *O*6-methylguanine adducts which produce mismatched base pairs resulting in apoptosis or point mutation (Kaina et al. 2010). A large number of studies have reported the use of temozolomide and dacarbazine for treatment of melanoma (Patel et al. 2011; Stacchiotti et al. 2013; Teimouri et al. 2013; Plummer et al. 2013). They are found to be effective in the treatment of metastatic colorectal cancer (Sartore-Bianchi et al. 2017) and metastatic medullary thyroid cancer (Lacin et al. 2015).

The ability of platinum complexes to suppress cell division of *Escherichia coli* at low concentration was discovered in 1965 (Rosenberg et al. 1965) and later utilized to establish their efficacy in chemotherapy (Desoize 2004). Afterward a huge number of platinum complexes have been synthesized (Boulikas et al. 2007). Currently used platinum compounds include cisplatin, carboplatin and oxaliplatin, nedaplatin and lobaplatin, and heptaplatin (Dilruba and Kalayda 2016; Torres et al. 2018). These compounds exert their cytotoxic effects by formation of platinum DNA adducts and affecting amine groups of DNA, RNA, and protein (Riddell and Lippard 2018).

Antimetabolites have been synthesized for so long and used as chemotherapeutics (P Burke et al. 2016). These are divided into analogs of folic acid, purine, pyrimidine, and cytidine (Jain et al. 2016). Their mode of action involves competitive binding at active sites of enzyme thus leading to incorporation into DNA and cell apoptosis (Mathews 2015). The role of microtubules in cellular processes like mitosis, maintenance of cell structure, and transportation of proteins makes them potential targets for anticancer drugs. These compounds are abundantly used for treatment of both solid and hematological malignancies (Katsetos et al. 2015; McCance and Huether 2018).

Various anthracycline antibiotics as chemotherapeutic drugs including doxorubicin, daunorubicin, epirubicin, and idarubicin have been used for cancer treatment (Rizvi 2018). Acute leukemia has been treated with daunorubicin and idarubicin while other two drugs are found to be more effective for treatment of solid tumors like those of lung, breast, thyroid, gastric, etc. (Moslehi 2016; Sethi et al. 2017). Despite their antitumor activity, these drugs also induce toxicity like irreversible cardio-toxicity (Vejpongsa and Yeh 2014; McGowan et al. 2017).

3.3.2 Hormonal Therapy

Considering the diverse roles of glucocorticoids (stress hormones) in intracellular functions, their use as anti-proliferative agents had been highly suggested. Glucocorticoids such as dexamethasone and prednisone are readily used in chemotherapy of leukemia and lymphoma. Additionally, they are also used for treatment of solid tumors (Schlossmacher et al. 2011). Other than its role in female reproductive system, progesterone is also involved in non-reproductive functions related to the heart, bone, and central nervous system. In case of hormone-dependent breast cancer, progesterone agonists such as PR megestrol and megestrol acetate have been used for treatment (Scarpin et al. 2009). Antiestrogens and antiendrogens inhibit the binding of estrogen and androgen to their respective receptors. Thus, they are used to block receptor signaling in cancer cells thus suppressing their division (Chen et al. 2009). Fulvestrant is an FDA-approved antiestrogen which completely blocks ER and is used for treatment of breast cancer in post-menopausal women (Finn et al. 2016; Friedman 2018).

3.3.3 Radiotherapy

Radiotherapy is considered as another option for cancer treatment as localized radiation induces DNA damage and inhibits the replication and division of tumor cells (Gotwals et al. 2017). However, to limit the possible side effects, the dose of radiation is restricted to a certain extent (Niemierko and Goitein 1993). Rate of application of this technique varies worldwide for cancer therapy. As previous research had suggested, the treatment of nearly 50% of cancer patients is with radiations. However, no valid evidence had been provided for this estimate (Delaney et al. 2005).

3.3.4 Combination Therapy

Combination therapy including both chemo- and radiotherapies had also been suggested for cervical cancer patients by US National Cancer Institute alert in February 1999. However, chemo-radiotherapy demands strict monitoring due to possibility of both acute and late toxicity, but survival rate without further progression has been improved by using this technique (Green et al. 2001). Likewise, radiotherapy can also be combined with surgery in variable ways such as during, before, and after surgical removal of the affected area. Owing to the advancements in engineering and technology, radiotherapy techniques have been significantly changed over the past few decades (Garibaldi et al. 2017). Some of the conventional anti-cancer therapies like chemotherapy, radiotherapy and resection are applied either alone or in combination for various cancer types and success rate depends upon different variables including stage and type of cancer, patient's overall health, treatment preferences and possible side effects (Smyth et al. 2016). Recent advancements in technology have enabled the researchers to upgrade these conventional therapies.

3.4 Adjuvant Therapy with Bacteria

More than a hundred years ago, it was suggested that infections may lead to cancer development (Torre et al. 2015); however, bacterial role in cancer remains unrecognized for many years within the cancer research community. Examples of mechanisms by which some antibiotics can cause apoptosis are illustrated in Fig. 3.2. Certain bacteria like *Salmonella typhi* and *Helicobacter pylori* were found to have strong association with cancer (Louis et al. 2014). Increasing evidence that infections are risk factors for carcinogenesis encourages antimicrobial therapy usage in certain pathogen-linked carcinomas. Antibiotics have potential to treat or reverse the cellular proliferation/infection induced by *Citrobacter rodentium*, *Lawsonia intracellularis*, and *Bartonella* species (Alibek et al. 2012). Antibiotic therapy is commonly used for lymphoma and cervical, gastric, bladder, breast, and lung cancer treatment (Tamim et al. 2011; Piccart-Gebhart et al. 2014; Joshi et al. 2015; Racioppi et al. 2018). Tamim et al. reported association between cervical cancer development and bacterial infections (*Chlamydia trachomatis* and *Neisseria gonorrhoeae*), and antibiotic therapy was proved to be effective from more than 15 years (Tamim et al. 2011). *Helicobacter pylori* is a well-known bacterial agent involved in gastric cancers and mucosa-associated lymphoid tissue (MALT) lymphoma (Wang et al. 2014) and is reported to be increasingly involved in breast, eye, lung, and bladder carcinomas (Codolo et al. 2015; Gurunathan et al. 2015; Gotoda et al. 2016; Van Der Post et al. 2018). Clarithromycin and amoxicillin in combination with proton pump blockers (lansoprazole/omeprazole) and muco-protectants (sucralfate) are recommended by the World Health Organization (WHO) and FDA for the treatment of *Helicobacter pylori* infection (Srikanta et al. 2011). It was reported that bacterial infection results in partial or complete remission of MALT lymphoma among 60–80% of cases (Moleiro et al. 2016; Teckie et al. 2017).

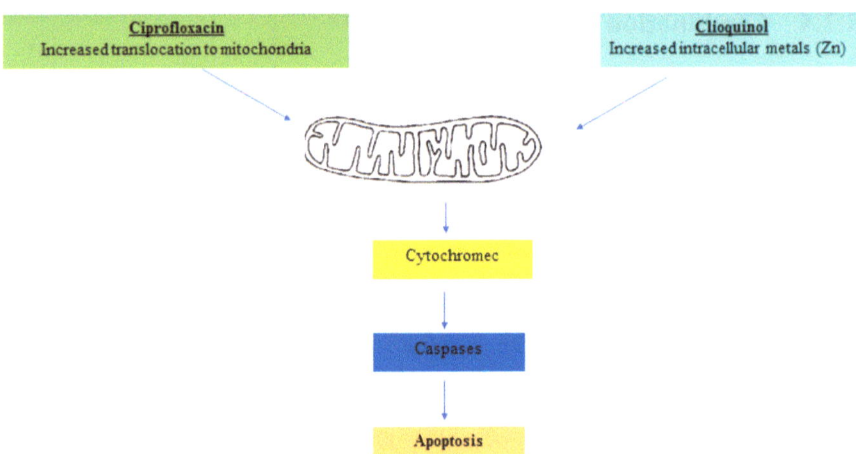

Fig 3.2 Schematic representation of antibiotics' effects on apoptosis

Andriani et al. conducted meta-analysis of 24 studies and reported complete respite of MALT lymphoma in 35–100% cases (Andriani et al. 2009). Therefore, antibiotic therapy is considered as safe and cost-effective first-line treatment for MALT lymphomas, whereas patients who are irresponsive to antibiotic therapy can ultimately be given other oncological treatments (Zullo et al. 2014; Foster and Portell 2015). Similarly, elimination of *Chlamydia psittaci* has been suggested to treat MALT-type eye cancers instead of opting for other aggressive treatment options (Alibek et al. 2012; Miguel 2016).

Use of bacterial extracts to treat bladder cancer leads to the approval of BCG (Bacillus Calmette-Guerin) therapy. Various studies have reported that use of BCG immunoprophylaxis after removal of tumor by surgery is linked with delayed recurrence time or decreased recurrence rate (Punj et al. 2004b; Türkeri et al. 2015; Miyazaki et al. 2018). BCG antitumor activity has been characterized due to its effect on the immune system particularly focusing on CD4 and CD8 T lymphocytes (Redelman-Sidi et al. 2014; Gotwals et al. 2017). In addition to bladder cancer (Kawai et al. 2013), BCG has been used in the treatment of central nervous system (Rosello et al. 2012; DeLance et al. 2013) and genitourinary malignancies (Gandhi et al. 2013). BCG in combination with CancerVax™ (vaccine) has been shown to improve or prolong overall survival among lymphoma patients (Danson and Lorigan 2005; Hoshimoto et al. 2012; Ogi and Aruga 2013).

Various human cancers have been treated with anthracycline antibiotics like daunomycin and doxorubicin (Yang et al. 2014). Even exact antitumor activity of anthracycline is debatable; certain suggested mechanisms are "intercalation into DNA, free radical production, DNA cross-linking, DNA binding and alkylation and topoisomerase II inhibition" (Alibek et al. 2012). However, doxorubicin concentration used in some researches was supra-clinical, intricating the anticipated mechanisms for antitumor activity. Anthracyclines are also responsible for inducing cardio-toxicity (Šimůnek et al. 2009; Ghigo et al. 2016). Researchers are continuously working to explore the anticancer potential and possible underlying mechanisms of other antibiotics. With the discovery that metal chelators, for example, desferrioxamine, have potential to inhibit tumor growth, researchers are looking forward to explore the antitumor potential of other antibiotic chelators (Jansson et al. 2015). Clioquinol is a zinc and copper chelator and supposed to have antitumor activity by inhibiting superoxide dismutase-1, a possible tumor target (Chen and Dou 2008). It was observed that clioquinol treatment results in apoptosis among human cancerous cells by caspase activation (Tardito et al. 2011; Tardito et al. 2012; Cater et al. 2013). Clioquinol is capable of transporting metals inside cell and behaves as transition metal ionophore (Weekley and He 2017); for example, increased metal concentration in mitochondria results in release of cytochrome c leading to cell death (Valko et al. 2016).

Several thiazole antibiotics can inhibit transcription factors, exhibit antiproliferative activity, and induce apoptosis (Amin et al. 2017). These antibiotics bind with larger subunit of ribosome resulting in protein synthesis inhibition among prokaryotes (Chellat et al. 2016). Thiazole antibiotics such as thiostrepton and siomycin A lead to inhibition of FoxM1/Forkhead box M1 transcriptional activity in many in vitro experiments (Kalinichenko and Kalin 2015). FoxM1 is an oncogenic transcription factor and potential target for tumor treatment (Wolter et al. 2017). Overexpression of FoxM1 was observed in various cancers including pancreatic (Cui et al. 2014), breast (O'Regan and Nahta 2018), lung (Liu et al. 2018a), colorectal (Liu et al. 2017), and hepatocellular carcinomas (Chen et al. 2015). Antiproliferative activity of thiazole antibiotics had been reported in xenograft model of human breast cancer as well (Safdari et al. 2015).

Several fluoroquinolones also depict antitumor activity (Sha et al. 2015) and are known to inhibit topoisomerase type II in bacterial cells (Redgrave et al. 2014; Sanderson et al. 2017). Ciprofloxacin inhibits the growth of colorectal cancer, transitional cell carcinoma (most prevalent type of bladder cancer), and prostatic, breast, and pancreatic carcinoma cells (Gurtowska et al. 2010; Yadav et al. 2015; Zhu et al. 2017; Beberok et al. 2018). It inhibits colorectal cancer cell's proliferation and induces apoptosis by blocking synthesis of mitochondrial DNA (Kozieł et al. 2006). Moxifloxacin is capable of inhibiting NFkB activation and synthesis of interleukin-8. It is involved in topoisomerase II inhibition by enhancing antitumor effects of etoposide (Fabian et al. 2006). Geldanamycin (benzoquinone ansamycin member) and its derivatives (17-AAG and 17-DMAG) bind with heat shock protein (Hsp90) and lead to the degradation of tumor-causing proteins (Hong et al. 2013). Hsp90 is responsible for promoting tumor cell survival, growth, and metastasis even in the absence of growth factors through continuous cellular proliferation and protein translation. Hsp-dependent kinases (responsible for Hsp90 inhibition) can be destabilized by 17-AAG (Li et al. 2012). They have cytotoxic and antiproliferative activity in human esophageal cancer cells and result in increased radiosensitization (Wu et al. 2009).

3.5 Role of Live, Attenuated, and Engineered Bacterial Strains in Cancer Treatment

Conventional and broadly available cancer treatments, for example, chemotherapy, surgery, and radiotherapy, fail to accomplish complete cancer remission (Urruticoechea et al. 2010). Furthermore, it is a well-known fact that chemotherapy and radiotherapy cause significant damaging effects on patient's health (Lyss et al. 2011; Monje and Dietrich 2012). Therefore, researchers started exploring less damaging and long-lasting treatment options for cancer. With the advancement in molecular biology, the scope of microbial-based cancer therapeutics gained much importance and opens new possibilities of using bacteria as sensitizing agent for chemotherapy, delivery agents for various drugs, and vectors for gene therapy (Lehouritis et al. 2013; Liu et al. 2018b). One of the most promising approaches is the use of live, attenuated, and engineered bacterial strains in cancer treatment (Patyar et al. 2010; Felgner et al. 2016). Live attenuated bacteria such as *Salmonella* and *Listeria* have great potential to deliver heterologous antigens as multivalent vaccine (Zhou et al. 2010; Hegazy and Hensel 2012; Hegazy et al. 2012; Lin et al. 2015). Many studies have reported that bacteria can serve as DNA vaccine delivery systems (Bermúdez-Humarán et al. 2011; Mura et al. 2013; Hosseinidoust et al. 2016; Gnopo et al. 2017). Literature has shown the incorporation of plasmids coding for antigens under the control of eukaryotic/viral promoter into *Listeria* or *Salmonella* (Miki et al. 2004; Sinha et al. 2017; Lubenau and Springer 2018; Wu and Hung 2018).

Genetically modified *Salmonella* strains have thousand times more potential to target cancerous cells as compared to normal ones and have been used as delivery agents for certain enzymes (cytosine deaminase) into cancerous cells (Hosseinidoust et al. 2016; Felgner et al. 2016; Alimoradi et al. 2016; Lubenau and Springer 2018). Cytosine deaminase is responsible for converting 5-fluorocytosine (nontoxic prodrug) into 5-fluorouracil (highly toxic anticancer drug) (Twitty et al. 2015). It was observed that mice bearing lung carcinoma showed significantly decreased tumor burden when injected with genetically engineered *Salmonella* strains (Zheng et al. 2017; Liang et al. 2018). In TAPET or "tumor amplified protein expression therapy," genetically altered *Salmonella* strains were used to deliver antitumor drug to treat solid tumors (Nemunaitis et al. 2003; Patyar et al. 2010; Lin et al. 2015; Yoon et al. 2018). Saltzman et al. reported that combination of low-dose chemotherapy and bacterial immunotherapy results in significantly decreased tumor burden in murine model harboring breast carcinoma (Saltzman et al. 2018). Various bacterial strains like *Salmonella*, *Clostridium*, and *Bifidobacterium* have potential to target solid tumors and be used for transporting and amplifying genes coding for factors like toxins, pro-drug-converting enzymes, cytokines, and angiogenesis inhibitors (Sun 2014; Chalkoo et al. 2016; Kucerova and Cervinkova 2016; Zhou et al. 2017; Ma et al. 2017). Mendes et al. reported potential chemopreventive effect of various probiotic strains such as *Clostridium*, *Lactobacillus*, and *Bifidobacterium* on colorectal cancer (Mendes et al. 2018). Probiotic supplementation results in altered microbial structures and is involved in the regulation of inflammatory responses preventing colorectal cancer prognosis (Malik et al. 2018).

Clostridium novyi-NT showed positive effects in the treatment of various carcinomas among animal models when given in combination with radiotherapy (Roberts et al. 2014). Radiotherapy targets highly oxygen-exposed and rapidly proliferating cells in solid tumors, and *C. novyi-NT* injections accelerate this process by hitting radiation-resistant cells (Staedtke et al. 2016; Dang et al. 2017). Combination of *C. novyi-NT* with radiotherapy was found to be beneficial for patients as it results in the reduced radiotherapeutic dosage (Bernardes et al. 2010). Cloned endostatin gene was transformed into *B. longum* and *B. adolescentis* and results in the inhibition of tumor growth (Lin et al. 2015; Routy et al. 2018a). It was found that systemic administration of *B. adolescentis* has more than 70% tumor inhibition potential in mice models as compared to the wild-type ones (Livanos et al. 2016). Oral administration of engineered *B. longum* results in the selective recognition of hepatocellular carcinoma (Yin et al. 2011). Anticancer delivery gene therapy was found to be safe and effective, and no biological effect/loss of plasmid stability occurred on selectively removing antibiotic resistance genes (Chatel et al. 2015; Hardee et al. 2017; Riglar and Silver 2018).

3.6 Bacteria-Derived Anticancer Agents

Worldwide billions of people suffer from cancer annually and have approximately 50% mortality rate. In advanced countries of the West, one-fifth of deaths are due to cancer, whereas in European countries, it ranges from 3% to 15 % depending upon age (Torre et al. 2015). The most widely accepted mode of treatment for cancer patients is the use of either chemotherapy or radiotherapy or surgery or combination of these, but these strategies often fail and if successful they pose serious side effects for patients making their lives miserable during and after treatment; therefore new strategies have been designed for treatment. One of the approaches is the use of microorganisms either live or purified products of them or attenuated forms.

However microorganisms like *Helicobacter pylori* and *Agrobacterium tumefaciens* are known to cause gastric cancer in humans and crown gall in plants, respectively. But at the same time, it is a known fact that usage of microorganisms for cancer treatment dates back to more than 100 years ago. In 1890, bacteria were identified first time as anticancer agents by Dr. Coley (Coley 1893). It was noted by him that the cancer cells regressed once infected with pathogenic bacteria and later *Clostridium histolyticum* was used to treat advanced stages of cancer. Later on few bacteria also caused appearance of cancer; therefore scientists used extracts of bacteria for treatment purposes (Wei et al. 2008). Some bacteria act as anticancer agents by targeting cancer cells only like *Bifidobacterium*, *Listeria*, *Bacillus*, *Mycobacterium*, *Salmonella*, and *Clostridium* which are used by oncologists in either engineered or attenuated or live forms. They have been most frequently used in patients where chemotherapy or radiotherapy stands unsuccessful. Their unique property is their growth in solid tumors in hypoxic core region with very less oxygen (Fialho et al. 2008). Due to restriction in their selectivity of cells, they act as desirable vector delivery vehicle for targeted gene therapy. Bladder cancer may be treated by activation of immune response by the use of *M. bovis*, and vectors of attenuated microorganisms used as vaccines (like *Listeria monocytogenes* and *Salmonella enterica*) attack antigen-presenting cells, or they act as strong innate immune system inducers or activate IL-12 (mediator of immune system); therefore they are used as anticancer agent and for prevention of cancer (Paglia and Guzman 1998).

Whenever we are dealing with the use of bacteria as a cancer treatment option, problems arise related to toxicity of agent, side effects associated with it, efficient immune response, and unpredictable efficacy. Numerous xenografted models were used by researchers in order to assess the ability of bacteria as treatment. Colorectal cancer was studied using such models with different species of *Bifidobacterium*, *Lactobacillus*, and *Clostridium*, and it was found that *Clostridium sordellii* and *Clostridium novyi* have strong colonization potential of tumor (Agrawal et al. 2004). In spite of all the literature, one-third of mice die within 16–18 h after spore injection, and it was found to be due to lethal toxin α produced by the strain. This toxin was removed by heating and new strain was named as *C. novyi-NT.* Later this strain was combined with ancient chemotherapy strategies named as COBALT (combined bacteriological therapy) (Parr et al. 1973). Another strain of *S. typhimurium* known as A1 is an auxotroph for arginine and leucine and receives these amino acids from tumor (Zhao et al. 2005). This potential was verified in mouse xenografted with prostate cancer of human. Based on different experiments with different microorganisms (live bacteria) and different tumors in animal models, it was found that *M. bovis* BCG (Bacillus Calmette-Guerin) is most successful in bladder cancer. The literature review shows that bacteria may act as an anticancer agent and their modified forms may be used as treatment option.

3.6.1 Bacterial Azurin (Promising Anticancer Agent)

Bacteria, plants, and animals have naturally occurring cupredoxins that has a family of compounds like halocyanins, amicyanins, auracyanins, rusticyanins, pseudoazurins, and azurins (De Rienzo et al. 2004). It is thought that these compounds have evolutionary similarity with immunoglobulins (Stevens 2008), and that is the reason why they can be used as therapeutic agent for cancer treatment. It has an extended α-helix domain for protein transduction (50–77) and at C terminal four loop regions known as GH, FG, EF, and CD loops that give it an immunoglobulin fold and β-sandwich core. This property of azurin allows it to mimic immune activity and act as anticancer agent when present in blood. One of the cupredoxins is produced by *Pseudomonas aeruginosa* known as azurin and is a copper-containing periplasmic protein that has scaffolding properties regulated by hydrophobic and electrostatic interaction and has potential to reduce molecular consequences associated with progression of cancer. Azurins may be extracted from bacteria very easily and it appeared to be associated with internalization of cholesterol-abundant microdomains. The domains are known as lipid rafts and are highly expressed in cancerous cells. Therefore, based on this information, azurins have the potential to act as anticancer agent for treatment.

Azurin interferes with growth of cells through many processes in cancer cells like formation of complex with p53 DNA binding domain and protects its functioning which leads to apoptosis (Apiyo and Wittung-Stafshede 2005). Mice that are xenografted with breast tumor injected with azurin exhibit regression of tumor with no side effect (Punj et al. 2004a). Azurin causes cell death and arrests growth through BCL2 and BAX imbalance created by higher mRNA expression of p53 (Yamada et al. 2005). Interestingly azurin also affects EphB2 (that is overexpressed on the surface of lung, prostate, and breast cancers) by binding to its receptor and in turn causes regression in proliferation and rapid growth of tumor due to its association with its ligand-associated receptor ephrin B2 (Chaudhari et al. 2007). Azurin binds to EphB2 through GH loop (88–113) and increases sensitivity to radiation therapy in lung cancer by attaching with radiosensitizer drug nicotinamide in oral SCC (Choi et al. 2011). Angiogenesis is also inhibited by azurin in cancer cells by decreasing the phosphorylation potential of Akt, FAK, and VEGF2 (Mehta et al. 2011). An important feature of azurin is that due to structural features it does not need to enter into the cancerous cells for complex formation with these genes. Most of the cancer proliferation is caused by the cell surface receptor dysregulation, and azurin targets those extracellular receptors as treatment strategy (Fig. 3.3).

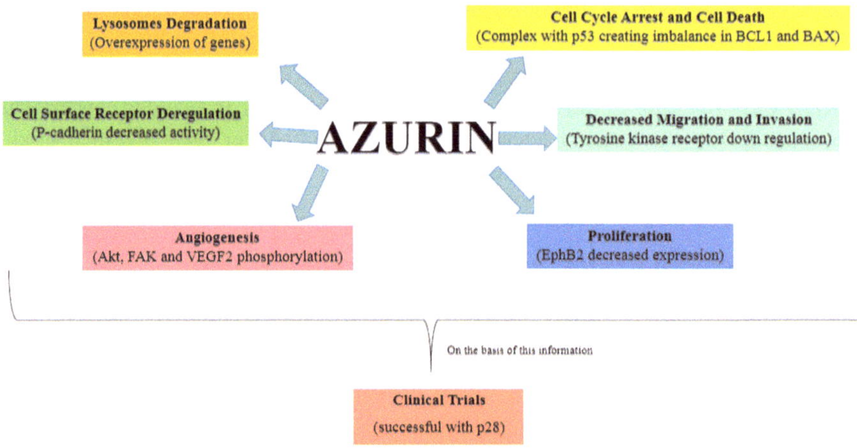

Fig 3.3 Multifunctional anticancer properties of bacterial azurin

Recently azurin has been found to exert its multivalent action in invasive breast tumors that have higher levels of P-cadherin which normally acts as a cell adhesion agent but in case of tumor leads to progression of malignant breast cancer (Albergaria et al. 2011). Azurin causing decreased activity of P-cadherin without affecting E-cadherin was an important discovery as it plays its role only in invasive forms of cancer as it reduces invasive capacity of tumor. Azurin and p28 (its peptide) are dependent on cholesterol for its penetration activity and are fast penetrators in tumor than normal cells (Taylor et al. 2009). They are nontoxic and nonimmunogenic and several patients have claimed for its use as cancer treatment drug solely or increasing the effect of other drugs without any cellular toxicity (Bernardes et al. 2016; Yamada et al. 2016). The success of azurin as therapeutic is due to its mode of entrance in the cell through caveolae-mediated endocytic pathway rather than usual clathrin or glycosaminoglycan use (Taylor et al. 2009).

FAK and src are involved in cross talks between cadherin- and integrin-associated adhesion in cell surface receptor and epithelial cells, whereas azurin and its peptide p28 are known to decrease the level of these two receptor tyrosine kinases resulting in decreased invasion and migration potential of cells (Goel et al. 2012). Azurin also overexpresses genes associated with degradation of lysosomes and endocytosis through vesicle-mediated transport by downregulation of cell surface receptors. Therefore, azurins have been found to act in multiple ways depending upon cancer type possibly by reducing receptor activity at membrane for those cells where active proliferation is going on. One of the factors contributing to resistance of cancer therapy is the receptors involved in different modes and azurin is best to counter this problem. Azurin may also act in combination with other drugs and reduce monoclonal antibodies or tyrosine kinase activity and decrease relapse of cancer in patients (Chakrabarty et al. 2014).

3.7 Clinical Trials

Cancer is a disease that is complicated due to complexity of pathways that promote growth or that regulate signaling networks or neuronal system involved in cancer development. Therefore in this era there is an urge to discover a drug which can target multiple sites and also has the least toxicity effects so that only cancer cells are targeted not the normal ones (Avner et al. 2012). Azurin is postulated to be one of the drugs which have multitarget capability and no toxicity so far; hence clinical trials are in progress. Azurin has not been directly used in any clinical trial but it has many domains in addition to p28 that have potential for anticancer effect generation due to its reduced resistance development, nontoxicity, and efficacy that can be proved through trials. In a phase I trial, 14 patients of different cancers (1 sarcoma, 1 prostate cancer, 1 pancreatic cancer, 4 colon cancer, and 7 melanoma patients) were given p28 as single agent in 5 dose levels. It was found that no toxicity was observed and survival rate was increased in p53-positive metastatic solid cancer patients (Richards et al. 2011). Another human trial conducted by Warso et al. involves 15 patients who were given p28 for 4 weeks and 3 times per week through short infusions. It was found that no patient revealed toxicity of drug and no immune response was generated against the drug; therefore it was concluded that p28 from azurin inhibited p53 ubiquitination (Warso et al. 2013). These trials revealed that azurin acts selectively on cancerous cells and not on normal cells and that may be the reason why no toxicity was detected on administration. Another important role of azurin may be attributed to its administration via twice a week injection in normal people who have a family history of breast cancer and are BRCA1 and BRCA2 polymorphic so that the onset of cancer may be reduced or prevented (Fialho and Chakrabarty 2012). In case of central nervous system malignancy, a clinical trial was conducted by Lulla et al. with 18 children patients, and they were given intravenous p28 injections for four weeks on the basis of 3 times a week. The dose was phase II recommended dose and results showed no toxicity and children tolerated it very well (Lulla et al. 2016). The success of every drug comes with its oral administration and it is now in progress alongside its clinical trials. If the phase I and II trials will be positive for its activities, then it is expected to be bio-encapsulated and delivered through oral route as illustrated in Fig. 3.4.

3.8 Hematopoietic Transplants, Complications, and the Microbiota

In the early 1960s cancer patients were treated with chemotherapy giving positive response, but later patients developed bacterial (mostly *Escherichia coli* and *Pseudomonas aeruginosa*) infections in the blood leading to increased rate of mortality and morbidity. Later it was found that these bacterial infections were originating from the gastrointestinal tract of the patient himself (Taur et al. 2012). Similarly, in hematopoietic cell transplants, the patients undergo reduced neutrophil count that is less than 100 cells/mm^3 (neutropenia) and fever that leads to higher number

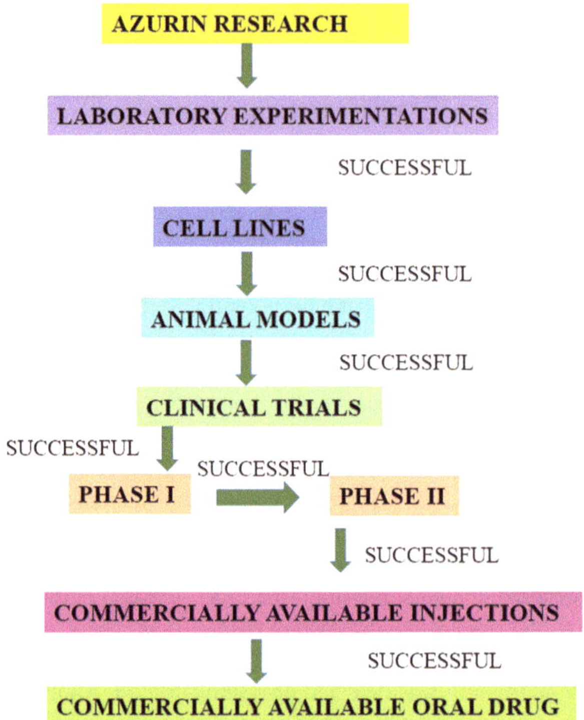

Fig 3.4 Azurin research: as promising anticancer agent

of infections by bacteria and fungi. Therefore cancer and hematopoietic cell transplant patients receive heavy doses of antibiotics for a long time and report no infections (Freifeld et al. 2011). Clinicians have now discovered that stem cell transplant and cancer patients have now developed antibiotic-resistant bacteria and also there is a lot of disturbance in the gut microbiota of these patients resulting in increased invasive infections as well as a post-transplant disease known as graft versus host disease (Taur et al. 2012; Bilinski et al. 2016; Holler et al. 2014; Simms-Waldrip et al. 2017).

It is a known fact that intestinal bacteria are commensals inhabiting our intestinal lining; they are known as intestinal microbiota. These bacteria play an important role when it comes to hematopoietic cell transplantation. Initially patient's intestinal microbiota was altered for hematopoietic cell transplantation but later the technique rendered less beneficial. The exact reason of the role of these bacteria was unknown until techniques of biotechnology like PCR, 16rRNA sequencing, and high-throughput sequencing were developed that helped researchers to get an insight into functioning of microbiota. Later many diseases were clinically evaluated for outcomes like asthma, allergies, cancer initiation, as well as hematopoietic cell transplantation (Docampo et al. 2015). One of the complications associated with hematopoietic cell transplantation is reduced IgA concentration after few months of transplant, and this has been observed in mouse as well as human. Although normal levels of IgA are maintained after some time, but if the patient develops graft versus host disease, then these levels are further reduced (Jenq et al. 2012). Although hematopoietic cell transplantation exerts little effect on microbiota, graft versus host disease results in decreased diversity of species and increased number of *Clostridiales*, *Streptococcus*, *Enterococcus*, *Lactobacillus*, *Enterobacter*, *Klebsiella*, and *Escherichia* (Eriguchi et al. 2012; Jenq et al. 2012; Heimesaat et al. 2010). In case of mice these changes are not more pronounced as they are not given antibiotics routinely like in humans. Sometimes during hospitalization patients of hematopoietic cell transplantation get reduced diversity of microflora due to more oligoclonal production of *Enterobacteriales*, *Streptococcus*, and *Enterococcus*. One of the reasons may be the use of drug metronidazole that harms the obligatory anaerobes leading to reduced suppression of these microorganisms (Taur et al. 2012).

The main problems of these bacterial transformations from the gastrointestinal tract are imbalance in microbiota (due to antibiotics), changed intestinal surface composition (mucositis), and reduced immunity of cells. It has been found that patients who suffer from gut microbiota misbalance developed more blood infections compared with others (Koh et al. 2008). One of the benefits of gut microbiota is that it prevents pathogenic bacteria from infecting by making colonies in the intestine. In mice these bacteria cause epithelial cells of gastrointestinal walls to form antimicrobial peptides that reduce the number of pathogenic bacteria in the intestine. But when antibiotics are administered, these commensal bacteria are depleted leading to reduced antimicrobial peptides which allows entry of pathogenic bacteria that cause infections (Buffie et al. 2015). In case of stem cell transplant, increased levels of gastrointestinal *Enterobacteriaceae* and *Enterococcus* spp. are observed resulting in increased blood infections. Therefore it was concluded that more perturbations of gut microbiota lead to more infections in the bloodstream of stem cell transplant patients, and one way to reduce infections is to limit the growth of bacteria in the gastrointestinal tract (Fan et al. 2015).

Antibiotic therapy in hematopoietic stem cell transplant is the treatment strategy that rescued the lives of these patients. Antibiotic resistance in these patients should be avoided by using broad-spectrum antibiotics and mostly eradicating gram-negative bacteria. Selective pathogen-targeting strategies are now being investigated like use of phages with CRISPR/cas9 carrying vectors that target specific pathogen or biotechnologically manipulated bacteria to outcompete bacteria that are pathogenic (Kommineni et al. 2015) or antibiotic conjugating a pathogen-specific antibody (Lehar et al. 2015).

3.9 Future Challenges, Recommendations, and Conclusion

Microbial-based anticancer therapy is often neglected and marginalized. Recent advancements have been made in the field of cancer drug discovery focusing on drug designing and targeting specific molecule/molecules or signaling pathway.

Novel synthetic chemical compounds with peculiar structural properties capable of blocking or inhibiting a reaction/signaling cascade responsible for tumor growth, proliferation, and survival are developed with the help of combinatorial chemistry and high-throughput assays. Currently developed anticancer drugs target cell division, DNA replication, angiogenesis, various signaling cascades, and apoptosis. Several commercially available therapeutic modalities in clinical practice include drugs and their combination with pro-drugs, small synthetic or natural molecules, and monoclonal antibodies. Cancer cells can acquire rapid mutations therefore resulting in therapeutic efficacy loss. To overcome this problem, various drugs have been used in combinations, but it was associated with increased cost, complexity, and problems like multi-drug resistance. This would lead to the development of novel, cost-effective, and targeted antitumor therapies.

Microbial-based anticancer therapy is an emerging field and some important advances have been made in the past few years regarding the use of live, attenuated, and engineered bacterial strains for cancer treatment. Several facultative and obligate anaerobic bacterial strains are used to target cancerous cell growth in hypoxic conditions particularly among solid tumors. A large number of studies are in progress to target multiple mechanisms responsible for inducing tumor growth with the help of engineered bacterial strains or proteins. Microbial-based cancer treatment focuses on the production of vaccines responsible for activating the immune system to fight against disease and vectors used for the transmission of antitumor drugs/therapeutics. Microbial-based cancer treatment is responsible for inhibiting tumor growth in patients who were unable to get any benefit from conventional cancer therapies. But like other treatments, this approach also has certain limitations like risk of infection which may lead to adverse conditions and death of the patient as well. In many experimental studies, bacterial strains result in tumor growth inhibition, but ultimately model animals died due to infection caused by various pathogens. Therefore, patient's safety must be ensured before using any attenuated bacteria by testing on animal models and sometimes in clinical trials as well. Proper anticancer treatment can only be achieved by finding or developing a perfect match between attenuated microorganism and its immunostimulation. Other constraints include high cost linked with clinical trials and launching a new drug/product in the market. Furthermore, it comes up with complex and strict legal regulations because the impact of microbes on carcinogenesis is not fully explored yet.

Another important factor is patient's current health and proper diagnosis. One of the best examples is the diagnosis, experimentation, and treatment implementation for *Plasmodium falciparum*. Now, second-generation drugs have gained researchers' attention as having greater efficacy and fewer side effects, and single drug can target multiple steps responsible for carcinogenesis. Azurin, protein produced by *P. aeruginosa*, is a point of focus by investigators because it is responsible for inducing apoptosis by forming complex with p53 gene or inhibiting tumor growth by angiogenesis prevention or tyrosine kinase-mediated cell signaling. With the advancement in knowledge that bacteria/bacterial products have immunostimulatory and antitumor activity and hit various cellular and molecular targets, new modalities have been opened for microbial-based cancer treatment.

References

Agrawal N, Bettegowda C, Cheong I, Geschwind J-F, Drake CG, Hipkiss EL, Tatsumi M, Dang LH, Diaz LA, Pomper M (2004) Bacteriolytic therapy can generate a potent immune response against experimental tumors. Proc Natl Acad Sci 101(42):15172–15177

Albergaria A, Ribeiro A-S, Vieira A-F, Sousa B, Nobre A-R, Seruca R, Schmitt F, Paredes J (2011) P-cadherin role in normal breast development and cancer. Int J Dev Biol 55(7–9):811–822

Alibek K, Bekmurzayeva A, Mussabekova A, Sultankulov B (2012) Using antimicrobial adjuvant therapy in cancer treatment: a review. Infect Agents Cancer 7(1):33

Alimoradi H, Matikonda SS, Gamble AB, Giles GI, Greish K (2016) Hypoxia responsive drug delivery systems in tumor therapy. Curr Pharm Des 22(19):2808–2820

Ambalam P, Raman M, Purama RK, Doble M (2016) Probiotics, prebiotics and colorectal cancer prevention. Best Pract Res Clin Gastroenterol 30(1):119–131

Amin HK, El-Araby AM, Eid S, Nasr T, Bondock S, Leheta O, Dawoud ME (2017) A thiazole analogue exhibits an anti-proliferative effect in different human carcinoma cell lines and its mechanism based on molecular modeling. Adv Biol Chem 7(01):76

Andriani A, Miedico A, Tedeschi L, Patti C, Di Raimondo F, Leone M, Schinocca L, Romanelli A, Bonanno G, Linea C (2009) Management and long-term follow-up of early stage *H. pylori*-associated gastric MALT-lymphoma in clinical practice: an Italian, multicentre study. Dig Liver Dis 41(7):467–473

Apiyo D, Wittung-Stafshede P (2005) Unique complex between bacterial azurin and tumor-suppressor protein p53. Biochem Biophys Res Commun 332(4):965–968

Avner BS, Fialho AM, Chakrabarty AM (2012) Overcoming drug resistance in multi-drug resistant cancers and microorganisms: a conceptual framework. Bioengineered 3(5):262–270

Beberok A, Wrześniok D, Rzepka Z, Respondek M, Buszman E (2018) Ciprofloxacin triggers the apoptosis of human triple-negative breast cancer MDA-MB-231 cells via the p53/Bax/Bcl-2 signaling pathway. Int J Oncol 52(5):1727–1737

Bermúdez-Humarán LG, Kharrat P, Chatel J-M, Langella P (2011) Lactococci and lactobacilli as mucosal delivery vectors for therapeutic proteins and DNA vaccines. In: Microbial cell factories, vol 1. BioMed Central, p S4

Bernardes N, Seruca R, Chakrabarty AM, Fialho AM (2010) Microbial-based therapy of cancer: current progress and future prospects. Bioeng Bugs 1(3):178–190. https://doi.org/10.4161/bbug.1.3.10903

Bernardes N, Abreu S, Carvalho FA, Fernandes F, Santos NC, Fialho AM (2016) Modulation of membrane properties of lung cancer cells by azurin enhances the sensitivity to EGFR-targeted therapy and decreased β1 integrin-mediated adhesion. Cell Cycle 15(11):1415–1424

Bharti AC, Vishnoi K, Singh SM, Aggarwal BB (2018) Pathways linked to cancer chemoresistance and their targeting by nutraceuticals. In: Role of nutraceuticals in cancer chemosensitization. Elsevier, London, pp 1–30

Bilinski J, Robak K, Peric Z, Marchel H, Karakulska-Prystupiuk E, Halaburda K, Rusicka P, Swoboda-Kopec E, Wroblewska M, Wiktor-Jedrzejczak W (2016) Impact of gut colonization by antibiotic-resistant bacteria on the outcomes of allogeneic hematopoietic stem cell transplantation: a retrospective, single-center study. Biol Blood Marrow Transplant 22(6):1087–1093

Boulikas T, Pantos A, Bellis E, Christofis P (2007) Designing platinum compounds in cancer: structures and mechanisms. Cancer Ther 5:537–583

Buffie CG, Bucci V, Stein RR, McKenney PT, Ling L, Gobourne A, No D, Liu H, Kinnebrew M, Viale A (2015) Precision microbiome reconstitution restores bile acid mediated resistance to *Clostridium difficile*. Nature 517(7533):205

Bultman SJ (2013) Emerging roles of the microbiome in cancer. Carcinogenesis 35(2):249–255

Bultman SJ (2016) The microbiome and its potential as a cancer preventive intervention. In: Seminars in oncology, vol 1. Elsevier, pp 97–106

Calinski DM (2013) Oxazaphosphorine metabolism by the human cytochrome P450s and some commonly expressed P450 polymorphic variants

Cann SH, Van Netten J, Van Netten C (2003) Dr. William Coley and tumour regression: a place in history or in the future. Postgrad Med J 79(938):672–680
Cater MA, Pearson HB, Wolyniec K, Klaver P, Bilandzic M, Paterson BM, Bush AI, Humbert PO, La Fontaine S, Donnelly PS (2013) Increasing intracellular bioavailable copper selectively targets prostate cancer cells. ACS Chem Biol 8(7):1621–1631
Chakrabarty AM, Bernardes N, Fialho AM (2014) Bacterial proteins and peptides in cancer therapy: today and tomorrow. Bioengineered 5(4):234–242
Chalkoo AH, Tariq S, Bhat ZA (2016) Gene therapy in the management of head and neck cancers: a review. Int J Health Sci Res 6(4):487–496
Chatel J-M, Breyner N, Gomes DLR, Azevedo V, Miyoshi A, Langella P (2015) Use of rec LABs: good bugs to deliver molecules of health interest: from mouse to man. Probiotics and Prebiotics:167
Chaudhari A, Mahfouz M, Fialho AM, Yamada T, Granja AT, Zhu Y, Hashimoto W, Schlarb-Ridley B, Cho W, Gupta TKD (2007) Cupredoxin– cancer interrelationship: azurin binding with EphB2, interference in EphB2 tyrosine phosphorylation, and inhibition of cancer growth. Biochemistry 46(7):1799–1810
Chellat MF, Raguž L, Riedl R (2016) Targeting antibiotic resistance. Angew Chem Int Ed 55(23):6600–6626
Chen D, Dou QP (2008) New uses for old copper-binding drugs: converting the pro-angiogenic copper to a specific cancer cell death inducer. Expert Opin Ther Targets 12(6):739–748
Chen Y, Clegg NJ, Scher HI (2009) Anti-androgens and androgen-depleting therapies in prostate cancer: new agents for an established target. Lancet Oncol 10(10):981–991
Chen J, Chen J-S, Zhang J, Gi Y, Phan LM, Muñoz NM, Katz LH, Shin J-H, Jeong YS, Menon VK (2015) Sa1851 prognostic significance of TGF-β signaling in human hepatocellular carcinoma. Gastroenterology 148(4):S-1024
Choi J-H, Lee M-H, Cho Y-J, Park B-S, Kim S, Kim G-C (2011) The bacterial protein azurin enhances sensitivity of oral squamous carcinoma cells to anticancer drugs. Yonsei Med J 52(5):773–778
Clowse ME, Behera MA, Anders CK, Copland S, Coffman CJ, Leppert PC, Bastian LA (2009) Ovarian preservation by GnRH agonists during chemotherapy: a meta-analysis. J Women's Health 18(3):311–319
Codolo G, Munari F, Fassan M, de Bernard M (2015) Evaluation of the efficacy of the *H. pylori* protein HP-NAP as a therapeutic tool for treatment of bladder cancer in an orthotopic murine model. J Vis Exp 99:e52743
Coley WB (1893) The treatment of malignant tumors by repeated inoculations of erysipelas: with a report of ten original cases. Am J Med Sci 105(6):487
Cui J, Shi M, Xie D, Wei D, Jia Z, Zheng S, Gao Y, Huang S, Xie K (2014) FOXM1 promotes the warburg effect and pancreatic cancer progression via transactivation of LDHA expression. Clin Cancer Res 20(10):2595–2606
Dang L, Bettegowda C, Kinzler KW, Vogelstein B (2017) Combination bacteriolytic therapy for the treatment of tumors. Google Patents
Danson S, Lorigan P (2005) Improving outcomes in advanced malignant melanoma. Drugs 65(6):733–743
De Martel C, Ferlay J, Franceschi S, Vignat J, Bray F, Forman D, Plummer M (2012) Global burden of cancers attributable to infections in 2008: a review and synthetic analysis. Lancet Oncol 13(6):607–615
De Rienzo F, Gabdoulline R, Wade R, Sola M, Menziani M (2004) Computational approaches to structural and functional analysis of plastocyanin and other blue copper proteins. Cell Mol Life Sci 61(10):1123–1142
DeLance AR, Safaee M, Oh MC, Clark AJ, Kaur G, Sun MZ, Bollen AW, Phillips JJ, Parsa AT (2013) Tuberculoma of the central nervous system. J Clin Neurosci 20(10):1333–1341
Delaney G, Jacob S, Featherstone C, Barton M (2005) The role of radiotherapy in cancer treatment: estimating optimal utilization from a review of evidence-based clinical guidelines. Cancer: Interdisciplinary Int J Am Cancer Soc 104(6):1129–1137

Desoize B (2004) Metals and metal compounds in cancer treatment. Anticancer Res 24(3A):1529–1544
Dickson RP, Erb-Downward JR, Huffnagle GB (2013) The role of the bacterial microbiome in lung disease. Expert Rev Respir Med 7(3):245–257
Dilruba S, Kalayda GV (2016) Platinum-based drugs: past, present and future. Cancer Chemother Pharmacol 77(6):1103–1124
Docampo MD, Auletta JJ, Jenq RR (2015) Emerging influence of the intestinal microbiota during allogeneic hematopoietic cell transplantation: control the gut and the body will follow. Biol Blood Marrow Transplant 21(8):1360–1366
dos Santos Guimarães I, Daltoé RD, Herlinger AL, Madeira KP, Ladislau T, Valadão IC, Junior PCML, Teixeira SF, Amorim GM, dos Santos DZ (2013) Conventional cancer treatment. In: Cancer treatment-conventional and innovative approaches. InTech, Rijeka
Eriguchi Y, Takashima S, Oka H, Shimoji S, Nakamura K, Uryu H, Shimoda S, Iwasaki H, Shimono N, Ayabe T (2012) Graft-versus-host disease disrupts intestinal microbial ecology by inhibiting Paneth cell production of α-defensins. Blood 120(1):223–231
Fabian I, Reuveni D, Levitov A, Halperin D, Priel E, Shalit I (2006) Moxifloxacin enhances antiproliferative and apoptotic effects of etoposide but inhibits its proinflammatory effects in THP-1 and Jurkat cells. Br J Cancer 95(8):1038
Fan D, Coughlin LA, Neubauer MM, Kim J, Kim MS, Zhan X, Simms-Waldrip TR, Xie Y, Hooper LV, Koh AY (2015) Activation of HIF-1α and LL-37 by commensal bacteria inhibits *Candida albicans* colonization. Nat Med 21(7):808
Felgner S, Kocijancic D, Frahm M, Weiss S (2016) Bacteria in cancer therapy: renaissance of an old concept. Int J Microbiol 2016:8451728
Fialho AM, Chakrabarty AM (2012) Patent controversies and court cases: cancer diagnosis, therapy and prevention. Cancer Biol Ther 13(13):1229–1234
Fialho AM, Das Gupta T, Chakrabarty AM, Sleator R, Hill C (2008) Promiscuous drugs from pathogenic bacteria in the post-antibiotics era. In: Patho-Biotechnology. Landes Bioscience, Austin, pp 145–162
Finn RS, Aleshin A, Slamon DJ (2016) Targeting the cyclin-dependent kinases (CDK) 4/6 in estrogen receptor-positive breast cancers. Breast Cancer Res 18(1):17
Foster LH, Portell CA (2015) The role of infectious agents, antibiotics, and antiviral therapy in the treatment of extranodal marginal zone lymphoma and other low-grade lymphomas. Curr Treat Options in Oncol 16(6):28
Freifeld AG, Bow EJ, Sepkowitz KA, Boeckh MJ, Ito JI, Mullen CA, Raad II, Rolston KV, Young J-AH, Wingard JR (2011) Clinical practice guideline for the use of antimicrobial agents in neutropenic patients with cancer: 2010 update by the Infectious Diseases Society of America. Clin Infect Dis 52(4):e56–e93
Fresneau B, Hackshaw A, Hawkins D, Paulussen M, Anderson J, Judson I, Litière S, Dirksen U, Lewis I, van den Berg H (2017) Investigating the heterogeneity of alkylating agents' efficacy and toxicity between sexes: a systematic review and meta-analysis of randomized trials comparing cyclophosphamide and ifosfamide (MAIAGE study). Pediatr Blood Cancer 64(8):e26457
Friedman C (2018) Endocrine therapy resistance in breast cancer: mechanisms and treatments. Stern college for women
Friedman AA, Letai A, Fisher DE, Flaherty KT (2015) Precision medicine for cancer with next-generation functional diagnostics. Nat Rev Cancer 15(12):747
Fulda S, Galluzzi L, Kroemer G (2010) Targeting mitochondria for cancer therapy. Nat Rev Drug Discov 9(6):447
Gandhi NM, Morales A, Lamm DL (2013) Bacillus Calmette-Guérin immunotherapy for genitourinary cancer. BJU Int 112(3):288–297
Garibaldi C, Jereczek-Fossa BA, Marvaso G, Dicuonzo S, Rojas DP, Cattani F, Starzyńska A, Ciardo D, Surgo A, Leonardi MC (2017) Recent advances in radiation oncology. Ecancermedicalscience 11
Garrett WS (2015) Cancer and the microbiota. Science 348(6230):80–86

Ghigo A, Li M, Hirsch E (2016) New signal transduction paradigms in anthracycline-induced cardiotoxicity. Biochim Biophys Acta (BBA)-Mol Cell Res 1863(7):1916–1925

Gilman A, Philips FS (1946) The biological actions and therapeutic applications of the B-chloroethyl amines and sulfides. Science 103(2675):409–436

Gnopo YM, Watkins HC, Stevenson TC, Delisa MP, Putnam D (2017) Designer outer membrane vesicles as immunomodulatory systems–reprogramming bacteria for vaccine delivery. Adv Drug Deliv Rev 114:132–142

Goel HL, Pursell B, Standley C, Fogarty K, Mercurio AM (2012) Neuropilin-2 regulates α6β1 integrin in the formation of focal adhesions and signaling. J Cell Sci 125(2):497–506

Gotoda T, Uedo N, Yoshinaga S, Tanuma T, Morita Y, Doyama H, Aso A, Hirasawa T, Yano T, Uchita K (2016) Basic principles and practice of gastric cancer screening using high-definition white-light gastroscopy: eyes can only see what the brain knows. Dig Endosc 28:2–15

Gotwals P, Cameron S, Cipolletta D, Cremasco V, Crystal A, Hewes B, Mueller B, Quaratino S, Sabatos-Peyton C, Petruzzelli L (2017) Prospects for combining targeted and conventional cancer therapy with immunotherapy. Nat Rev Cancer 17(5):286

Green JA, Kirwan JM, Tierney JF, Symonds P, Fresco L, Collingwood M, Williams CJ (2001) Survival and recurrence after concomitant chemotherapy and radiotherapy for cancer of the uterine cervix: a systematic review and meta-analysis. Lancet 358(9284):781–786

Gurtowska N, Kloskowski T, Drewa T (2010) Ciprofloxacin criteria in antimicrobial prophylaxis and bladder cancer recurrence. Med Sci Monit 16(10):RA218–RA223

Gurunathan S, Jeong J-K, Han JW, Zhang X-F, Park JH, Kim J-H (2015) Multidimensional effects of biologically synthesized silver nanoparticles in *Helicobacter pylori*, *Helicobacter felis*, and human lung (L132) and lung carcinoma A549 cells. Nanoscale Res Lett 10(1):35

Hardee CL, Arévalo-Soliz LM, Hornstein BD, Zechiedrich L (2017) Advances in non-viral DNA vectors for gene therapy. Genes 8(2):65

Hegazy WAH, Hensel M (2012) *Salmonella enterica* as a vaccine carrier. Future Microbiol 7(1):111–127

Hegazy WAH, Xu X, Metelitsa L, Hensel M (2012) Evaluation of Salmonella enterica type III secretion system effector proteins as carriers for heterologous vaccine antigens. Infect Immun:06056–06011

Heimesaat MM, Nogai A, Bereswill S, Plickert R, Fischer A, Loddenkemper C, Steinhoff U, Tchaptchet S, Thiel E, Freudenberg MA (2010) MyD88/TLR9 mediated immunopathology and gut microbiota dynamics in a novel murine model of intestinal graft-versus-host disease. Gut 59(8):1079–1087

Hillmen P (2011) Using the biology of chronic lymphocytic leukemia to choose treatment. ASH Educ Program Book 2011(1):104–109

Holler E, Butzhammer P, Schmid K, Hundsrucker C, Koestler J, Peter K, Zhu W, Sporrer D, Hehlgans T, Kreutz M (2014) Metagenomic analysis of the stool microbiome in patients receiving allogeneic stem cell transplantation: loss of diversity is associated with use of systemic antibiotics and more pronounced in gastrointestinal graft-versus-host disease. Biol Blood Marrow Transplant 20(5):640–645

Hong DS, Banerji U, Tavana B, George GC, Aaron J, Kurzrock R (2013) Targeting the molecular chaperone heat shock protein 90 (HSP90): lessons learned and future directions. Cancer Treat Rev 39(4):375–387

Hoshimoto S, Shingai T, Morton DL, Kuo C, Faries MB, Chong K, Elashoff D, Wang H-J, Elashoff RM, Hoon DS (2012) Association between circulating tumor cells and prognosis in patients with stage III melanoma with sentinel lymph node metastasis in a phase III international multicenter trial. J Clin Oncol 30(31):3819

Hosseinidoust Z, Mostaghaci B, Yasa O, Park B-W, Singh AV, Sitti M (2016) Bioengineered and biohybrid bacteria-based systems for drug delivery. Adv Drug Deliv Rev 106:27–44

Irrazábal T, Belcheva A, Girardin SE, Martin A, Philpott DJ (2014) The multifaceted role of the intestinal microbiota in colon cancer. Mol Cell 54(2):309–320

Jain K, Arya N, Inamdar N, Auti P, Unawane S, Puranik H, Sanap M, Inamke A, Mahale V, Prajapati C (2016) The chemistry and bio-medicinal significance of pyrimidines & condensed pyrimidines. Curr Top Med Chem 16(28):3133–3174

Jansson PJ, Kalinowski DS, Lane DJ, Kovacevic Z, Seebacher NA, Fouani L, Sahni S, Merlot AM, Richardson DR (2015) The renaissance of polypharmacology in the development of anti-cancer therapeutics: Inhibition of the "Triad of Death" in cancer by Di-2-pyridylketone thiosemicarbazones. Pharmacol Res 100:255–260

Jenq RR, Ubeda C, Taur Y, Menezes CC, Khanin R, Dudakov JA, Liu C, West ML, Singer NV, Equinda MJ (2012) Regulation of intestinal inflammation by microbiota following allogeneic bone marrow transplantation. J Exp Med 209(5):903–911

Joshi M, Rizvi SM, Belani CP (2015) Afatinib for the treatment of metastatic non-small cell lung cancer. Cancer Manag Res 7:75

Kaimala S, Al Sbiei A, Cabral-Marques O, Fernandez-Cabezudo MJ, Al-Ramadi BK (2018) Attenuated bacteria as immunotherapeutic tools for cancer treatment. Front Oncol 8:136

Kaina B, Margison GP, Christmann M (2010) Targeting O6-methylguanine-DNA methyltransferase with specific inhibitors as a strategy in cancer therapy. Cell Mol Life Sci 67(21):3663–3681

Kalinichenko VV, Kalin TV (2015) Is there potential to target FOXM1 for 'undruggable'lung cancers? Taylor & Francis

Katsetos CD, Reginato MJ, Baas PW, D'Agostino L, Legido A, Tuszyn JA, Dráberová E, Dráber P (2015) Emerging microtubule targets in glioma therapy. In: Seminars in pediatric neurology, vol 1. Elsevier, pp 49–72

Kawai K, Miyazaki J, Joraku A, Nishiyama H, Akaza H (2013) Bacillus Calmette–Guerin (BCG) immunotherapy for bladder cancer: current understanding and perspectives on engineered BCG vaccine. Cancer Sci 104(1):22–27

Kensler TW, Spira A, Garber JE, Szabo E, Lee JJ, Dong Z, Dannenberg AJ, Hait WN, Blackburn E, Davidson NE (2016) Transforming cancer prevention through precision medicine and immune-oncology. Cancer Prev Res 9(1):2–10

Koh AY, Köhler JR, Coggshall KT, Van Rooijen N, Pier GB (2008) Mucosal damage and neutropenia are required for *Candida albicans* dissemination. PLoS Pathog 4(2):e35

Kommineni S, Bretl DJ, Lam V, Chakraborty R, Hayward M, Simpson P, Cao Y, Bousounis P, Kristich CJ, Salzman NH (2015) Bacteriocin production augments niche competition by enterococci in the mammalian gastrointestinal tract. Nature 526(7575):719

Kozieł R, Zabłocki K, Duszyński J (2006) Calcium signals are affected by ciprofloxacin as a consequence of reduction of mitochondrial DNA content in Jurkat cells. Antimicrob Agents Chemother 50(5):1664–1671

Kucerova P, Cervinkova M (2016) Spontaneous regression of tumour and the role of microbial infection – possibilities for cancer treatment. Anti-Cancer Drugs 27(4):269–277. https://doi.org/10.1097/CAD.0000000000000337

Lacin S, Esin E, Karakas Y, Yalcin S (2015) Metastatic medullary thyroid cancer: a dramatic response to a systemic chemotherapy (temozolomide and capecitabine) regimen. OncoTargets Ther 8:1039–1042. https://doi.org/10.2147/OTT.S82906

Laport GG, Wu J, Logan B, Bachanova V, Hosing C, Fenske T, Longo W, Devine SM, Nademanee A, Gersten I (2016) Reduced-intensity conditioning with fludarabine, cyclophosphamide, and high-dose rituximab for allogeneic hematopoietic cell transplantation for follicular lymphoma: a phase two multicenter trial from the blood and marrow transplant clinical trials network. Biol Blood Marrow Transplant 22(8):1440–1448

Lehar SM, Pillow T, Xu M, Staben L, Kajihara KK, Vandlen R, DePalatis L, Raab H, Hazenbos WL, Morisaki JH (2015) Novel antibody–antibiotic conjugate eliminates intracellular S. aureus. Nature 527(7578):323

Lehouritis P, Springer C, Tangney M (2013) Bacterial-directed enzyme prodrug therapy. J Control Release 170(1):120–131

Li W, Sahu D, Tsen F (2012) Secreted heat shock protein-90 (Hsp90) in wound healing and cancer. Biochim Biophys Acta (BBA)-Mol Cell Res 1823(3):730–741

Liang K, Liu Q, Li P, Han Y, Bian X, Tang Y, Kong Q (2018) Endostatin gene therapy delivered by attenuated *Salmonella typhimurium* in murine tumor models. Cancer Gene Ther 25(7–8):167–183

Lin IY, Van TTH, Smooker PM (2015) Live-attenuated bacterial vectors: tools for vaccine and therapeutic agent delivery. Vaccines 3(4):940–972

Liu K, Fan J, Wu J (2017) Forkhead box protein J1 (FOXJ1) is overexpressed in colorectal cancer and promotes nuclear translocation of β-catenin in SW620 cells. Med Sci Monit 23:856
Liu B, Su F, Lin R, Teng H, Ju Y (2018a) Overexpression of forkhead box M1 is associated poor survival in patients with nonsmall cell lung cancer. J Cancer Res Ther 14:1121–1123
Liu J, Song L, Liu S, Jiang Q, Liu Q, Li N, Wang Z-G, Ding B (2018b) A DNA-based nanocarrier for efficient gene delivery and combined cancer therapy. Nano Lett 18(6):3328–3334
Livanos AE, Greiner TU, Vangay P, Pathmasiri W, Stewart D, McRitchie S, Li H, Chung J, Sohn J, Kim S (2016) Antibiotic-mediated gut microbiome perturbation accelerates development of type 1 diabetes in mice. Nat Microbiol 1(11):16140
López-Tarruella S, Martín M (2009) Recent advances in systemic therapy. Advances in adjuvant systemic chemotherapy of early breast cancer. Breast Cancer Res 11(2):204
Louis P, Hold GL, Flint HJ (2014) The gut microbiota, bacterial metabolites and colorectal cancer. Nat Rev Microbiol 12(10):661
Lubenau H, Springer M (2018) *Salmonella*-based vectors for cancer immunotherapy targeting wilms' tumor gene WT1. Google Patents
Luginbuehl V, Meier N, Kovar K, Rohrer J (2018) Intracellular drug delivery: potential usefulness of engineered Shiga toxin subunit B for targeted cancer therapy. Biotechnol Adv 36:613–623
Lukasiewicz K, Fol M (2018) Microorganisms in the treatment of cancer: advantages and limitations. J Immunol Res 2018:2397808. https://doi.org/10.1155/2018/2397808
Lulla RR, Goldman S, Yamada T, Beattie CW, Bressler L, Pacini M, Pollack IF, Fisher PG, Packer RJ, Dunkel IJ (2016) Phase I trial of p28 (NSC745104), a non-HDM2-mediated peptide inhibitor of p53 ubiquitination in pediatric patients with recurrent or progressive central nervous system tumors: a Pediatric Brain Tumor Consortium Study. Neuro-Oncology 18(9):1319–1325
Lyss AP, Fagundes H, Corrigan P (2011) Chemotherapy and radiation For dummies. Wiley, Somerset
Ma YT, Leonard SM, Gordon N, Anderton J, James C, Huen D, Woodman CB, Palmer DH (2017) Use of a genome-wide haploid genetic screen to identify treatment predicting factors: a proof-of-principle study in pancreatic cancer. Oncotarget 8(38):63635
Madkour LH (2017) Biotechnology of nucleic acids medicines as gene therapeutics and their drug complexes. Chron Pharm Sci 1:204–253
Malik SS, Saeed A, Baig M, Asif N, Masood N, Yasmin A (2018) Anticarcinogenecity of microbiota and probiotics in breast cancer. Int J Food Prop 21(1):655–666
Mao Q, Jiang F, Yin R, Wang J, Xia W, Dong G, Ma W, Yang Y, Xu L, Hu J (2018) Interplay between the lung microbiome and lung cancer. Cancer Lett 415:40–48
Marchesi JR, Dutilh BE, Hall N, Peters WH, Roelofs R, Boleij A, Tjalsma H (2011) Towards the human colorectal cancer microbiome. PLoS One 6(5):e20447
Marchesi JR, Adams DH, Fava F, Hermes GD, Hirschfield GM, Hold G, Quraishi MN, Kinross J, Smidt H, Tuohy KM (2015) The gut microbiota and host health: a new clinical frontier. Gut 65:330–339
Mathews CK (2015) Deoxyribonucleotide metabolism, mutagenesis and cancer. Nat Rev Cancer 15(9):528
McCance KL, Huether SE (2018) Pathophysiology-E-book: the biologic basis for disease in adults and children. Elsevier Health Sciences
McGowan JV, Chung R, Maulik A, Piotrowska I, Walker JM, Yellon DM (2017) Anthracycline chemotherapy and cardiotoxicity. Cardiovasc Drugs Ther 31(1):63–75
Mehta RR, Yamada T, Taylor BN, Christov K, King ML, Majumdar D, Lekmine F, Tiruppathi C, Shilkaitis A, Bratescu L (2011) A cell penetrating peptide derived from azurin inhibits angiogenesis and tumor growth by inhibiting phosphorylation of VEGFR-2, FAK and Akt. Angiogenesis 14(3):355–369
Mendes MCS, Paulino DSM, Brambilla SR, Camargo JA, Persinoti GF, Carvalheira JBC (2018) Microbiota modification by probiotic supplementation reduces colitis associated colon cancer in mice. World J Gastroenterol 24(18):1995–2008. https://doi.org/10.3748/wjg.v24.i18.1995
Mey U, Hitz F, Lohri A, Pederiva S, Taverna C, Tzankov A, Meier O, Yeow K, Renner C (2012) Diagnosis and treatment of diffuse large B-cell lymphoma. Swiss Med Wkly 142:w13511

Miguel N (2016) Hematological extradigestive manifestations of helicobacter pylori infection in childhood. In: Extradigestive manifestations of helicobacter pylori infection-an overview. InTech

Miki K, Nagata T, Tanaka T, Kim Y-H, Uchijima M, Ohara N, Nakamura S, Okada M, Koide Y (2004) Induction of protective cellular immunity against Mycobacterium tuberculosis by recombinant attenuated self-destructing Listeria monocytogenes strains harboring eukaryotic expression plasmids for antigen 85 complex and MPB/MPT51. Infect Immun 72(4):2014–2021

Miller KD, Siegel RL, Lin CC, Mariotto AB, Kramer JL, Rowland JH, Stein KD, Alteri R, Jemal A (2016) Cancer treatment and survivorship statistics, 2016. CA Cancer J Clin 66(4):271–289

Miyazaki J, Onozawa M, Takaoka E, Yano I (2018) Bacillus Calmette–Guérin strain differences as the basis for immunotherapies against bladder cancer. Int J Urol 25(5):405–413

Moleiro J, Ferreira S, Lage P, Dias Pereira A (2016) Gastric malt lymphoma: analysis of a series of consecutive patients over 20 years. United European Gastroenterol J 4(3):395–402

Monje M, Dietrich J (2012) Cognitive side effects of cancer therapy demonstrate a functional role for adult neurogenesis. Behav Brain Res 227(2):376–379

Moslehi JJ (2016) Cardiovascular toxic effects of targeted cancer therapies. N Engl J Med 375(15):1457–1467

Mura S, Nicolas J, Couvreur P (2013) Stimuli-responsive nanocarriers for drug delivery. Nat Mater 12(11):991

Nemunaitis J, Cunningham C, Senzer N, Kuhn J, Cramm J, Litz C, Cavagnolo R, Cahill A, Clairmont C, Sznol M (2003) Pilot trial of genetically modified, attenuated Salmonella expressing the E. coli cytosine deaminase gene in refractory cancer patients. Cancer Gene Ther 10(10):737

Niemierko A, Goitein M (1993) Modeling of normal tissue response to radiation: the critical volume model. Int J Radiat Oncol Biol Phys 25(1):135–145

O'Regan RM, Nahta R (2018) Targeting forkhead box M1 transcription factor in breast cancer. Biochem Pharmacol 154:407–413

Ogi C, Aruga A (2013) Immunological monitoring of anticancer vaccines in clinical trials. Oncoimmunology 2(8):e26012

P Burke M, M Borland K, A Litosh V (2016) Base-modified nucleosides as chemotherapeutic agents: past and future. Curr Top Med Chem 16(11):1231–1241

Paglia P, Guzman CA (1998) Keeping the immune system alerted against cancer. Cancer Immunol Immunother 46(2):88–92

Parr I, Wheeler E, Alexander P (1973) Similarities of the anti-tumour actions of endotoxin, lipid A and double-stranded RNA. Br J Cancer 27(5):370

Pasquini MC, Logan B, Wu J, Devine S, Porter DL, Maziarz RT, Warlick E, Fernandez HF, Alyea EP, Hamadani M (2015) Results of a phase III randomized, multi-center study of allogeneic stem cell transplantation after high versus reduced intensity conditioning in patients with myelodysplastic syndrome (MDS) or acute myeloid leukemia (AML): Blood and Marrow Transplant Clinical Trials Network (BMT CTN) 0901. Am Soc Hematol

Patel PM, Suciu S, Mortier L, Kruit WH, Robert C, Schadendorf D, Trefzer U, Punt CJ, Dummer R, Davidson N (2011) Extended schedule, escalated dose temozolomide versus dacarbazine in stage IV melanoma: final results of a randomised phase III study (EORTC 18032). Eur J Cancer 47(10):1476–1483

Patyar S, Joshi R, Byrav DP, Prakash A, Medhi B, Das B (2010) Bacteria in cancer therapy: a novel experimental strategy. J Biomed Sci 17(1):21

Paul A, Bhattacharya S (2012) Chemistry and biology of DNA-binding small molecules. Curr Sci 102:212–231

Piccart-Gebhart MJ, Holmes AP, Baselga J, De Azambuja E, Dueck AC, Viale G, Zujewski JA, Goldhirsch A, Santillana S, Pritchard KI (2014) First results from the phase III ALTTO trial (BIG 2-06; NCCTG [Alliance] N063D) comparing one year of anti-HER2 therapy with lapatinib alone (L), trastuzumab alone (T), their sequence (T→ L), or their combination (T+ L) in the adjuvant treatment of HER2-positive early breast cancer (EBC). J Clin Oncol 32(15 suppl):LBA4

Plummer R, Lorigan P, Steven N, Scott L, Middleton MR, Wilson RH, Mulligan E, Curtin N, Wang D, Dewji R (2013) A phase II study of the potent PARP inhibitor, Rucaparib (PF-01367338, AG014699), with temozolomide in patients with metastatic melanoma demonstrating evidence of chemopotentiation. Cancer Chemother Pharmacol 71(5):1191–1199

Punj V, Bhattacharyya S, Saint-Dic D, Vasu C, Cunningham EA, Graves J, Yamada T, Constantinou AI, Christov K, White B (2004a) Bacterial cupredoxin azurin as an inducer of apoptosis and regression in human breast cancer. Oncogene 23(13):2367

Punj V, Saint-Dic D, Daghfal S, Kanwar JR (2004b) Microbial based therapy of cancer: a new twist to an age old practice. Cancer Biol Ther 3(8):708–714

Pushalkar S, Hundeyin M, Daley D, Zambirinis CP, Kurz E, Mishra A, Mohan N, Aykut B, Usyk M, Torres LE (2018) The pancreatic cancer microbiome promotes oncogenesis by induction of innate and adaptive immune suppression. Cancer Discov 8(4):403–416

Puyo S, Montaudon D, Pourquier P (2014) From old alkylating agents to new minor groove binders. Crit Rev Oncol Hematol 89(1):43–61

Racioppi M, Di Gianfrancesco L, Ragonese M, Palermo G, Sacco E, Bassi P (2018) The challenges of Bacillus of Calmette-Guerin (BCG) therapy for high risk non muscle invasive bladder cancer treatment in older patients. J Geriatr Oncol 9(5):507–512

Rau K-M, Lin Y-C, Chen Y-Y, Chen J-S, Lee K-D, Wang C-H, Chang H-K (2015) Pegylated liposomal doxorubicin (Lipo-Dox®) combined with cyclophosphamide and 5-fluorouracil is effective and safe as salvage chemotherapy in taxane-treated metastatic breast cancer: an open-label, multi-center, non-comparative phase II study. BMC Cancer 15(1):423

Redelman-Sidi G, Glickman MS, Bochner BH (2014) The mechanism of action of BCG therapy for bladder cancer—a current perspective. Nat Rev Urol 11(3):153

Redgrave LS, Sutton SB, Webber MA, Piddock LJ (2014) Fluoroquinolone resistance: mechanisms, impact on bacteria, and role in evolutionary success. Trends Microbiol 22(8):438–445

Richards J, Warso M, Mehta D, Christov K, Schaeffer C, Yamada T, Beattie C, Bressler L, Das Gupta T (2011) A first-in-class, first-in-human phase I trial of p28, a non-HDM2-mediated peptide inhibitor of p53 ubiquitination in patients with metastatic refractory solid tumors. J Clin Oncol 29(15_suppl):2511–2511

Riddell IA, Lippard SJ (2018) Cisplatin and oxaliplatin: our current understanding of their actions. Metallo-drugs: development and action of anticancer agents. Dev Action Anticancer Agents 18:1

Riglar DT, Silver PA (2018) Engineering bacteria for diagnostic and therapeutic applications. Nat Rev Microbiol 16(4):214

Rizvi SFA (2018) Anthracyclines: mechanism of action, classification, pharmacokinetics and future–a mini review. Int J Biotech Bioeng 4:4–81

Roberts NJ, Zhang L, Janku F, Collins A, Bai R-Y, Staedtke V, Rusk AW, Tung D, Miller M, Roix J (2014) Intratumoral injection of Clostridium novyi-NT spores induces antitumor responses. Sci Transl Med 6(249):249ra111

Rosello A, Warnes G, Meier UC (2012) Cell death pathways and autophagy in the central nervous system and its involvement in neurodegeneration, immunity and central nervous system infection: to die or not to die–that is the question. Clin Exp Immunol 168(1):52–57

Rosenberg B, Van Camp L, Krigas T (1965) Inhibition of cell division in Escherichia coli by electrolysis products from a platinum electrode. Nature 205(4972):698–699

Routy B, Gopalakrishnan V, Daillère R, Zitvogel L, Wargo JA, Kroemer G (2018a) The gut microbiota influences anticancer immunosurveillance and general health. Nat Rev Clin Oncol 15(6):382–396

Routy B, Le Chatelier E, Derosa L, Duong CP, Alou MT, Daillère R, Fluckiger A, Messaoudene M, Rauber C, Roberti MP (2018b) Gut microbiome influences efficacy of PD-1–based immunotherapy against epithelial tumors. Science 359(6371):91–97

Safdari Y, Khalili M, Ebrahimzadeh MA, Yazdani Y, Farajnia S (2015) Natural inhibitors of PI3K/AKT signaling in breast cancer: emphasis on newly-discovered molecular mechanisms of action. Pharmacol Res 93:1–10

Saltzman D, Augustin L, Leonard A, Mertensotto M, Schottel J (2018) Low dose chemotherapy combined with attenuated *Salmonella* decreases tumor burden and is less toxic than high dose chemotherapy in an autochthonous murine model of breast cancer. Surgery 163(3): 509–514

Sanderson MR, Laponogov I, Pan X, Veselkov DA, Crevel IM, Selvarajah J, Branstrom A, Cirz R, Moser HE, Fisher LM (2017) Studies of prokaryotic Type II topoisomerase drug inhibition. Found Crystallogr 70:C633

Sartore-Bianchi A, Pietrantonio F, Amatu A, Milione M, Cassingena A, Ghezzi S, Caporale M, Berenato R, Falcomatà C, Pellegrinelli A (2017) Digital PCR assessment of MGMT promoter methylation coupled with reduced protein expression optimises prediction of response to alkylating agents in metastatic colorectal cancer patients. Eur J Cancer 71:43–50

Scarpin KM, Graham JD, Mote PA, Clarke CL (2009) Progesterone action in human tissues: regulation by progesterone receptor (PR) isoform expression, nuclear positioning and coregulator expression. Nucl Recept Signal 7:e009

Schlossmacher GA, Stevens A, White A (2011) Glucocorticoid receptor mediated apoptosis: mechanisms of resistance in cancer cells. J Endocrinol 211:17–25

Schwabe RF, Jobin C (2013) The microbiome and cancer. Nat Rev Cancer 13(11):800

Sears CL, Garrett WS (2014) Microbes, microbiota, and colon cancer. Cell Host Microbe 15(3):317–328

Sethi TK, Basdag B, Bhatia N, Moslehi J, Reddy NM (2017) Beyond anthracyclines: preemptive management of cardiovascular toxicity in the era of targeted agents for hematologic malignancies. Curr Hematol Malig Rep 12(3):257–267

Sha S, Han H-W, Gao F, Liu T-B, Li Z, Xu C, Zhong W-Q, Zhu H-L (2015) Discovery of fluoroquinolone derivatives as potent, selective inhibitors of PI3Kγ. Med Chem Commun 6(11):2029–2035

Shaked Y, Emmenegger U, Francia G, Chen L, Lee CR, Man S, Paraghamian A, Ben-David Y, Kerbel RS (2005) Low-dose metronomic combined with intermittent bolus-dose cyclophosphamide is an effective long-term chemotherapy treatment strategy. Cancer Res 65(16): 7045–7051

Sharma P, Allison JP (2015) Immune checkpoint targeting in cancer therapy: toward combination strategies with curative potential. Cell 161(2):205–214

Siegel RL, Miller KD, Jemal A (2018) Cancer statistics, 2018. CA Cancer J Clin 68(1):7–30. https://doi.org/10.3322/caac.21442

Sikov WM, Berry DA, Perou CM, Singh B, Cirrincione CT, Tolaney SM, Kuzma CS, Pluard TJ, Somlo G, Port ER (2015) Impact of the addition of carboplatin and/or bevacizumab to neoadjuvant once-per-week paclitaxel followed by dose-dense doxorubicin and cyclophosphamide on pathologic complete response rates in stage II to III triple-negative breast cancer: CALGB 40603 (Alliance). J Clin Oncol 33(1):13

Simms-Waldrip TR, Sunkersett G, Coughlin LA, Savani MR, Arana C, Kim J, Kim M, Zhan X, Greenberg DE, Xie Y (2017) Antibiotic-induced depletion of anti-inflammatory *Clostridia* is associated with the development of graft-versus-host disease in pediatric stem cell transplantation patients. Biol Blood Marrow Transplant 23(5):820–829

Šimůnek T, Štěrba M, Popelová O, Adamcová M, Hrdina R, Geršl V (2009) Anthracycline-induced cardiotoxicity: overview of studies examining the roles of oxidative stress and free cellular iron. Pharmacol Rep 61(1):154–171

Sinha S, Kuo C-Y, Ho JK, White PJ, Jazayeri JA, Pouton CW (2017) A suicidal strain of *Listeria monocytogenes* is effective as a DNA vaccine delivery system for oral administration. Vaccine 35(38):5115–5122

Smyth MJ, Ngiow SF, Ribas A, Teng MW (2016) Combination cancer immunotherapies tailored to the tumour microenvironment. Nat Rev Clin Oncol 13(3):143

Srikanta BM, Nayaka MAH, Dharmesh SM (2011) Inhibition of *Helicobacter pylori* growth and its cytotoxicity by 2-hydroxy 4-methoxy benzaldehyde of Decalepis hamiltonii (Wight & Arn); a new functional attribute. Biochimie 93(4):678–688

Stacchiotti S, Tortoreto M, Bozzi F, Tamborini E, Morosi C, Messina A, Libertini M, Palassini E, Cominetti D, Negri T (2013) Dacarbazine in solitary fibrous tumor: a case series analysis and preclinical evidence vis-a-vis temozolomide and antiangiogenics. Clin Cancer Res 19:5192–5201

Staedtke V, Roberts NJ, Bai R-Y, Zhou S (2016) *Clostridium novyi-NT* in cancer therapy. Genes Dis 3(2):144–152

Stankiewicz-Kranc A, Miltyk W, Skrzydlewska E (2010) Comparison of influence of carmustine and new proline analog of nitrosourea on antioxidant system in breast carcinoma cells (MCF-7). Drug Chem Toxicol 33(1):55–63

Stevens FJ (2008) Homology versus analogy: possible evolutionary relationship of immunoglobulins, cupredoxins, and Cu, Zn-superoxide dismutase. J Mol Recognit 21(1):20–29

Sun Z (2014) Development of gene expression systems in *Bifidobacterium bifidum* S17 and their application for tumor therapy. Universität Ulm

Tamim HM, Musallam KM, Al Kadri HM, Boivin J-F, Collet J-P (2011) Antibiotic use and risk of gynecological cancer. Eur J Obstet Gynecol Reprod Biol 159(2):388–393

Tardito S, Bassanetti I, Bignardi C, Elviri L, Tegoni M, Mucchino C, Bussolati O, Franchi-Gazzola R, Marchiò L (2011) Copper binding agents acting as copper ionophores lead to caspase inhibition and paraptotic cell death in human cancer cells. J Am Chem Soc 133(16):6235–6242

Tardito S, Barilli A, Bassanetti I, Tegoni M, Bussolati O, Franchi-Gazzola R, Mucchino C, Marchiò L (2012) Copper-dependent cytotoxicity of 8-hydroxyquinoline derivatives correlates with their hydrophobicity and does not require caspase activation. J Med Chem 55(23):10448–10459

Taur Y, Xavier JB, Lipuma L, Ubeda C, Goldberg J, Gobourne A, Lee YJ, Dubin KA, Socci ND, Viale A (2012) Intestinal domination and the risk of bacteremia in patients undergoing allogeneic hematopoietic stem cell transplantation. Clin Infect Dis 55(7):905–914

Taylor BN, Mehta RR, Yamada T, Lekmine F, Christov K, Chakrabarty AM, Green A, Bratescu L, Shilkaitis A, Beattie CW (2009) Noncationic peptides obtained from azurin preferentially enter cancer cells. Cancer Res 69(2):537–546

Teckie S, Qi S, Chelius M, Lovie S, Hsu M, Noy A, Portlock C, Yahalom J (2017) Long-term outcome of 487 patients with early-stage extra-nodal marginal zone lymphoma. Ann Oncol 28(5):1064–1069

Teimouri F, Nikfar S, Abdollahi M (2013) Efficacy and side effects of dacarbazine in comparison with temozolomide in the treatment of malignant melanoma: a meta-analysis consisting of 1314 patients. Melanoma Res 23(5):381–389

Torre LA, Bray F, Siegel RL, Ferlay J, Lortet-Tieulent J, Jemal A (2015) Global cancer statistics, 2012. CA Cancer J Clin 65(2):87–108

Torres M, Khan S, Duplanty M, Lozano HC, Morris TJ, Nguyen T, Rostovtsev YV, DeYonker NJ, Mirsaleh-Kohan N (2018) Raman and infrared studies of platinum-based drugs: cisplatin, carboplatin, oxaliplatin, nedaplatin and heptaplatin. J Phys Chem A 122:6934–6952

Tripathi K (2013) Essentials of medical pharmacology. JP Medical Ltd, New Delhi

Türkeri LN, Wientjes MG, Au JL (2015) Intravesical chemotherapy bladder cancer: diagnosis and clinical management:64

Twitty CG, Diago OR, Hogan DJ, Burrascano C, Ibanez CE, Jolly DJ, Ostertag D (2015) Retroviral replicating vectors deliver cytosine deaminase leading to targeted 5-fluorouracil-mediated cytotoxicity in multiple human cancer types. Hum Gene Ther Methods 27(1):17–31

Urruticoechea A, Alemany R, Balart J, Villanueva A, Vinals F, Capella G (2010) Recent advances in cancer therapy: an overview. Curr Pharm Des 16(1):3–10

Valko M, Jomova K, Rhodes CJ, Kuča K, Musílek K (2016) Redox-and non-redox-metal-induced formation of free radicals and their role in human disease. Arch Toxicol 90(1):1–37

Van Der Post RS, van Dieren J, Grelack A, Hoogerbrugge N, van der Kolk LE, Snaebjornsson P, Lansdorp-Vogelaar I, Van Krieken JH, Bisseling TM, Cats A (2018) Outcomes of screening gastroscopy in first-degree relatives of patients fulfilling hereditary diffuse gastric cancer criteria. Gastrointest Endosc 87(2):397–404. e392

van Laar JM, Farge D, Sont JK, Naraghi K, Marjanovic Z, Larghero J, Schuerwegh AJ, Marijt EW, Vonk MC, Schattenberg AV (2014) Autologous hematopoietic stem cell transplantation

vs intravenous pulse cyclophosphamide in diffuse cutaneous systemic sclerosis: a randomized clinical trial. JAMA 311(24):2490–2498
Vejpongsa P, Yeh ET (2014) Prevention of anthracycline-induced cardiotoxicity: challenges and opportunities. J Am Coll Cardiol 64(9):938–945
Vogtmann E, Goedert JJ (2016) Epidemiologic studies of the human microbiome and cancer. Br J Cancer 114(3):237
Wang F, Meng W, Wang B, Qiao L (2014) Helicobacter pylori-induced gastric inflammation and gastric cancer. Cancer Lett 345(2):196–202
Warso M, Richards J, Mehta D, Christov K, Schaeffer C, Bressler LR, Yamada T, Majumdar D, Kennedy S, Beattie C (2013) A first-in-class, first-in-human, phase I trial of p28, a non-HDM2-mediated peptide inhibitor of p53 ubiquitination in patients with advanced solid tumours. Br J Cancer 108(5):1061
Weekley CM, He C (2017) Developing drugs targeting transition metal homeostasis. Curr Opin Chem Biol 37:26–32
Wei MQ, Mengesha A, Good D, Anné J (2008) Bacterial targeted tumour therapy-dawn of a new era. Cancer Lett 259(1):16–27
White SC, Pharoah MJ (2014) Oral radiology-E-Book: principles and interpretation. Elsevier Health Sciences
Wishart DS (2016) Emerging applications of metabolomics in drug discovery and precision medicine. Nat Rev Drug Discov 15(7):473
Wolter P, Hanselmann S, Pattschull G, Schruf E, Gaubatz S (2017) Central spindle proteins and mitotic kinesins are direct transcriptional targets of MuvB, B-MYB and FOXM1 in breast cancer cell lines and are potential targets for therapy. Oncotarget 8(7):11160
Wu T-C, Hung C-F (2018) Anti-cancer DNA vaccine employing plasmids encoding signal sequence, mutant oncoprotein antigen, and heat shock protein. Google Patents
Wu X, Wanders A, Wardega P, Tinge B, Gedda L, Bergstrom S, Sooman L, Gullbo J, Bergqvist M, Hesselius P (2009) Hsp90 is expressed and represents a therapeutic target in human oesophageal cancer using the inhibitor 17-allylamino-17-demethoxygeldanamycin. Br J Cancer 100(2):334
Yadav V, Varshney P, Sultana S, Yadav J, Saini N (2015) Moxifloxacin and ciprofloxacin induces S-phase arrest and augments apoptotic effects of cisplatin in human pancreatic cancer cells via ERK activation. BMC Cancer 15(1):581
Yamada T, Fialho AM, Punj V, Bratescu L, Gupta TKD, Chakrabarty AM (2005) Internalization of bacterial redox protein azurin in mammalian cells: entry domain and specificity. Cell Microbiol 7(10):1418–1431
Yamada T, Gupta TKD, Beattie CW (2016) p28-mediated activation of p53 in G2/M phase of the cell cycle enhances the efficacy of DNA damaging and antimitotic chemotherapy. Cancer Res 76:2355
Yang F, Teves SS, Kemp CJ, Henikoff S (2014) Doxorubicin, DNA torsion, and chromatin dynamics. Biochim Biophys Acta (BBA)-Rev Cancer 1845(1):84–89
Yin Y, Wang Q-X, Chen X, Xing J, Fan Y-R, Wu Z-W, Wang J-J, Xu G-X (2011) Antitumor efficacy of *Bifidobacterium longum* carrying endostatin gene enriched with selenium and the distribution of selenium. Afr J Microbiol Res 5(31):5615–5621
Yoon W, Yoo Y, Chae Y, Kee S, Kim B (2018) Therapeutic advantage of genetically engineered *Salmonella typhimurium* carrying short hairpin RNA against inhibin alpha subunit in cancer treatment. Ann Oncol 29:2010–2017
Zhao M, Yang M, Li X-M, Jiang P, Baranov E, Li S, Xu M, Penman S, Hoffman RM (2005) Tumor-targeting bacterial therapy with amino acid auxotrophs of GFP-expressing *Salmonella typhimurium*. Proc Natl Acad Sci 102(3):755–760
Zheng JH, Nguyen VH, Jiang S-N, Park S-H, Tan W, Hong SH, Shin MG, Chung I-J, Hong Y, Bom H-S (2017) Two-step enhanced cancer immunotherapy with engineered *Salmonella typhimurium* secreting heterologous flagellin. Sci Transl Med 9(376):eaak9537
Zhou L, Wang X, Liu Q, Wang Q, Zhao Y, Zhang Y (2010) A novel multivalent vaccine based on secretary antigen-delivery induces protective immunity against *Vibrio anguillarum* and *Aeromonas hydrophila*. J Biotechnol 146(1–2):25–30

Zhou S, Vogelstein B, Kinzler K, Kim K, Saha S (2017) Use of bacteria, bacterial products, and other immunoregulatory entities in combination with anti-ctla-4 and/or anti-pd-1 antibodies to treat solid tumor malignancies. Google Patents

Zhu H, Xu W-Y, Hu Z, Zhang H, Shen Y, Lu S, Wei C, Wang Z-G (2017) RNA virus receptor Rig-I monitors gut microbiota and inhibits colitis-associated colorectal cancer. J Exp Clin Cancer Res 36(1):2

Zitvogel L, Ayyoub M, Routy B, Kroemer G (2016) Microbiome and anticancer immunosurveillance. Cell 165(2):276–287

Zitvogel L, Daillère R, Roberti MP, Routy B, Kroemer G (2017) Anticancer effects of the microbiome and its products. Nat Rev Microbiol 15(8):465

Zullo A, Hassan C, Ridola L, Repici A, Manta R, Andriani A (2014) Gastric MALT lymphoma: old and new insights. Ann Gastroenterol 27(1):27

Biodiesel and the Potential Role of Microbial Lipases in Its Production

4

Abhishek Sharma, Shadiya, Tanvi Sharma, Rakesh Kumar, Khemraj Meena, and Shamsher Singh Kanwar

Abstract

Biodiesel is an alternative energy source to conventional fuel and has environmental acceptance due to its biodegradability, low toxicity and renewability. Biodiesel consists chemically of fatty acid alkyl esters and had immense prospect as a substitute liquid fuel. Biodiesel is primarily produced by trans-esterification of oils or fats with chemical catalysts or microbial lipases. The best thing about using microbial lipases as catalysts is that they have the capability to use all types of glycerides as well as free fatty acids. The main characteristics of microbial lipases are high activity, less production inhibition, less reaction time, thermostability, reusability of the immobilized enzyme and alcohol resistance. Using immobilized lipase trans-esterification of oil feedstock becomes quite feasible for biodiesel production. Various immobilization processes have to be optimized to enhance biodiesel production. The present chapter significantly deals with the basis and history of biodiesel as well as different sources of oil feedstock for biodiesel production and how microbial lipases act on these oil feedstocks for biodiesel production.

Keywords

Biodiesel · Fatty acid alkyl esters · Microbial lipases · Immobilization

A. Sharma (✉) · Shadiya · R. Kumar · K. Meena · S. S. Kanwar
Department of Biotechnology, Himachal Pradesh University, Shimla, India

T. Sharma
Department of Biotechnology and Bioinformatics, Jaypee University of Information Technology, Solan, India

P. K. Arora (ed.), *Microbial Technology for the Welfare of Society*, Microorganisms for Sustainability 17,
https://doi.org/10.1007/978-981-13-8844-6_4

4.1 Introduction

Fatty acid alkyl esters are also commonly known as biodiesel and show great potential as an alternative liquid fuel. Biodiesel is usually produced by trans-esterification of oils or fats with chemical catalysts or lipase. Biodiesel is an alternative clean-burning renewable fuel similar to conventional 'fossil' diesel. It is made using natural vegetable oils, animal oil/fats or bio-lipids, tallow and waste cooking oil. As it is biodegradable in nature, it is anticipated to be used as a replacement for fossil diesel fuel. It can also be mixed with petroleum diesel fuel in any proportion. Immobilized lipase as the biocatalyst for biodiesel production always draws attention because trans-esterification process by immobilized lipases is 'greener'. The development of green replaceable energy which is renewable and clean has attracted more and more attention due to the exhaustion of fossil fuel and problems faced by the ecological environment generated by the use of fossil fuel in the world today (Verma and Sharma 2016; Manurung et al. 2017; Taher et al. 2017; Irimie et al. 2018; Vakros 2018). The necessity of searching for alternative energy sources that combine environmental friendliness with biodegradability, low toxicity, renewability and less dependence on petroleum products has never been greater. Enzymatic reactions involving lipases can be an excellent alternative to produce biodiesel through a process commonly referred to as alcoholysis, a form of trans-esterification reaction, or through an esterification reaction. Enzymatic trans-esterification has certain advantages over the chemical catalysis of trans-esterification, as it is less energy intensive and allows easy recovery of glycerol and the trans-esterification of glycerides with high free fatty acid contents (Baskar and Aiswarya 2016; Sharma et al. 2016; Sood et al. 2016; Pedro et al. 2018; Sharma et al. 2017a).

Due to the increasing awareness of the depletion of fossil fuel resources and environmental issues, biodiesel became more and more attractive in the recent years. Biodiesel production is a promising and important field of research because of the relevance it gains from the rising petroleum prices and its environmental advantages (Haziratul et al. 2017). Biodiesel is an attractive alternative source of biofuel amongst all available renewable energies. Chemical-based conversion of feedstock into fatty acid methyl esters (FAMEs), a non-purified form of biodiesel, is non-environment-friendly as it generates higher amounts of waste and requires large quantities of water to clean up the fuel. Microbial lipases have the ability to convert oil/fat into FAMEs at ambient conditions and generate less waste with respect to the chemical-based methods, thereby decreasing overall process cost (Baadhe et al. 2014; Budžaki et al. 2017). However, lipase-catalysed biodiesel production is still an immature process which requires detailed understanding of the process at various levels. Overall this chapter details various factors, sources and steps involved in lipase-catalysed biodiesel production and the challenges involved in developing a successful process.

4.1.1 Basis of Biodiesel

Biodiesel is a renewable, biodegradable fuel manufactured domestically from vegetable oils, animal fats or recycled restaurant grease. It is a cleaner-burning replacement for petroleum diesel fuel. Biodiesel meets both the biomass-based diesel and in general advanced biofuel requirement of the renewable fuel standard. Similar to petroleum diesel, biodiesel is used to fuel compression-ignition engines and normally made by chemically reacting lipids like vegetable oil, soybean oil and animal fat (tallow) with alcohol-producing fatty acid esters (Huang et al. 2015; Verma and Sharma 2016; Manurung et al. 2017; Taher et al. 2017; Irimie et al. 2018). Biodiesel is meant to be used in standard diesel engines and is thus distinct from the vegetable and waste oils used to fuel converted diesel engines. Biodiesel can be used alone or blended with petro-diesel in any proportion and has many other physical characteristics (Table 4.1). The structural features of individual fatty esters determine the physical properties of biodiesel including ignition quality, heat of combustion, exhaust emissions, oxidative stability, cold flow, viscosity, density and lubricity. Biodiesel blends can also be used as heating oil. The National Biodiesel Board (United States) also has a technical definition of 'biodiesel' as a mono-alkyl ester. The fuel can be produced domestically, from seed to pump, and is non-toxic and biodegradable. Biodiesel typically produces about 60% less net carbon dioxide emissions than petroleum-based diesel, as it is itself (partially, at least) produced from atmospheric carbon dioxide via photosynthesis in plants.

Table 4.1 Physical Features of Biodiesel. (Joshi and Pegg 2007; Kumar and Kant 2013; Budžaki et al. 2017)

Salient features	Values
Specific gravity	0.88
Kinematic viscosity	4.0 to 6.0 (at 40 °C)
Cetane number	47 to 65
Higher heating value	~127,960 Btu/gal
Lower heating value	~119,550
Density	7.3 lb/gal (15.5 °C)
Hydrogen	12 (wt %)
Oxygen	11 (wt %)
Boiling point	315 to 350 (°C)
Flash point	100 to 170 (°C)
Sulphur	0.0 to 0.0015 (wt %)
Cloud point, °C	−3 to 15 (°C)
Pour point, °C	−5 to 10 (°C)

4.1.2 Historical Aspects of Biodiesel

J. Patrick and E. Duffy conducted trans-esterification reaction in 1853 and four decades later the first diesel engine became functional. On 10 August 1893 Rudolf Diesel's prime model ran on its own power for the first time in Germany, running with only peanut oil. In memory of this incident, 10 August has been declared as 'International Biodiesel Day' (IEA 2007). Regardless of the prevalent utilization of petroleum-derived diesel fuels, curiosity in vegetable oils as fuels for internal combustion engines was reported in a number of countries like Belgium, the United States, France, Italy, the United Kingdom, Portugal, Germany, Brazil, Argentina, Japan, China, etc. during the 1920s–30s and later during World War II. Several operational inconveniences were reported due to the high viscosity of vegetable oils compared to petroleum diesel fuel, which results in pitiable atomization of the fuel in the fuel spray and frequently leads to deposits and coking of the injectors, combustion chamber and valves. Several attempts to prevail over these troubles included heating of the vegetable oil, blending it with petroleum-derived diesel fuel or ethanol, pyrolysis and cracking of the oils (Zhang et al. 2013; Budžaki et al. 2017; Irimie et al. 2018). The Austrian Biofuels Institute had recognized 21 countries with business-related biodiesel projects. At present biodiesel (100%) is available at various normal service stations across Europe. A Belgian inventor in 1937 proposed that by using trans-esterification vegetable oils can be converted into fatty acid alkyl esters and used as a diesel fuel (Khan et al. 2009).

4.2 Some Common Sources of Oil for Biodiesel Production

The production of biodiesel is commonly done using vegetable oil of soybean, sunflower, palm, rapeseed canola, cotton seed, *Jatropha*, microalgae and other plants (Fig. 4.1) (Irimie et al. 2018;Venkatesagowda et al. 2018; Sharma et al. 2018b; Gutierrez-Lazaro et al. 2018; Liu et al. 2018). Waste vegetable oils and non-edible vegetable oils are economical and hence are predominantly chosen for biodiesel production (Udeh 2017). On the other hand are animal fats, which cannot be used in a similar scope due to differences in the natural properties with the vegetable oils like presence of high saturated fatty acid (Fangrui 1999; Pedro et al. 2018). Generally the production source of oil for biodiesel relies on the climate of that area where the crop desires to be fully fledged grown for biodiesel production.

Foremost challenges of the modern world are energy protection, oil price, resources depletion and climate change, and these challenges have incited significant advances in research and development of biomass-derived energy and fuels. Algal biofuels are generally seen as one of the most promising solutions to global energy catastrophe and climate. The most imperative reward of algae is potentially high yield and no rivalry with food crops (Taher et al. 2011).

Algae mainly grow in aquatic environments and use light and CO_2 to generate biomass. Numerous types of microalgae have long been renowned as potentially good-quality sources for biofuel production because of their relatively high oil

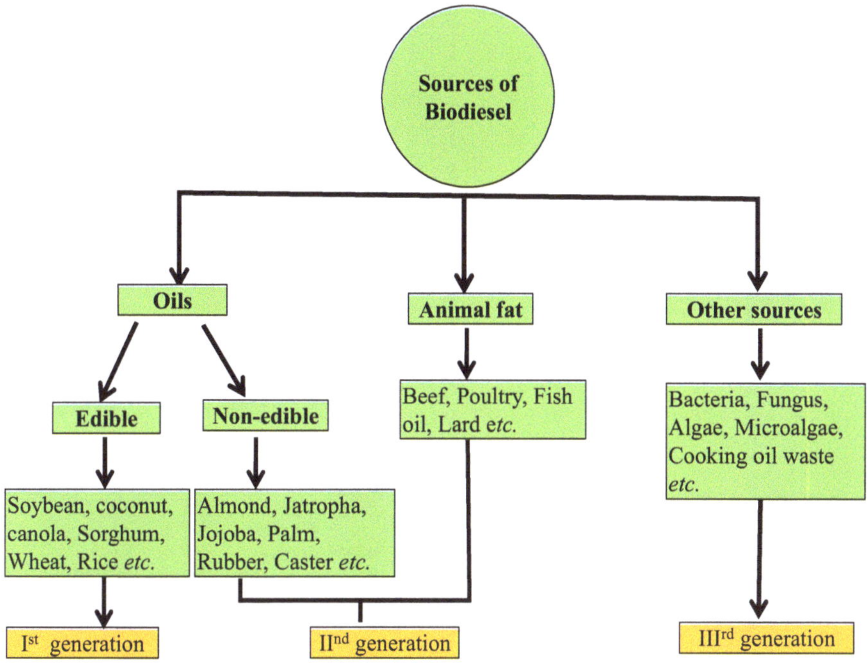

Fig. 4.1 Different sources of oil for biodiesel production. (Irimie et al. 2018)

content and speedy biomass production (Table 4.2) (Ayhan and Fatih 2011). Microalgae grow exceptionally fast compared to terrestrial crops, and also, algal mass culturing can be performed on non-arable lands with non-potable saline water and waste water. Algae and non-edible seeds in contrast to refined or recycled oils have cheaper production cost and have more availability. Hence, use of microalgae as a substitute biodiesel/biofuel feedstock is gaining ever-increasing curiosity from researchers and entrepreneurs (Fig. 4.2; Fatih 2010). Amongst all the feedstock, oleaginous microalgae, when it is not possible for plants, offer another significant advantage.

4.3 Various Production Methods of Biodiesel

Direct use of vegetable oils in engine leads to various inconveniences like fuel filter clogging, poor atomization and incomplete combustion because of high viscosity, high density and poor non-volatility. But transformation of vegetable oils into biodiesel can overcome these problems. These transformations can be performed using four technologies: (i) heating/pyrolysis, (ii) dilution/blending, (iii) micro-emulsion and (iv) trans-esterification (Fig. 4.3). Amongst all these techniques, trans-esterification is an extensive and convenient and the most promising method for the reduction of viscosity, density and other properties of the straight vegetable oils.

Table 4.2 Different algae with oil content involved in biodiesel production. (Ayhan and Fatih 2011)

S. no.	Algae	Oil content
1.	*Botryococcus braunii*	25–75
2.	*Chlorella* sp.	28–32
3.	*Schizochytrium* sp.	50–77
4.	*Crypthecodinium cohnii*	20
5.	*Phaeodactylum tricornutum*	20–30
6.	*Nitzschia* sp.	45–47
7.	*Cylindrotheca* sp.	16–37
8.	*Nannochloris* sp.	20–35
9.	*Nannochloropsis* sp.	31–68
10.	*Tetraselmis suecica*	15–23
11.	*Dunaliella* sp.	23–25

Fig. 4.2 Schematic representation of biodiesel production through algae. (Fatih 2010)

4.3.1 Biodiesel Production Through Microbial Lipases

Biocatalysis is the key feature of microbial enzymes in synthetic chemistry at the present time. Principally it refers to the employment of natural catalysts for novel purposes (Sharma et al. 2016; Sood et al. 2016; Jamwal et al. 2017; Sharma et al. 2017a; Sharma et al. 2018c; Thakur et al. 2018). Moreover, for the synthesis of an extensive variety of commodity chemicals, enzymes are indispensable in bioremediation and waste management because the growing urbanization and industrialization sector had resulted in addition of xenobiotic compounds in nature (Sood et al. 2016; Sharma et al. 2017a, b, c, d, e).

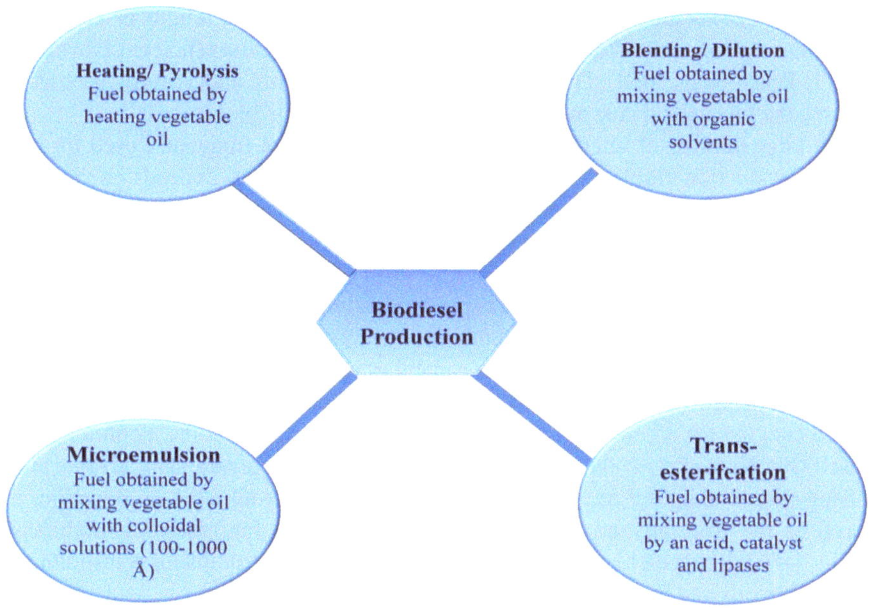

Fig. 4.3 Some common methods of production of biodiesel

In conventional chemical process for biodiesel production, alkali or acid is frequently adopted as the catalyst. Nevertheless, there are a number of problems aligned to chemical processes such as difficulty in glycerol recovery, excessive energy cost and the need for removal of catalyst from the product (Talha and Sulaiman 2016; Vakros 2018). Enzymatic methods *via* lipases can triumph over these troubles which allow mild reaction conditions and no chemical waste is produced. The use of microbial lipase as a biocatalyst for the trans-esterification reaction step in biodiesel production has been extensively investigated (Verma and Sharma 2016). Lipase is produced by all living organisms and can be used intracellularly or extracellularly. The newest drift in microbial lipase study is the expansion of novel and superior lipases through molecular approaches like directed evolution and meta-genomic approach (Gupta et al. 2004). Lipases desire water-insoluble substrates, characteristically triglycerides composed of long-chain fatty acids, whereas esterases preferentially hydrolyse simple esters (Sharma et al. 2016; Sood et al. 2016; Sharma et al. 2017a, b, c, d). Both enzymes have been revealed to be stable and active in organic solvents, although this attribute is extra-prominent with lipases (Fojan et al. 2000; Sharma et al. 2018a, b).

Despite the fact that enzymatic approaches for biodiesel production have become more and more striking, still these approaches have not been realized for industrialization of biodiesel due to the relatively high price of lipase and its short operational life caused by the negative effects of excessive methanol and by-product glycerol (Bajaj et al. 2010; Saifuddin et al. 2015; Pollardo et al. 2017; Taher et al. 2017; Vakros 2018; Pedro et al. 2018). Generally lipases that are used for biodiesel production are acquired from microbial sources like bacteria, fungus or yeast, and enzymatic purification from these sources is cost effective and efficient (Taher et al. 2011; Pedro et al. 2018). Different strategies are presently under study in order to lessen the biocatalyst cost and to achieve a more dynamic biocatalyst.

A large number of microbial lipase variants have been frequently used for biodiesel production (Table 4.3). The choice of microbial lipases depends on their origin and formulation, and in order to transform the glycerides and free fatty acids into alkyl esters, microbial lipases must be nonspecific in nature (Wang et al. 2009; Bhangu et al. 2017; Makareviciene et al. 2017). The most important features of microbial lipase are high activity, less production inhibition, less reaction time, thermostability, reusability of the immobilized enzyme and alcohol resistance. The most excellent microbial lipases have capability to produce conversions >90% with moderate temperature (30–50 °C) and for 7–8-h reaction times while for immobilized enzyme for 85–90 h for the same free enzymes based on the alcohol and oil used (Fjerbaek et al. 2009; Hama and Kondo 2013). The preeminent advantage of using microbial lipases as catalysts is their potential to use all types of glycerides as well as the free fatty acids from different oil sources for biodiesel production (Fig. 4.4). It has been confirmed that >1/2 molar equivalent methanol is insoluble in vegetable oils and the immobilized lipases are easily inactivated by contacting with insoluble methanol existing as drops in the oils (Shimada et al. 1999; Su and Wei 2008). Stepwise addition of methanol (Shimada et al. 1999; Shimada et al. 2002;

Table 4.3 Production of biodiesel by different microbial lipases

Enzyme	References
Candida rugosa lipase and Novozym 435	Watanabe et al. (2007)
Thermomyces lanuginosus	Yucel (2011)
Novozym 435 and Lipozyme TL IM	Wang et al. (2006)
Candida cylindracea	Park et al. (2008)
Immobilized *Rhizopus oryzae*	Li et al. (2012)
Penicillium expansum	Lai et al. (2012)
Burkholderia sp.	Tran et al. (2012)
Immobilized lipase NS40044 and NS40042	Zhao et al. (2012)
Thermomyces lanuginosus, *Candida antarctica* B and *Mucor miehei*	Bautista et al. (2015)
Bacillus aerius	Narwal et al. (2015)
Novozym 435	Manurung et al. (2017)
Candida antarctica	Pollardo et al. (2017)
Novozym 435	Taher et al. (2017)
Novozym 435, Lipozyme RM IM and Lipozyme TL IM	Pedro et al. (2018)
Lasiodiplodia theobromae	Venkatesagowda et al. (2018)
Rhizopus arrhizus	Sharma et al. (2018b)
Candida antarctica (CALB)	Gutierrez-Lazaro et al. (2018)
Penicillium cyclopium	Liu et al. (2018)
Thermomyces lanuginosus	Wancura et al. (2018)

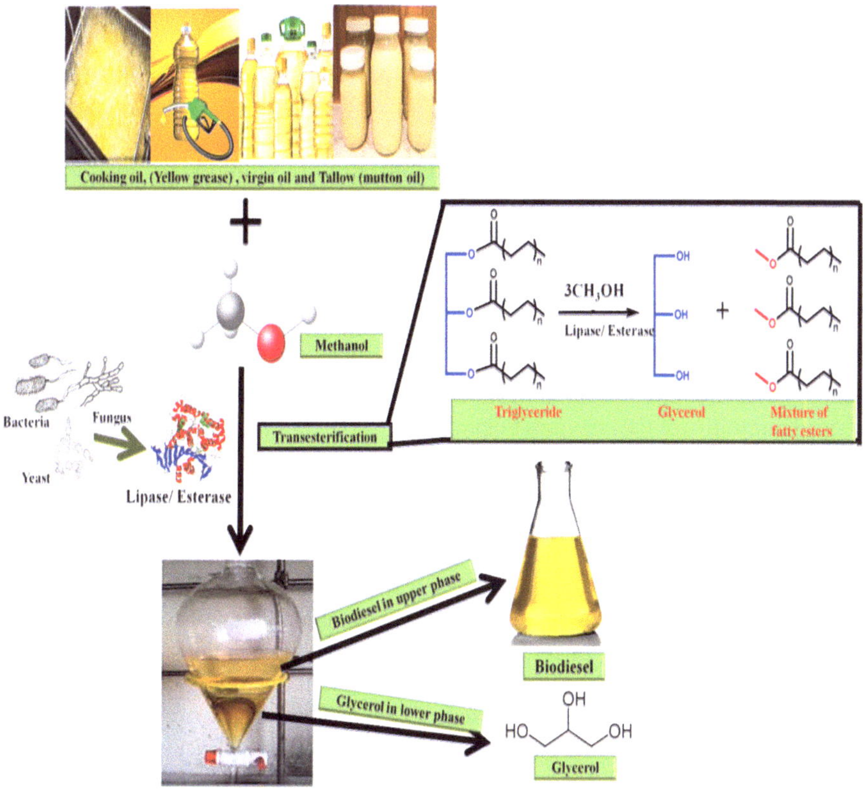

Fig.4.4 A schematic approach for biodiesel production through microbial lipases

Fukuda et al. 2009) or using hydrophobic solvent like n-hexane or petroleum ether as reaction media (Nelson et al. 1996; Lara and Park 2004; Soumanou and Bornscheuer 2003) has been proposed to decrease the negative effect of methanol on lipase activity. By-product, i.e. glycerol, is hydrophilic and insoluble in the oil, which is quite frequently adsorbed onto the surface of the immobilized lipase leading to negative effect on lipase activity and operational stability (Soumanou and Bornscheuer 2003; Dossat et al. 2002; Su et al. 2007). Numerous methods have also been proposed to eradicate the negative effect caused by glycerol. The important one is addition of silica gel into the reaction cocktail to absorb the glycerol (Sharma and Kanwar 2014) or washing the lipase with some organic solvents regularly to eliminate glycerol (Kumar et al. 2016). Microbial enzymes often show higher activity in comparatively hydrophobic organic solvents with higher log P >2 like n-hexane and petroleum ether, and hence these hydrophobic organic solvents should also be tried as reaction media for biodiesel production (Sharma and Kanwar 2014; Kumar et al. 2016; Sharma et al. 2018d). Noticeably, microbial lipases from different sources with different properties are suitably selected for biodiesel production.

At present, trans-esterification using plant oils, animal fats or lipids from microalgae is the major method of biodiesel production.

Competency of reuse and recycling of lipase is a crucial factor in enzymatic biodiesel production as the high price of microbial lipase enzymes is one of the constraints. In order to decrease the cost, enzymes must be reused while maintaining a high level of activity. Enzyme immobilization is an important approach that could be used as a tool to improve and optimize operation stability, activity and selectivity which allows the enzymes to be studied under harsher environmental conditions and also provides their separation from the reaction mixture without filtration in case of packed bed reactor (Fernandez-Lafuente et al. 1998; Bhushan et al. 2008; Gao et al. 2006; Sharma et al. 2017a; Sharma et al. 2018a, b, c) and, hence, could lead to more favourable economical benefits. It is the cultivation method and strength of immobilization matrix which ultimately decide the longevity and durability of the enzyme (Fukuda et al. 2009; Sharma et al. 2017b, c; Sharma et al. 2018a, b, c). Numerous immobilization techniques have been discovered in recent decades. The methods for immobilization can be categorized into reversible and irreversible immobilization based on the interactions between lipase and the carrier. The common techniques available for enzyme immobilization are carrier binding, cross linking, entrapment and encapsulation (Fig. 4.5). Amongst all the methods available, adsorption is the most favourable, as it is a simple and cost-effective approach. Each of the techniques has its own advantages and disadvantages based on the operating conditions; thus different operating conditions require different immobilized lipase techniques. The major drawback of the immobilized lipase, when compared with the free one, would be the possibility of creating mass

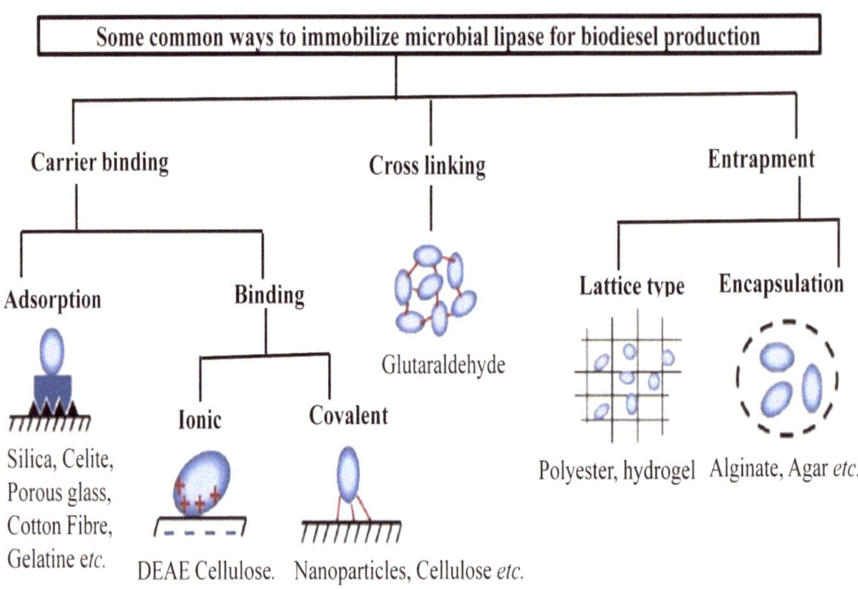

Fig. 4.5 A schematic approach for biodiesel production through microbial lipases

transfer barriers due to enzyme entrapment/encapsulation or deactivation of enzymes when binding them to the surface of the carriers.

4.4 Present Scenario of Biodiesel Production in the World

The United States and Brazil were amongst the largest biodiesel producers in the world, totalling some 5.5 and 3.8 billion litres, respectively, in 2016. The United States is projected to reach production levels of over 1 billion gallons of biodiesel by 2025. After the implementation of the Energy Policy Act of 2005 which provided tax incentives for certain types of energy, biodiesel production in the United States began to increase (Fig. 4.6). The Volumetric Ethanol Excise Tax Credit is currently one of the main sources of financial support for biofuels in the United States. In 2010, the United States exported about 85 million gallons of their biodiesel products. Comparatively, Argentina accounted for over half of the world's total exports. The United States has one of the highest bioenergy capacities in the world, totalling 13,764 megawatts in 2015 (Araújo et al. 2017).

4.5 Significance of Biodiesel

The use of biodiesel helps reduce dependence on finite fossil fuel reserves. As an alternative energy source, it is relatively easy to process for all communities from rural communities in developing nations to urban ones in developed countries. Scientific research confirms that biodiesel exhaust has a less harmful impact on human health than petroleum diesel fuel. Hence, due to the various well-promised advantages of biodiesel, it imparts a valuable role to human beings in the modern era (Fig. 4.7). The advancement of alternative energy is primarily catalysed by negative environmental impacts and energy depletion caused by the excessive usage of

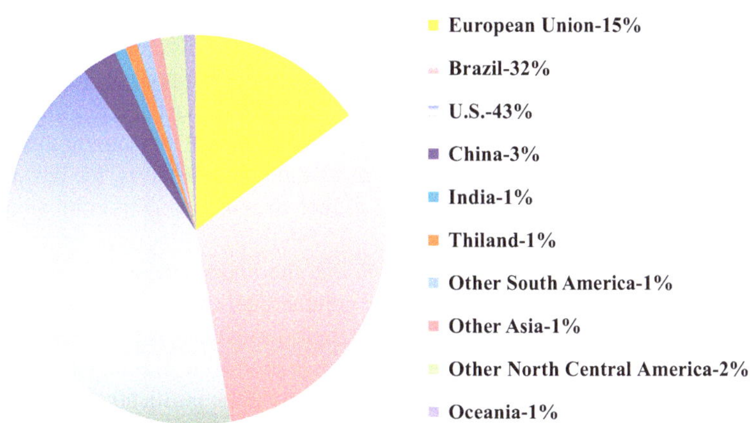

Fig. 4.6 Current status of biodiesel production in the world. (Araújo et al. 2017)

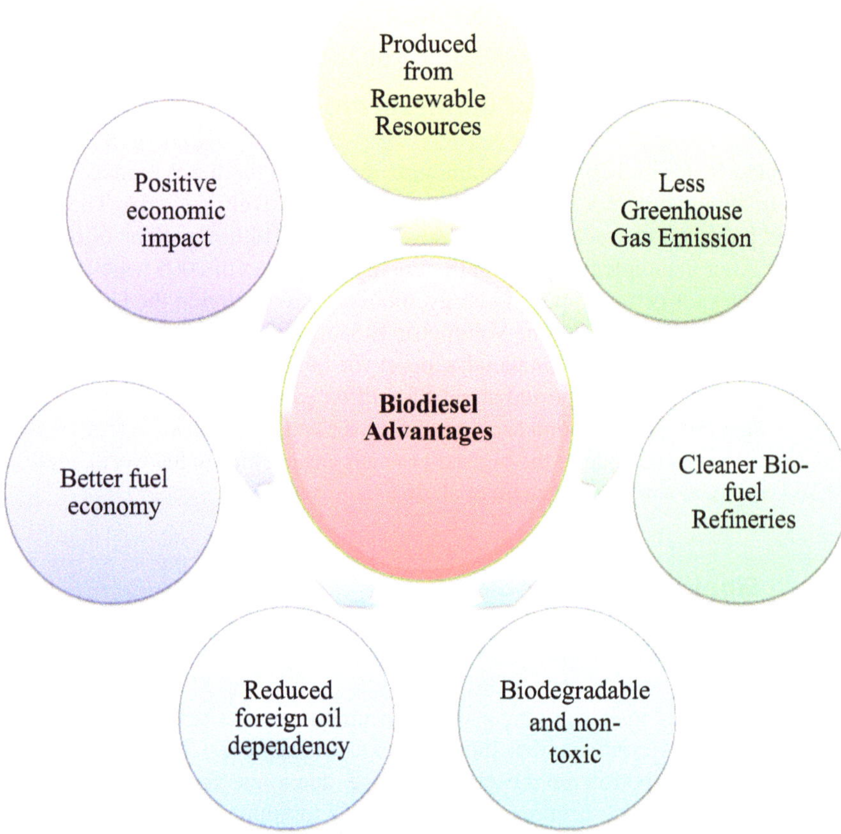

Fig. 4.7 Advantages of the biodiesel. (Babu et al. 2012; Tenzin and Krishna 2014)

fossil fuels. Biodiesel has emerged as a promising substitute to petro-diesel because it is biodegradable and less toxic and reduces greenhouse gas emission. Apart from that, biodiesel can be used as blending component or direct replacement for diesel fuel in automotive engines. A diverse range of methods have been reported for the conversion of renewable feedstocks (vegetable oil or animal fat) into biodiesel with trans-esterification being the most preferred method. Nevertheless, the cost of producing biodiesel is higher compared to fossil fuel, thus impeding its commercialization potentials.

The limited source of reliable feedstock and the underdeveloped biodiesel production route have prevented the full-scale commercialization of biodiesel in many parts of the world and, hence, overall impart their disadvantages to human beings in the modern era (Fig. 4.8). Although biodiesel is a promising alternative fuel, it still has many challenging issues. One of the problems is its purification process. The traditional wet washing method can increase the production costs so it is important to investigate other reusable and inexpensive purification methods. In addition, the

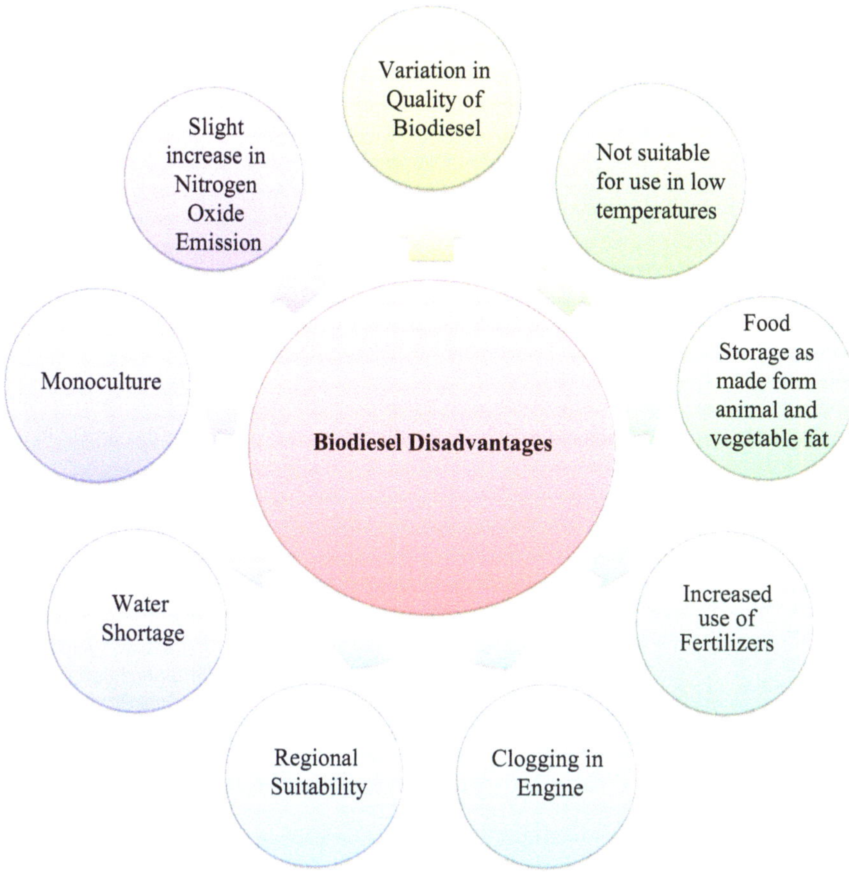

Fig. 4.8 Disadvantages of biodiesel. (Babu et al. 2012; Tenzin and Krishna 2014)

application of biodiesel is limited due to its poor cold weather properties. In order to improve performance in the cold climate, additives are typically added to biodiesel.

4.6 Conclusion

Continuously increasing use of fossil fuels to fulfil our daily requirements of energy will soon lead to depletion in the availability of these fuels. A clear indication of this situation can be observed with the increase in fuel prices in recent years. The need to look for alternative fuel sources to meet the energy requirements has increased manifold. Out of the many alternative energy sources known, biodiesels have been attracting attention as liquid transportation fuels. This is because biodiesel is produced from biomass sources which can provide a renewable carbon feedstock. The biodiesel from refined vegetable oils meets the Indian requirements of high-speed diesel oil. But the production of biodiesel from edible oils is currently much more expensive than diesel fuels due to relatively high cost of edible oils. However, there remains a huge scope for exploration of more oil-rich sources. This is because long-term utilization of one or two species may affect the ecological balance. Moreover, the animal sources for biodiesel production are still underexplored. In the present piece of work, an attempt was made to explore new oil sources for biodiesel and also to develop eco-friendly and economic catalytic methods for conversion of the oil to biodiesel.

Acknowledgements The financial support provided by DST-SRF (DST/INSPIRE Fellowship/2013/1036; IF 131077) to one of the authors (AS) in the form of a junior research fellowship by the Department of Science and Technology, New Delhi (India), is thankfully acknowledged.

Conflict of Interest The authors declare no conflict of interest publishing this article in this book.

References

Araújo K, Mahajan D, Kerr R, Silva MD (2017) Global biofuels at the crossroads: an overview of technical, policy, and investment complexities in the sustainability of biofuel development. Agriculture 7:1–22

Ayhan D, Fatih DM (2011) Importance of algae oil as a source of biodiesel. Energy Convers Manag 52:163–170

Baadhe RR, Potumarthi R, Gupta VK (2014) Lipase-catalysed biodiesel production: technical challenges. Bioenergy Res Adv Appl 2014:119–129

Babu S, Penugonda, Venkata M (2012) Significance of biodiesel use as I.C. Engine Fuels 1:40–43

Bajaj A, Purva L, Jha P, Mehrotra R (2010) Biodiesel production through lipase catalyzed transesterification: an overview. J Mol Catal B Enzym 62:9–14. https://doi.org/10.1016/j.molcatb.2009.09.018

Baskar G, Aiswarya R (2016) Trends in catalytic production of biodiesel from various feedstocks. Renew Sust Energ Rev 57:496–504

Bautista LF, Vicente G, Mendoza Á, González S, Morales V (2015) Enzymatic production of biodiesel from *Nannochloropsis gaditana* microalgae using immobilized lipases in mesoporous materials. Energy Fuel 29(8):4981–4989

Bhangu SK, Gupta S, Ashok KM (2017) Ultrasonic enhancement of lipase-catalysed transesterification for biodiesel synthesis. Ultrason Sonochem 34:305–309

Bhushan I, Parshad R, Gazi G, Gupta VK (2008) Immobilization of lipase by entrapment in ca-alginate beads. J Bioact Compat Polym 23:552–562

Budžaki S, Miljić G, Tišma M, Sundaram S, Hessel V (2017) Is there a future for enzymatic biodiesel industrial production in microreactors? Appl Energy 201:124–134

Dossat V, Combes D, Marty A (2002) Efficient lipase catalysed production of a lubricant and surfactant formulation using a continuous solvent-free process. J Biotechnol 97:117–124

Fangrui M (1999) Hanna MA. 'Biodiesel production: A review'. Bio Source Technol 70:1–15

Fatih DM (2010) Microalgae as a feedstock for biodiesel. Energ Educ Sci Technol A Energ Sci Res 25:31–43

Fernandez-Lafuente R, Armisen P, Sabuquillo P, Fernandez-Lorente G, Guisan JM (1998) Immobilization of lipases by selective adsorption on hydrophobic supports. Chem Phys Lipids 93:185–197

Fjerbaek L, Christensen KV, Norddahl B (2009) A review of the current state of biodiesel production using enzymatic transesterification. Biotechnol Bioeng 102(5):1298–1315
Fojan P, Jonson PH, Petersen MT, Petersen SB (2000) What distinguishes an esterase from a lipase: a novel structural approach. Biochimie 82:1033–1041
Fukuda H, Akihiko K, Sriappareddy T (2009) Bioenergy: sustainable fuels from biomass by yeast and fungal whole-cell biocatalysts. Biochem Eng J 44:2–12
Gao Y, Tan T, Nie K, Wang F (2006) Immobilization of lipase on macroporous resin and its application in synthesis of biodiesel in low aqueous media. Chin J Biotechnol 22:114–118
Gupta R, Gupta N, Rathi P (2004) Bacterial lipases: an overview of production, purification and biochemical properties. Appl Microbiol Biotechnol 64:763–781
Gutierrez-Lazaro A, Velasco D, Boldrini DE, Yustos P, Esteban J, Ladero M (2018) Effect of operating variables and kinetics of the lipase catalyzed transesterification of ethylene carbonate and glycerol. Fermentation 4:1–14
Hama S, Kondo A (2013) Enzymatic biodiesel production: an overview of potential feedstocks and process development. Bioresour Technol 135:386–395
Haziratul HM, Ong HW, Masjuki HH, Lim S, Lee HV (2017) A review on latest developments and future prospects of heterogeneous catalyst in biodiesel production from non-edible oils. Renew Sustain Energ Rev 67:1225–1236
Huang D, Zhou H, Lin L (2015) Biodiesel: an alternative to conventional fuel. Energy Procedia 16:1874–1885
IEA (2007) World energy outlook 2007. International Energy Agency Paris. http://www.mybiodiesel.com/biodiesel-history.php
Irimie FD, Paizs C, Tosa MI, Bencze LC (2018) Biodiesel, a green fuel obtained through enzymatic catalysis. Biomass Renewable Raw Material Obtain Bioproducts of High-tech Value 191–234. https://doi.org/10.1016/B978-0-444-63774-1.00006-5
Jamwal S, Kumar R, Sharma A, Kanwar SS (2017) Response surface methodology (RSM) approach for improved extracellular RNase production *by a Bacillus* sp. J Adv Microbiol 3:131–144
Joshi RM, Pegg MJ (2007) Flow properties of biodiesel fuel blends at low temperatures. Fuel 86:143–151
Khan SA, Mir ZR, Prasad HS, Banerjee UC (2009) Prospects of biodiesel production from microalgae in India. Renew Sustain Energ Rev 13:2361–2372
Kumar V, Kant P (2013) Study of physical and chemical properties of biodiesel from sorghum oil. Res J Chem Sci 3:64–68
Kumar A, Dhar K, Kanwar SS, Arora PK (2016) Lipase catalysis in organic solvents: advantages and applications. Biol Procedures Online 2016:1–18
Lai J-Q, Hu Z-L, Wang P-W, Yang Z (2012) Enzymatic production of microalgal biodiesel in ionic liquid [BMIm][PF 6]. Fuel 95:329–333
Lara PV, Park EY (2004) Potential application of waste activated bleaching earth on the production of fatty acid alkyl esters using *Candida cylindracea* lipase in organic solvent system. Enzym Microb Technol 34:270–277
Li X, He X-Y, Li Z-L, Wang Y-D, Wang C-Y, Shi H, Wang F (2012) Enzymatic production of biodiesel from *Pistacia chinensis* bge seed oil using immobilized lipase. Fuel 92(1):89–93
Liu YH, Lin Z, Dong F, Yu S, Mengying X, Zehua M, Jieying LF (2018) Development of a *Pichia pastoris* whole-cell biocatalyst with overexpression of mutant lipase I PCL G47I from *Penicillium cyclopium* for biodiesel production. RSC Adv 8:26161–26168
Makareviciene V, Gumbyte M, Virginija S, Sendzikiene E (2017) Biodiesel fuel production by enzymatic microalgae oil transesterification with ethanol. J Renewable Sustain Energy 9:1–13
Manurung R, Wage S, Taslim T (2017) Optimization of enzymatic process in biodiesel production for green development. J Eng Appl Sci 12(9):2485–2491
Narwal SK, Saun NK, Dogra P, Chaihan G, Gupta R (2015) Production and characterization of biodiesel using nonedible castor oil by immobilized lipase from *Bacillus aerius*. Biomed Res Int 2015:1–6
Nelson LA, Foglia TA, Marmer WN (1996) Lipase-catalyzed production of biodiesel. J Am Oil Chem Soc 73:1191–1195

Park EY, Sato M, Kojima S (2008) Lipase-catalyzed biodiesel production from waste activated bleaching earth as raw material in a pilot plant. Bioresour Technol 99(8):3130–3135
Pedro KCNR, Parreiraa JM, Correiaa IN, Henriquesa CA, Langone MAP (2018) Enzymatic biodiesel synthesis from acid oil using a lipase mixture. Quim Nova 41:284–291
Pollardo AA, Lee H, Lee D, Sangyong K, Jaehoon K (2017) Effect of supercritical carbon dioxide on the enzymatic production of biodiesel from waste animal fat using immobilized *Candida antarctica* lipase B variant. BMC Biotechnol 2017:1–6
Saifuddin N, Samiuddin A, Kumaran P (2015) A review on processing technology for biodiesel production. Trends Appl Sci Res 10:1–37
Sharma S, Kanwar SS (2014) Organic solvent tolerant lipases and applications. Sci World J 2014:1–15
Sharma A, Sharma T, Kanwar SS (2018a) Entrapment of *Streptomyces* sp. lipase in cupric-beads for methyl salicylate synthesis. Trends Carbo Res 10:21–29
Sharma T, Sharma A, Kanwar SS (2016) Purification and characterization of an esterase from *Bacillus pumilus*. J Adv Biotechnol Bioeng 4:9–16
Sharma A, Meena KR, Kanwar SS (2017a) Molecular characterization and bioinformatics studies of a lipase from *Bacillus thermoamylovorans* BHK67. Int J Biol Macromol 107:2131–2140
Sharma A, Meena KR, Rana S, Singh M, Kanwar SS (2017b) Fabrication and functionalization of magnesium nanoparticle for lipase immobilization in *n*-propyl gallate synthesis. J King Saud Univ Sci 29:536–546
Sharma A, Sharma T, Meena KR, Kanwar SS (2017c) Physical adsorption of lipase onto mesoporous silica. Int J Curr Adv Res 6:3836–3841
Sharma A, Sharma T, Gupta R, Kanwar SS (2017d) Correlation of enzyme activity and stabilities with extreme environments: complexities and flexibilities in adaptations. In: Dr VR (ed) Recent advances in biotechnology, vol 1. Shree Publishers, New Delhi, pp 17–31
Sharma T, Sharma A, Kanwar SS (2017e) An overview on esterases: structure, classification, sources and their application. In Rai V (Ed.) Recent advances in biotechnology, Shree Publishers New Delhi. 2:216-229.
Sharma RK, Saxena M, O'Neill CA, Ramos HAR, Griebenow K (2018b) Synthesis of *Rhizopus arrhizus* lipase nanoparticles for biodiesel production. ACS Omega 3:18203–18213
Sharma A, Sharma T, Meena KR, Kumar A, Kanwar SS (2018c) High throughput synthesis of ethyl pyruvate by employing superparamagnetic iron nanoparticles-bound esterase. Process Biochem 71:109–117
Sharma A, Meena KR, Kanwar SS (2018d) Microbial thermo-tolerant and solvent-tolerant lipases and their applications. Curr Res Microbiol 2:1–15
Shimada Y, Watanabe Y, Samukawa T (1999) Conversion of vegetable oil to biodiesel using immobilized *Candida antarctica* lipase. J Am Oil Chem Soc 76:789–793
Shimada Y, Watanabe Y, Sugihara A, Tominaga Y (2002) Enzymatic alcoholysis for biodiesel fuel production and application of the reaction to oil processing. J Mol Catal B Enzym 17:133–142
Sood S, Sharma A, Sharma N, Kanwar SS (2016) Carboxylesterases: sources, characterization and broader applications. Insight Enzym Res 1:1–11
Soumanou MM, Bornscheuer UT (2003) Improvement in lipase-catalyzed synthesis of fatty acid methyl esters from sunflower oil. Enzym Microb Technol 33:97–103
Su E, Wei D (2008) Improvement in lipase-catalyzed methanolysis of triacylglycerols for biodiesel production using a solvent engineering method. J Mol Catal B Enzym 55:118–125
Su EZ, Zhang MJ, Zhang JG, Gao JF, Wei DZ (2007) Lipase-catalyzed irreversible transesterification of vegetable oils for fatty acid methyl esters production with dimethyl carbonate as the acyl acceptor. Biochem Eng J 36:167–173
Taher H, Al-Zuhair S, Al-Marzouqi AH, Haik Y, Farid MM (2011) A review of enzymatic transesterification of microalgal oil-based biodiesel using supercritical technology. Enzym Res 468292:25
Taher H, Nashefa E, Anvarb N, Al-Zuhair S (2017) Enzymatic production of biodiesel from waste oil in ionic liquid medium. Biofuels 0:1–10

Talha NS, Sulaiman S (2016) Overview of catalysts in biodiesel production. ARPN J Eng Appl Sci 11:439–448

Tenzin T, Krishna M (2014) Advances and biotechnological applications in biofuel production: a review. Open J Renew Sustain Energy 2014:29–34

Thakur N, Kumar A, Sharma A, Bhalla TC, Kumar D (2018) Purification and characterization of alkaline, thermostable and organic solvent stable protease from a mutant of *Bacillus* sp. Biocatal Agric Biotechnol 16:217–224

Tran D-T, Yeh K-L, Chen C-L, Chang J-S (2012) Enzymatic transesterification of microalgal oil from *Chlorella vulgaris* ESP-31 for biodiesel synthesis using immobilized *Burkholderia* lipase. Bioresour Technol 108:119–127

Udeh BA (2017) Biodiesel production from waste vegetable oil (Sunflower) obtained from fried chicken and plant. J Pet Environ Biotechnol 8:1–4

Vakros J (2018) Biochars and their use as transesterification catalysts for biodiesel production: a short review. Catalysts *8*(11):562

Venkatesagowda B, Ponugupaty E, Barbosa-Dekker AM et al (2018) The purification and characterization of lipases from *Lasiodiplodia theobromae*, and their immobilization and use for biodiesel production from coconut oil. Appl Biochem Biotechnol 185:619–640

Verma P, Sharma MP (2016) Review of process parameters for biodiesel production from different feedstocks. Renew Sustain Energ Rev 62:1063–1071

Wancura JHC, Rosset DV, Tres MV, Oliveira JV, Mazutti MA, Jahn SL (2018) production of biodiesel catalyzed by lipase from *Thermomyces lanuginosus* in its soluble form. Can J Chem Eng 96:2361–2368

Wang L, Du W, Liu D, Li L, Dai N (2006) Lipase-catalyzed biodiesel production from soybean oil deodorizer distillate with absorbent present in tert-butanol system. J Mol Catal B Enzym 43:29–32

Wang Z-G, Wan L-S, Liu Z-M, Huang X-J, Xu Z-K (2009) Enzyme immobilization on electrospun polymer nanofibers: an overview. J Mol Catal B Enzym 56:189–195

Watanabe Y, Nagao T, Nishida Y, Takagi Y, Shimada Y (2007) Enzymatic production of fatty acid methyl esters by hydrolysis of acid oil followed by esterification. J Am Oil Chem Soc 84:1015–1021

Yücel Y (2011) Biodiesel production from pomace oil by using lipase immobilized onto olive pomace. Bioresour Technol 102:3977–3980

Zhang XL, Yan S, Surampalli RY, Tyagi RD (2013) Biodiesel production from heterotrophic microalgae through transesterification and nanotechnology application in the production. Renew Sustain Energ Rev 26:216–223

Zhao X, Peng F, Du W, Liu C, Liu D (2012) Effects of some inhibitors on the growth and lipid accumulation of oleaginous yeast *Rhodosporidium toruloides* and preparation of biodiesel by enzymatic transesterification of the lipid. Bioprocess Biosyst Eng 35:993–1004

Light Mediation as a Strategy to Induce Production of Valuable Microbial Compounds

5

Peck Ting Gan and Adeline Su Yien Ting

Abstract

Microorganisms are highly valued for their production of valuable natural compounds. These microbial-based compounds are often with multiple bioactivities, demonstrating antimicrobial, anti-tumour, and antioxidant attributes, among others. The yield of the compounds is, however, typically low and insufficient to supplement large-scale extraction and purification. To address this limitation, several strategies have been attempted to enhance the production of microbial-based compounds. These include optimization of media composition, genetic level modifications, and modifications to external stimuli. This review emphasizes on the role of light, an external stimuli, in enhancing the production of microbial-based compounds. The use of light as a regulating factor is clearly a more environmental-friendly and low-cost approach than the modification of other regulating factors. In this chapter, the major types of microbial compounds, factors regulating the production of microbial compounds, influence of light on the production of microbial compounds, mechanisms regulating light mediation, and applications for light mediation are discussed.

Keywords

Bacteria · Bioactivities · Fungi · Light mediation · Mechanisms of light regulation · Microbial-based compounds

---

P. T. Gan · A. S. Y. Ting (✉)
School of Science, Monash University Malaysia,
Bandar Sunway, Selangor Darul Ehsan, Malaysia
e-mail: adeline.ting@monash.edu

P. K. Arora (ed.), *Microbial Technology for the Welfare of Society*,
Microorganisms for Sustainability 17,
https://doi.org/10.1007/978-981-13-8844-6_5

5.1 Introduction

Microbial-based compounds are natural compounds produced or isolated from microorganisms (Mishra and Tiwari 2011). The synthesis of these compounds is typically via their metabolic pathway; hence, the compounds are classified as primary or secondary compounds. Primary compounds such as polysaccharides, amino acids, and carbohydrates are produced as part of the natural physiological requirement to maintain cell viability and are often without pharmacological properties. On the contrary, secondary compounds (or often referred to as secondary metabolites) are low-molecular-weight organic compounds that are not involved in growth, development, or reproduction of the organisms. Instead, secondary compounds are usually produced in response to stress, to enhance survival (Li et al. 2016). The structures of the secondary compounds are widely diverse and unique as they are produced typically in response to various biotic and abiotic stresses. Hence, secondary metabolites typically demonstrate antimicrobial, antioxidant, anti-tumour, or antiviral properties (Rao et al. 2017). These compounds often have the potential as lead compounds for the synthesis of useful drugs (Seca and Pinto 2018). In fact, from 1981 to 2006, 52% of the Food and Drug Administration (FDA)-approved drugs comprised of natural compounds. And the World Health Organization (WHO) has estimated that 65% of the natural compounds used are predominantly of plant origin (Cragg and Newman 2013).

Although plant-derived natural products show promising bioactivities, the collection of plant species is limited due to environmental factors, anthropogenic pressure, and the distinct plant growth cycles of various plants (Atanasov et al. 2015). Valuable plants also risk the likelihood of becoming endangered or extinct due to poor harvesting approaches (Cordell 2011; David et al. 2015). Based on a report by the UK-based Botanic Gardens Conservation International, approximately 15,000 medicinal plants are endangered or threatened with extinction (Brower 2008). The high demand of valuable plants led to rampant wildcrafting. One of the most well-known examples of plants 'wildcrafted' is plants with the taxol attribute, in response to the 'taxol supply crisis'. The demand for taxol, an important anticancer compound, increased tremendously after the compound was proven to be effective in treating ovarian and breast cancers. Taxol was derived from the bark of the western yew tree (*Taxus brevifolia* L.), and it remained the only known source for taxol for many years. Unfortunately, yew trees grow slowly and were too small for extraction (Blume 1989; Cragg et al. 1993; Ji et al. 2006). Hence, this results in the extensive wildcrafting of western yew, which subsequently resulted in the rapid increase of yew tree collection from 60,000 pounds in 1987 to 1.6 million pounds in 1992 (Kate and Laird 2000).

The wildcrafting of plants is a temporal solution to the high demands of valuable natural compounds. Alternatives using microorganisms are sought, as microorganisms are easier to cultivate in large quantities and at a lesser cost (Yuan et al. 2016). The use of microorganisms as one of the sources in drug production began with the discovery of penicillin from the filamentous fungus *Penicillium notatum* by Alexander Fleming in the 1930s. And this is followed by discoveries of other new bioactive compounds such as griseofulvin and gramicidin, increasing the rate of new bioactive compounds from microbes by 50% (Penesyan et al. 2015; Gaynes 2017; Lobanovska and Pilla 2017). To date, there are approximately 22,500 bioactive microbial-derived compounds in the world. From this, 17% of the compounds are from prokaryotic sources (e.g. *Bacillus* and *Pseudomonas*), 45% of the compounds are produced by filamentous actinomycetes, and the remaining 38% are derived from eukaryotic fungi (Berdy 2005; Amedei and D'Elios 2012; Shanthakumar et al. 2015). In spite of the high number of available bioactive microbial-based compounds, only less than 1% (~150 compounds) of these compounds can be used directly for human therapy (Raja and Prabakarana 2011).

This review will discuss firstly the types of valuable microbial compounds produced and factors that regulate the production of these compounds and gradually emphasize on the influence of light on production of microbial compounds. A brief description on the mechanisms involved in light regulation will also be presented. The mechanisms to how microbes respond to light, and the influence of light with the microbial compounds produced, are then explored as a strategy to induce production of beneficial microbial compounds.

5.2 Major Types of Microbial Compounds

Microbial compounds typically exhibit more than one specific biological activity due to their diverse and complex functional groups (Cragg and Newman 2013). As such, they find multipurpose use in the pharmaceutical, agricultural, and healthcare industries, primarily as antibiotics against bacteria and fungi, anti-tumour drugs, immunosuppressants, and enzyme inhibitors (Luo et al. 2016). Some examples of the commonly used microbial compounds are summarized in Table 5.1.

Generally, the microbial compounds can be classified into several classes based on their structural differences. And each of these classes of compounds has valuable bioactivities that allow for their multipurpose use in various applications.

Terpenoids, also known as isoprenoids, are a large class of natural organic compounds present in all classes of living organisms. To date, there are 40,000 molecules classified as terpenoids. In general, all microbial terpenoids consist of the basic five-carbon isopentenyl diphosphate (IDP) and the isomer dimethylallyl pyrophosphate (DMAPP) as their building blocks (Pattanaik and Lindberg 2015). Terpenoids can be classified into different groups such as monoterpenoids, diterpenoids, and triterpenoids, according to the number of their basic building blocks (Ali et al. 2017). Terpenoids exhibit a wide range of biological functions, from having anticancer and antimalarial activities to functions as flavouring or fragrance (Mikami 1988). Terpenoids also have a key role in biological processes required for the growth and survival of prokaryotic organism such as cell wall biosynthesis, electron transport, and conversion of light into chemical energy. One of the well-known examples of terpenoids is paclitaxel, which is also known as taxol. Taxol is a diterpenoid, which showed good bioactivity in treating breast, ovarian, and lung cancers (Pattanaik and Lindberg 2015). Besides taxol, another terpenoid, D-limonene (a monoterpenoid), is

Table 5.1 Common microbial compounds and their origin and applications in various industries

Compounds	Produced by	Applications	Reference
Sacrosidase	*Saccharomyces cerevisiae* (baker's yeast)	Yeast enzyme to treat sucrase-isomaltase deficiency in human	Silva et al. (2014)
Lipase	*Aspergillus niger*; *A. oryzae*; *A. flavus*	Accelerate cheese ripening; catalyst in processing biodiesel	Huode et al. (2004)
Erythromycin	*Saccharopolyspora erythraea*	Active against pathogenic Gram-positive bacteria	Wu et al. (2011)
Tetracycline	*Streptomycetes*	Active against Gram-positive and Gram-negative bacteria	Chopra and Roberts (2001)
Melanin	*Alternaria alternata*; *Colletotrichum lagenarium*	Biomaterials for construction and design of biodegradable and biocompatible immune tolerance devices	Solano (2017)
Penicillin	*Penicillium chrysogenum*	Active against Gram-positive bacteria	Ziemons et al. (2017)
Cyclosporin	*Tolypocladium inflatum*	Immunosuppressant for rheumatoid arthritis, Crohn's disease; prevents organ transplant rejection	Colombo and Ammirati (2011); Dehesa et al. (2012)
Daptomycin	*Streptomyces roseosporus*	Active against Gram-positive infections	Steenbergen et al. (2005)
Gibberellin	*Gibberella fujikuroi*	Growth hormones that stimulate cell elongation and fruit ripening	Tudzynski et al. (1998)
Carotenoids	*Zygomycetes*, *Ascomycetes*, *Basidiomycetes*, *Neurospora crassa*	Prevents cancer, antimicrobial agent	Kirti et al. (2014)
Lovastatin	*Aspergillus terreus*	Reduces risk of cardiovascular disease	Luo et al. (2016)
Aclacinomycin	*Streptomyces galilaeus*	Treatment for lung cancer	Singh et al. (2017)
Taxol	*Taxomyces andreanae*; *Nodulisporium sylviforme*	Treatment for breast cancer	Singh et al. (2017)
Lipstatin	*Streptomyces toxytricini*	Pancreatic lipase inhibitor; obstructs gastrointestinal absorption of fat	Singh et al. (2017)

also valued for its anti-tumour activity including towards kidney, skin, and lung cancers (Jongedijk et al. 2016). In recent years, some terpenoids were produced in engineered microbes. These engineered microbes are purposefully manipulated to express production of favoured compounds. Chambon et al. (1991) achieved this by blocking the mevalonate diphosphate decarboxylase (*erg19*) in farnesyl pyrophosphate (FPP) synthase (*erg 20*) in yeast, resulting in the production of two monoterpenoids geraniol and linalool that are widely used as floral fragrance.

Flavonoids are another class of natural compounds that are primarily found in plants. Flavonoids are characterized by the presence of at least one hydroxylated aromatic ring (Halbwirth 2010). Flavonoids can be further classified into flavanones, flavones, isoflavones, and flavonols, based on the carbon (C) ring, which the B ring is attached to, and the degree of unsaturation and oxidation of the C ring (Panche et al. 2016). Flavonoids are recovered in very low yield from plants; thus, they are usually extracted from genetically engineered (recombinant) microorganisms. For example, when the flavonoid specific gene, chalcone isomerases (CHIs) from soybeans, was cloned into *Saccharomyces cerevisiae*, the isoflavone genistein (angiogenesis inhibitor) was produced abundantly (Siddiqui et al. 2011). Flavonoids have also been synthesized by engineered *Escherichia coli*. To date, there are more than 20 pathways from fungus and plants that have been incorporated successfully into *S. cerevisiae* to enhance flavonoid production (Luo et al. 2016).

Alkaloid compounds are nitrogenous compounds, which represent a prominent class of compounds elicited as a defence response to stress factors. Alkaloids exhibit more than one biological function, attributed to their chemical structure of having more than one active functional group. Indole alkaloids are the largest class of alkaloids, which contain more than 4100 known different compounds (Heravi et al. 2018). Novel flavonoids are known to be generated via hydroxylation, dehydroxylation, methylation, glycosylation, and hydrogenation. These biotransformation processes can be performed by fungal species such as *Cunninghamella*, *Penicillium*, and *Aspergillus* sp. One such example is the biotransformation of flavanone to 6-hydroxyflavanone by *Aspergillus niger* and *Penicillium chermesinum*, which is a novel compound exhibiting strong cytotoxic and apoptotic activities against cancer cells (Mikell et al. 2015).

Polyketides are another group of microbial compounds that are of interest to the pharmaceutical industry. Aromatic polyketides exhibit diverse structures between strains and species; hence, they have a broad spectrum of bioactivities, i.e. anticancer, antibacterial, antifungal, and antiviral activities (Das and Khosla 2009). This family includes a number of well-known pharmaceutical compounds such as tetracycline, geldanamycin, doxycyclin, and erythromycin. For example, erythromycin A is a well-known polyketide produced by *Saccharopolyspora erythraea*, which is a common antibiotic to treat various diseases such as respiratory infections, acne, and whooping cough and also as a substitute of penicillin.

5.3 Factors Regulating Production of Microbial Compounds

The production of valuable compounds by microorganisms is dependent on several factors: firstly, the genetic make-up of the microbes themselves and, secondly, their capacity to produce these compounds in response to the stress stimuli. By nature, microorganisms have very low yield of such compounds, and this limits their application in the pharmaceutical, agricultural, and food industries. The microbial compounds are not only produced in low quantities, but also in mixtures with other compounds (non-targeted compounds) in the crude yield (Salas et al. 2017). It is therefore the general aim of many industries to have superior microorganisms, which respond to optimum stimuli, so that high yield can be recovered in the most feasible and economical manner. At such, several factors are extensively studied as means to optimize the conditions, leading to high production and recovery of the targeted microbial compounds.

5.3.1 Genetic Level Regulations

With the advancement of biotechnology and bioinformatics, much can be achieved through genetic manipulation. Metabolic engineering, random mutagenesis, and proteomics are all useful strategies in enhancing the yield of metabolites (Baltz 2001). At present, metabolic engineering is the most popular genetic technology adopted to improve the production of bioactive compounds. This technology uses recombinant DNA technology to modify the expression (in some cases, over-expression) and production of specific compounds (Kumar and Prasad 2011). Haleem et al. (2007) discovered that the genes of *Ralstonia eutropha* that encode for the synthesis of polyhydroxybutyrate (PHB), a polymer used in the synthesis of biodegradable plastic, can be transferred into *Saccharomyces cerevisiae*. The resulting recombinant yeast was able to produce 0.44% polymer of cell dry weight, a significant increase from the 0.08% polymer of cell dry weight produced by the wild strain. Recombining citramalate synthase (CimA) genes (from *Methanococcus jannaschii*, *Leptospira interrogans*, or *Geobacter sulfurreducens*) into *S. cerevisiae* also led to the production of n-butanol, which was seven times higher than the wild strain (Shi et al. 2016).

5.3.2 Carbon Sources

The synthesis of compounds in microorganisms is also affected by the medium composition (Ruiz et al. 2010). Carbon source is one of the most common parameters in regulating the growth of microorganisms and their subsequent synthesis of compounds (Sanchez and Demain 2008). Simple sugars such as glucose, fructose, and sucrose usually elicit better growth and metabolite production than complex carbon sources such as starch, galactose, and mannitols. In most cases, glucose is usually the preferred source for growth and production of metabolites (Bren et al. 2016). Srinubabu et al. (2006) also discovered that production of protease by *Aspergillus oryzae 637* increased significantly from 520 to 530 U/mL when 0.6% (w/v) of glucose was added into the culture medium. Although glucose was the most-studied carbon source, some species have demonstrated preference for other carbon sources such as citrate, sucrose, and aromatic organic acids (Ng and Dawes 1973; Basu et al. 2006; Wang et al. 2015).

5.3.3 Nitrogen Sources

Nitrogen can be obtained through inorganic or organic sources such as ammonia, amino acids, uric acid, nitrate, or nucleobases. Among these, ammonia is usually the preferred nitrogen source for most of the microbes (Tudzynski 2014; Kutvonen et al. 2015; Wang et al. 2016a, b). Studies have revealed that 30 of 45 putative secondary metabolite gene clusters of fungi (e.g. polyketide synthase (PKS) gene, diterpene cyclase gene, and non-ribosomal peptide synthetase gene) depend on nitrogen sources (Tudzynski 2014). Therefore, with optimum nitrogen source supplied, most microorganisms respond positively and production of compounds are enhanced.

Often, a good combination of carbon and nitrogen source (multiple substrates) could render synergistic effect on microorganisms (Egli and Quayle 1986; Wawrik et al. 2005). As such, a balance carbon:nitrogen (C:N) ratio is usually designed to promote optimal growth. Kapoor et al. (2011) discovered that optimizing C:N ratio to 1:1 encouraged the highest production of α-amylase in *Streptomyces* spp. MSC702 (807.64 U/mL). In the absence of carbon sources, production of antimicrobial compounds by *Bacillus* sp. was eight times lower compared to cultures with both nitrogen and carbon sources (Sreerag et al. 2014).

5.3.4 Temperature

Temperature is one of the most common environmental factors that can regulate the production of microbial compounds. The optimum temperature for the production of compounds usually depends on the nature of the microorganisms and the compounds. For example, the mesophilic microorganisms usually preferred the growing temperature of 20–45 °C. Pudi et al. (2016) reaffirmed that growth and alkaloid production of the mesophilic actinomycetes increased when incubated in temperatures from 10 to 30°C, with optimum growth and alkaloid production observed at 30°C (50% improved growth and 190 μg/mL alkaloid production). Similarly, the influence of temperature was also observed for fungi where mesophilic *Aspergillus terreus* achieved optimum growth and production of antimicrobial compounds at 25°C (Mathan et al. 2013). As for thermophiles, the optimum temperature for growth and production of compounds was between 45 and 122 °C. This was revealed by the thermophilic bacteria *Bacillus* spp. LBN 2, which had maximum lipase production at 50 °C (Bora and Bora 2012). On the contrary, low temperatures between −15 and 10°C were only favourable towards psychrophiles (Struvay and Feller 2012). In short, temperature is an important factor as when the temperature exceeds the optimum range, it implicates the ability of the microbes to uptake substrate, repressing microbial growth and metabolism (Nedwell 1999).

5.3.5 pH

The pH factor influences the growth of microbial cells as proton translocation and amino acid degradation may occur in response to adaptation to various pH conditions (Raghavulu et al. 2009). Production of microbial compounds is affected by pH especially via the fermentation process, as fermentation at any pH other than the optimum range will impair the process (Nakashimada et al. 2002; Yu and Fang 2003; Kawagoshi et al. 2005). The impact of pH is evident in the production of several compounds such as acids. One example is the production of itaconic acid by *Aspergillus terreus*. Production of itaconic acids was more favourable under acidic conditions (pH 1.6 to 5.9) than at any other pH range (Boruta and Bizukojc 2017). The precursor of itaconic acids, the cis-aconitic acid, originates from the citric acid cycle, in which citric acids can only be produced under acidic conditions (pH 2.5) (Klement and Buchs 2013). As such, low pH is a requirement for the formation of citric acid and subsequently the itaconic acid. Hence, the manipulation of pH is a primary factor in the formation of itaconic acid.

5.3.6 Light

Light as a factor influencing growth and production of compounds is lesser studied than any other factors (genetic, carbon sources, nitrogen sources, temperature, pH). The influence of light towards microbial cells was attributed to the discovery of phytochromes (a light sensor protein that responds to far red and red light) in cyanobacteria *Synechocystis PCC6803* and *Fremyella diplosiphon*. This came many years after the observation on how plants were affected by the presence of red and far red light (600–700 nm and above 700 nm, respectively) (Butler et al. 1959). Thereafter, researchers discovered that light sensor proteins do not only exist in phototropic cyanobacteria but also in heterotrophic bacteria such as *Deinococcus radiodurans* and *Pseudomonas aeruginosa* and in fungi (Romero et al. 2010).

Microbial cells typically respond and react to specific wavelengths in the light spectrum, rather than to the full spectrum of the white light (390–700 nm). Certain wavelengths in visible light such as blue (400–460 nm) and red light (600–700 nm) are known to be effective regulators in microbial metabolite production. On the contrary, ultraviolet light such as UV-A (315–400 nm), UV-B (280–315 nm), and UV-C (100–280 nm) are detrimental and usually result in cell death (Gharaie et al. 2017). Hence, the earlier studies were primarily focussed on the effects of light on microbial growth and reproduction (Parkin and Brock 1980; Saez et al. 2006; Kuo et al. 2012) rather than on regulation of compound production. It was only in recent years that it was revealed that light may be involved in the regulation and enhanced production of bioactive compounds (Calvo et al. 2002). Upon stimulation by different light wavelengths, certain genes including silent genes in the gene clusters can be activated or regulated. Otherwise, these genes are not expressed under typical white light spectrum, to prevent simultaneous gene expression and the overproduction of compounds. One of the more established genes known to respond to light is the gene responsible for the production of carotenoids. In the presence of blue light (400–460 nm), the carotenoid biosynthesis gene cluster was activated (Imbert and Blondeau 1999; Takano et al. 2005), resulting in light-induced carotenogenesis in *Streptomyces* sp. Similar observations were reported by Rubio et al. (2001) where the biosynthesis of carotenoid in *Myxococcus xanthus* was induced upon the presence of blue light.

The role of light in regulating the production of microbial-based compounds is greatly underestimated. However, based on the few available literature, light can be proposed as a promising approach in regulating production of beneficial microbial compounds. One of the most prominent use of light is to induce the production of carotenoids. Carotenoids are key compounds used widely in the industries as food colourant or feed additives in poultry farming. To date, there are more than 600 different carotenoids that can be produced by plants, algae, bacteria, and fungi, but only a handful can be produced in sufficient quantities for the extraction (Chandi and Gill 2011). In addition, microbial carotenoids have greater stability against external stress such as pH and temperature compared to plant-sourced carotenoids. Studies also revealed that under typical conditions, microorganisms may not produce carotenoids or produce them in very low quantities. However, with the presence of blue light, it was shown that most of the carotenoid-producing microorganisms such as *Neurospora crassa*, *Dunaliella* spp., and *Fusarium fujikuroi* showed a higher yield of carotenoids (Gmoser et al. 2017). As such, the following sections will further discuss the influence of light and the mechanisms of light regulation in microbial cells.

5.4 Influence of Light on Production of Microbial Compounds

5.4.1 Influence of Light Mediation on Fungi

Light is a significant stimulator for fungi. Although fungi do not use light as a source of energy, it triggers other responses in the fungi. Fungi respond to light signals ranging from 450 nm (blue light) to 700 nm (red light), depending on the presence of the various photoreceptors in the fungi. The most common responses by fungi to light are demonstrated by changes to pigmentation and biomass (growth). The effects of light on pigmentation had been widely studied in *Monascus* sp. and *Fusarium* sp. (Velmurugan et al. 2010; Feng et al. 2012; Gmoser et al. 2017). For *Monascus* sp., the biomass increased significantly when exposed to yellow (590–595 nm), red (620–625 nm), and blue (465–479 nm) light. Additionally, in the presence of blue light, production of red pigment was 50% higher compared to cultures incubated in the dark (Wang et al. 2016a, b). In a separate study by Buhler et al. (2015), both red light and dark conditions were able to induce higher production of red pigments (8.36 UA and 7.21 UA, respectively) in *Monascus ruber* compared to cultures incubated under blue, yellow, and green light (6.25 UA, 6.17 UA, and 6.17 UA, respectively). Incubation under normal white light condition revealed the lowest pigment production (4.48 UA).

Lights also influence the developmental stages of the fungi. By exposing *Monascus pilosus* to red (635 nm) and blue light (470 nm), mycelial growth was greater and it forms longer mycelial structures in the latter than in cultures incubated under red light. Nevertheless, red light (1247 ± 175 spores μL^{-1}) was able to induce sporulation in *M. pilosus* compared to blue light (1030 ± 121 μL^{-1}) and dark conditions (643 ± 126 μL^{-1}) (Miyake et al. 2005). Prub et al. (2014) reported similar observations where red light (680 nm) increased sporulation in *Alternaria alternata*, while blue light (450 nm) suppressed sporulation. Based on these studies, it is generally assumed that both blue and red light influence fungal growth, with the former more likely to trigger negative sporulation behaviour while the latter stimulates sporulation. And, more importantly, the response of fungi to light is most likely strain dependent. Cheong et al. (2016) further illustrated the strain-dependency factor when several species of *Aspergillus* responded differently to red light. In their study, irradiation with red and far red light (625 and 740 nm) significantly enhanced the production of ochratoxin A (mycotoxin) in *Aspergillus westerdijkiae* (*A. westerdijkiae*), but a contrasting effect was observed in *Aspergillus carbonarius* (*A. carbonarius*). This hypothesis was further supported by Schmidt-Heydt et al. (2011) where red (627 nm) and blue light (455–470 nm) reduced or inhibited the production of ochratoxin A in *Penicillium verrucosum*, *Aspergillus steynii*, *A. carbonarius*, and *P. nordicum*.

For green light (520–560 nm), only several species such as *Neurospora sp.*, *Fusarium fujikuroi*, *Cryptococcus neoformans*, *Lentinula edodes*, and *Phanerochaete chrysosporium* have been tested for their responses (Idnurm and Heitman 2005; Chen et al. 2009; Estrada and Avalos 2008; Ramirez et al. 2010; Glukhova et al. 2014). Green light did not benefit *Lentinula edodes* as growth (biomass) and carotenoid production was reduced substantially in the presence of green light (Glukhova et al. 2014). However, green light was an excellent stimulus for higher production of lignin peroxidase (LiP) in *P. chrysosporium*, increasing LiP activities by 20% as compared to cultures incubated in the dark.

5.4.2 Influence of Light Mediation on Bacteria

Unlike fungi, only a few bacterial groups are influenced by light, such as Cyanobacteria, Proteobacteria, Chlorobi, Chloroflexi, and Heliobacteria. And, in contrast with fungi, bacteria use light as a source of energy rather than as stimuli. To date, most of the published studies focused on the behavioural responses of bacteria to light, and very little on their role in regulating metabolite production (Armitage and Hellingwerf 2003; Bhaya 2004; Kim 2017). Bacteria have been hypothesized to have poorer defence system against light irradiation, particularly at wavelengths from 260 to 400 nm (ultraviolet range), which is detrimental to bacterial DNA. This is further complicated by the fact that bacteria have large DNA material within their small cell volume; thus, they are extremely susceptible to harmful ultraviolet range (Gonzalez et al. 2013). UV-A (315–400 nm) has been shown to slow bacterial growth (Sieracki and Sierburth 1986; Hortnagl et al. 2011). On the other hand,

UV-B (280–315 nm) not only damages DNA but also decreases production of extracellular enzymes (β-glucosidase, lipase, and leucine aminopeptidase) (Herndll et al. 1993). Other light wavelengths beyond the ultraviolet range are not known to have any significant impact on bacteria. This was evident in a study by Nagendran and Lee (2014), where exposure of both tomato plant and bacterial strain *Pseudomonas cichorii* JBC1 to red, blue, and green lights (630–655 nm, 448–475 nm, and 497–544 nm, respectively) revealed impact on plant growth but not the bacteria.

Some bacteria, however, respond favourably to light. Visible light (400–700 nm) can bring positive effects to the photoheterotrophic bacteria that have high resistance to solar and UV light. Cooney et al. (2006) discovered that production of caloxanthin sulfate (a photoprotective pigment) was highest when the photoheterotrophic bacteria *Erythrobacter longus* strain NK3Y was incubated under fluorescent light. By alternating the 12-h light and dark cycles, higher production of caloxanthin (4811 nmol g^{-1}) was derived compared to continuous incubation either in dark (3961 nmol g^{-1}) or light conditions (2741 nmol g^{-1}). Other studies have documented the positive influence of light on the growth and biomass of the aerobic anoxygenic photosynthetic bacteria (Harashima et al. 1987; Yurkov and Gemerden 1993; Kolber et al. 2001). Metagenomic studies have further revealed that there are several photoresponsive proteins present in various bacteria (Singh et al. 2017). This has suggested that light could influence bacterial physiology in many ways other than growth. Hence, more studies should follow on the functions of these photoresponsive proteins in bacteria.

5.4.3 Influence of Light Mediation on Algae

The impacts of light on algae are of greater significance compared to bacteria or fungi, as the growth of all algal species is highly dependent on solar radiation. As in the case with fungi, pigmentation is one of the most common responses in algae towards light. The effects of light on algae are influenced by light intensity, light wavelengths (colour), and the corresponding species of algae (Singh and Singh 2015). In algae, light regulates the synthesis of pigments such as chlorophyll a, b, c and auxiliary pigments (Lopez-Figueroa et al. 2003; Boss et al. 2004; Mouget et al. 2005; Kula et al. 2014). For example, production of chlorophyll a, chlorophyll c, and fucoxanthin (colour pigments) in the microalgae *Laminaria hyperborea* is influence by the blue and red lights (Dring 1986). The response to light is also highly species dependent. In a study by Wu (2016) on the algae *Pyropia haitanensis*, irradiation of fluorescent and blue light (420–460 nm) yielded higher concentrations of chlorophyll a, while red light (640–680 nm) decreased the production of chlorophyll a. Green light (490–530 nm) did not have any impact on the production of chlorophyll a. Light irradiation of any wavelength also showed no influence on carotenoid production. Nevertheless, higher levels of phycoerythrin and phycocyanin were detected in cultures exposed to blue and green light compared to others. In addition to light spectrum, light intensity is also known to influence the growth and lipid content in algae as well. It was reported that the biomass and lipid production in *Isochrysis galbana* was the highest (8.35 g/L and 0.45 g l^{-1} day^{-1}) when incubated at light intensity of 100 μmol m^{-2} s^{-1} for 18 h (Babuskin et al. 2014). Similar observations were reported in which high light intensity increased the biomass and lipid content (Beatriz et al. 2006; Sousa et al. 2007; Singh and Singh 2015).

5.5 Mechanisms Regulating Light Mediation

5.5.1 Mechanisms of Light Mediation in Fungi

The capability of fungi in sensing light is genetically modulated and expressed by the photo-responsive proteins present in the fungi. These photo-responsive proteins are diverse and found to be highly conserved across the entire fungal kingdom (Idnurm and Heitman 2005). In the presence of light, biochemical reactions occur in response to light, ranging from the synthesis of photolyases to the activation of more complicated signalling cascades, which then causes a change in gene expression that triggers the production of enzymes and chemicals to protect the fungi (Fuller et al. 2015). Some of the more commonly known responses to light are the production of protective pigments such as carotenoids or melanin, which are typically initiated to protect the fungi from harmful UV lights (Esbelin et al. 2012). Other than carotenoids and melanin, synthesis of photolyases (photolyase encoding gene *cryA*) is also activated as this enzyme repairs the damaged DNA. The various responses of fungi towards light depend largely on the presence of the various photo-responsive proteins present in the fungi. The mechanisms of how these proteins respond to light may not be well established, but it provides an early indication on the possible compounds produced as defence mechanisms towards light and how some of these responsive proteins can be further manipulated (via light mediation) to enhance or suppress production of certain important compounds. Three major photo-responsive proteins (blue-light, red-light, and green-light receptors) have been identified, and these will be discussed in the following sections.

5.5.1.1 Blue-Light Receptors

The first identified photo-responsive proteins are the white collar (*wc*) orthologs, which are located on the specific sequences (light response elements) in the light regulated genes. It is a heterodimer formed by two proteins: white collar-1 protein (*wc-1*) and white collar-2 protein (*wc-2*). The *wc-1* and *wc-2* proteins are commonly found in most fungi, but some species such as *Saccharomycotina* yeasts (*Saccharomyces* and *Candida)* and dermatophytic fungi (*Malassezia* and *Microsporum*) lack *wc-1*. These *wc* proteins are sensitive to light signals to the blue spectrum (~ 400–495 nm) (Chen and Loros 2009). The first cloned photo-responsive gene (*wc-1*) was isolated from *Neurospora crassa* (Ballario et al. 1996; Froehlich et al. 2002). The *wc-1* proteins are transcription factors, which consist of the zinc finger DNA-binding domains (GATA) and the protein domain (Per-Ant-Sim, PAS), which allows for binding with other proteins. The *wc-1* proteins also consist of the light-oxygen-voltage-sensing (LOV) domain, which enables direct light sensing that is responsible for adjusting the circadian temporal organization of the fungi to allow adaptation to the environment (Brych et al. 2016). Similarly, *wc-2* proteins also have the Zn-finger domain, but not the LOV domain; hence, direct light sensing is not possible. However, in theory, *wc-1* alone is sufficient as a photoreceptor, attributed to the presence of both PAS and LOV domain in the *wc-1* proteins.

The *wc-1* and *wc-2* proteins can either function independently, and/or both could interact with one another via the PAS domain to form the transcriptionally and photoactive white collar complex (*wcc*) (Idnurm and Heitman 2005). The *wcc* is of interest as it serves as an activator of the frequency (frq) gene that encodes the protein function in the circadian clock (Froehlich et al. 2002; He et al. 2002; Zoltowshi et al. 2007). After light exposure, the *wcc* will bind to the promoter of light responsive genes to initiate photo-protection such as induction of the synthesis of photoprotection pigments and sporulation. Nevertheless, the function of *wcc* can be repressed by VIVID (vvd), another blue light receptor which acts as the negative feedback for *wcc*.

Other blue-light receptors include cryptochromes and photolyases (Yu et al. 2010). Unlike *wcc*, both cryptochromes and photolyases contain an N-terminal domain (photolyase-related (PHR) region) that allows for the binding of FAD and 5,10 methyltetrahydrofolate (MTHF) (Losi 2007). Both receptors are widely distributed among species of *Ascomycota*, *Basidiomycota*, and *Mucoromycotina* (Coesel et al. 2009; Kim et al. 2014; Mei and Dvornyk 2015). It was hypothesized that blue light can excite the electrons of the flavin molecules in the cryptochromes, leading to changes of the photoreceptors to interact with light signalling proteins and altering gene expressions (Yu et al. 2010). In some cases, deletion of certain blue-light receptors can result in up-regulation or down-regulation of certain proteins. As in the case of *Trichoderma atroviride*, the deletion of *wc-1* proteins leads to the down-regulation of proteins in the dark, while the deletion of *wc-2* leads to the up-regulation of proteins (Sanchez-Arreguin et al. 2012).

5.5.1.2 Red-Light Receptors

Besides blue light, far red and red-light spectrums (approximately 600–850 nm) trigger responses in fungi as well. The first red-light receptor, known as phytochrome (Fph), was discovered in *Aspergillus nidulans* and is responsive to far red and red-light spectrums (Fuller et al. 2015). This receptor is widely distributed in species of *Ascomycota* and *Basidiomycota*. The structures of the receptor typically consist of an N-terminal photo-sensory (HY) domain, with C-terminal domains differing among species, but always consisting of a histidine kinase domain (Idnurm and Heitman 2005). Interestingly, the structure of phytochromes is reversible, such that they exist as two distinct forms, the red-light absorbing form (Pr) and the far red-light absorbing form (Pfr), depending on the red and far red light ratio in the light signal (Brandt et al. 2008; Sharrock 2008). The presence of the phytochrome genes, however, does not confirm absolute response to red and far red lights. In some fungi such as *Cryptococcus neoformans* and *Neurospora* sp., there was no response to red light despite the presence of the phytochrome genes (Idnurm and Heitman 2005; Froehlich et al. 2005). They observed that the transcription of phytochrome genes *phy-1* and *phy-2* is not light regulated; rather the circadian clock influences the abundance of *phy-1* mRNA. Hence, molecular evidence is needed to explain how the phytochromes react to light and regulate the production of metabolites.

Several light regulator proteins such as *VeA*, *VelB*, and *LaeA* have been reported to have important roles in the signalling pathway of the phytochromes. It has been hypothesized that the phytochromes cooperate with light regulator proteins to form a complex homologous to the *wcc* to respond to the light signals. The role of these light regulator proteins extends beyond their responses to light, as consequently, sexual and asexual development of the fungi is also impacted. For example, *VeA* is known to positively regulate sexual development in fungi but inhibits the asexual reproduction upon light stimulation. This fungal development is then closely associated to the production rate of metabolites, as most of the metabolites are produced upon the completion of the initial growth phase (Calvo et al. 2002). Hence, *VeA* influences not only the reproductive cycles but also the secondary metabolite production such as the production of mycotoxins (e.g. trichothecenes, fusarins, cyclopiazonic acid, and aflatrem) (Estiarte et al. 2016). Interestingly, *VeA* genes also regulate the production of antimicrobial compounds such as penicillins and cephalosporin C (Kato et al. 2003; Dreyer et al. 2007). Clearly, red-light receptors and their interaction with red and far red light could be manipulated to produce various compounds for multi-applications.

5.5.1.3 Green-Light Receptors

To date, only few microorganisms are known to respond to green light. They include the fungi *Allomyces reticulatus* and *Blastocladiella emersonii* and several species of Archaea and Eubacteria. Responses to green light are mediated by a photoreceptor called rhodopsin. Rhodopsin is formed by seven-trans-membrane domain opsin bound to the chemophore retinal via a conserved lysine (Heintzen 2012). In general, there are four opsin-like proteins identified, which are the heat shock protein 30 (HSP30), opsin-related protein 1 (ORP-1), *Neurospora* opsin protein 1 (NOP-1), and auxiliary ORP-like phodopsins (CarO). The exact mechanisms of the green-light receptors are still not fully understood, although it is generally accepted that light causes photoexcitation that leads to the isomerization of the chromophore and changes to the protein backbone. These changes then influence the ion transportation across membranes and the activation of the signal transduction pathway (Adam et al. 2018). Preliminary studies found contrasting observations, where the expression of rhodopsin gene in *Neurospora, Fusarium fujikuroi*, and *Cryptococcus neoformans* was said to be regulated by white light and not green light. Several ascomycete and basidiomycete species also did not show response towards green light, despite harbouring rhodopsin genes (Idnurm and Heitman 2005; Chen et al. 2009, Estrada and Avalos 2008). In the study by Bieszke et al. (2007), the result revealed that the transcription of *nop-1* was not affected by light irradiation but reproduction (conidiation) was affected instead. Hence, it can be suggested that similar to the phytochromes, the rhodopsins might react to the light indirectly via a series of cascade reactions.

5.5.2 Mechanisms of Light Mediation in Bacteria

There are five classes of photoreceptor proteins for bacteria, namely, blue-light sensing using FAD (BLUF), bacteriophytochromes, light-oxygen-voltage-sensing (LOV) domain, photoactive yellow protein (PYP), and rhodopsins (Gomelsky and Hoff 2011). These light-photoreceptor proteins are similar to fungi, but are not well understood, as bacteria are generally considered less sensitive to light, except for photosynthetic bacteria. Early studies would indicate that the light receptors are mainly derived from phototrophic bacteria. It was in 1992 that researchers found red photo-responsive receptors (identified as phytochromes) in non-phototrophic *Deinococcus radiodurans* and *Pseudomonas aeruginosa.* This revealed that the photoreceptors exist in both phototrophic and non-phototrophic bacteria (Davis et al. 1999). Among the five photoreceptor proteins, bacteriophytochromes are the most abundant photoreceptors in bacteria, as 17% of the bacteria harbour the phytochrome genes, followed by BLUF and LOV (Gomelsky and Hoff 2011). The chromophores of phytochrome vary among the species, with phytochromes in cyanobacteria consisting of phytochromobilin and phycocyanobilin, while anoxygenic phototrophic and non-photosynthetic bacteria have phytochromes with biliverdin IXα (BV). The phytochromes in bacteria regulate light similarly to the phytochromes in fungi, in which light will cause conformational changes to the protein. The changes in protein conformation then provide for transduction of a light-induced signal from the effector domains, most commonly with a histidine kinase, which then activates the effector domains and enables the transduction of light signal to biochemical signalling cascades (Fraikin et al. 2015).

The LOV photosensory proteins in bacteria consist of short β-module domain bound to the flavin chromophore. Upon elicitation by the blue-light signal, the flavin in the protein will bind to the cysteine residue located near the chromophore. This binding then induces transformation in protein structures, which then triggers a series of light signalling responses. The LOV proteins have been shown to regulate intracellular proliferation of *Brucella abortus*, induce autophosphorylation in *Caulobacter crescentus*, and regulate phosphodiesterase activity in *Synechococcus elongatus* (Ricciuti and Lubin 1976; Swartz et al. 2007; Wright et al. 2008; Herrou and Crosson 2012). The exact mechanisms leading to the expressions, however, remained unclear. Unlike LOV, BLUF consists of short α-module domain that binds to the flavin chromophores. BLUF interacts with the signalling proteins to form a complex that transmits light signals and regulates physiological responses. Recent studies suggested that the photoexcited BLUF domains can regulate the catalytic activity of enzymatic effectors responsible for the synthesis of secondary messenger molecules such as blue-light-regulated phosphodiesterase and cellular cAMP (Fraikin et al. 2015).

The rhodopsins are another class of bacteria photoreceptors, which is more well-studied compared to others. Rhodopsins act as photo-sensory receptors to enable the microbes to sense the changes of light in the environment. There are several types of rhodopsins identified in bacteria including halorhodopsins (HR) which function as transmembrane chloride pumps, sensory rhodopsins (SRI and SRII) as effective light sensors, and bacteriorhodopsin (BR), proteorhodopsin (PR), and xanthorhodopsins which function as energy-conserving transmembrane proton pumps (Sineshchekov et al. 2002; Spudich and Leucke 2002). The mechanisms involving rhodopsin are relatively simple, where a proton motive force will be created upon the presence of light. The absorbed light energy will then directly drive proton expulsion from cell via BR and PR. With the created proton motive force, the synthesis of ATP is promoted, and various secondary transport processes will be activated (Bryant and Frigaard 2006). In addition to BR and HR, bacteria such as *Halobacterium* sp. can also react to green, orange, and blue-light signals via SRI and SRII (Hoff et al. 1997; Cercignani et al. 2000; Spudich and Leucke 2002). To date, the photoreceptors of bacteria are not well understood; hence, there is room for further studies. This is especially in the downstream analyses of how these light receptor proteins interact with light and their subsequent influence on the production of compounds/metabolites.

5.6 Conclusions and Future Prospects

Light is clearly an effective, environmental-friendly, clean-energy, and low-cost approach in regulating the production of natural compounds of microbes. The use of light not only enhances the production of natural compounds such as pigments, antibiotics, and lipids but also enhances or activates some silent gene clusters that are not expressed under normal growing conditions, such as in the case of cre-1 gene cluster regulating the production of dihydrotrichotetronin in *Trichoderma reesei*. As such, there is potential in using light to mediate the expression of a variety of genes to produce various microbial-based compounds that may be beneficial. Light-mediation appears to have a more significant influence on fungi and algae compared to bacteria, perhaps due to the nature of the eukaryotes being less susceptible to light than prokaryotes. To date, most studies reported the influence of light on growth and the expression of certain genes, and these are typically carried on model fungal species such as *Neurospora crassa* and *Aspergillus nidulans*. As such, further studies can be conducted in exploring the effects of light on other beneficial fungal species, as well as the exploration on how to use light mediation to control the production of important compounds. There is also potential to further explore and understand the mechanisms of photoreceptors in various microorganisms and their response to light in regulating the production of microbial-based compounds.

Acknowledgements The authors express their gratitude to Monash University Malaysia for the financial support and facilities that enable the endeavour into light mediation research.

References

Adam A, Deimel S, Medina JP, Martinez JG, Konte T, Limon MC, Avalos J, Terpitz U (2018) Protein activity of the *Fusarium fujikuroi* rhodopsins CarO and OpsA and their relation to fungus-plant interaction. Int J Mol Sci 19(1):215. https://doi.org/10.3390/ijms19010215

Ali M, Li PH, She GB, Chen DF, Wan XC, Zhao J (2017) Transcriptome and metabolite analyses reveal the complex metabolic genes involved in volatile terpenoid biosynthesis in garden sage (*Salvia officinalis*). Sci Rep 7. https://doi.org/10.1038/s41598-017-15478-3

Amedei A, D'Elios M (2012) New therapeutic approaches by using microorganism-derived compounds. Curr Med Chem 19:3822–3840

Atanasov AG, Waltenberger B, Wenzig EMP, Linder T, Wawrosch C, Uhrin P, Temml V, Wang LM, Schwaiger S, Heiss EH, Rollinger JM, Schuster D, Breuss JM, Bochkov V, Mihovilovic MD, Kopp B, Bauer R, Dirsch VM, Stuppner H (2015) Discovery and resupply of pharmacologically active plant-derived natural products: a review. Biotechnol Adv 33(8):1582–1614

Armitage JP, Hellingwerf KJ (2003) Light induced behavioural responses (phototaxis) in prokaryotes. Photosynth Res 76:145–155

Babuskin S, Radhakrishnan K, Baby PAS, Sivarajan M, Sukumar M (2014) Effect of photoperiod, light intensity and carbon sources on biomass and lipid productivities of *Isochrysis galbana*. Biotechnol Lett 36:1653–1660

Ballario P, Vittorioso P, Magrelli A, Talora C, Cabibbo A, Macino G (1996) White collar-1, a central regulator of blue light responses in *Neurospora*, is a zinc finger protein. EMBO J 15:1650–1657

Baltz RH (2001) Genetic methods and strategies for secondary metabolite yield improvement in actinomycetes. Antonie Van Leeuwenhoek 79:251–259

Basu A, Apte SK, Phale PS (2006) Preferential utilization of aromatic compounds over glucose by *Pseudomonas putida* CSV86. Appl Environ Microbiol 72(3):2226–2230

Beatriz P, Jose V, Pilar R, Jose F (2006) Study of the effect of temperature, irradiance and salinity on growth and yessotoxin production by the dinoflagellate *Protoceratium reticulatum* in culture by using a kinetic and factorial approach. Mar Environ Res 62:286–300

Berdy J (2005) Bioactive microbial metabolites. J Antibiot 58(1):1–26

Bhaya D (2004) Light matters : phototaxis and signal transduction in unicellular cyanobacteria. Mol Microbiol 53(3):745–754

Bieszke JA, Li L, Borkovich KA (2007) The fungal opsin gene nop-1 negatively regulated by a component of the blue light sensing pathway and influences conidiation specific gene expression in *Neurospora crassa*. Curr Genet 52(3-4):149–157

Blume E (1989) Investigators seek to increase taxol supply. J Natl Cancer Inst 81(15):1122–1123

Bora L, Bora M (2012) Optimization of extracellular thermophilic highly alkaline lipase from thermophilic *Bacillus sp.* isolated from hotspring of Arunachal Pradesh India. Braz J Microbiol 43(1):30–42

Boruta T, Bizukojc M (2017) Production of lovastatin and itaconic acid by *Aspergillus terreus*: a comparative perspective. World J Microbiol Biotechnol 33(2). https://doi.org/10.1007/s11274-017-2206-9

Boss PK, Bastow RM, Mylne JS, Dean C (2004) Multiple pathways in the decision to flower : enabling, promoting and resetting. Plant Cell 16:18–31

Brandt S, Stetten DV, Gunther M, Hildebrandt P, Frankerberg-Dinkel N (2008) The fungal phytochrome FphA from Aspergillus nidulans. J Biol Chem 283(50):34605–34614

Bren A, Park JY, Towbin BD, Dekel E, Rabinowitz JD, Alon U (2016) Glucose becomes one of the worst carbon sources for *E.coli* on poor nitrogen sources due to suboptimal levels of cAMP. Sci Rep 6. https://doi.org/10.1038/srep24834

Brower V (2008) Back to nature: extinction of medicinal plant threatens drug discovery. J Natl Cancer Inst 100(12):838–839

Bryant DA, Frigaard NU (2006) Prokaryotic photosynthesis and phototrophy illuminated. Trends Microbiol 14(11):488–496

Brych A, Mascarenhas J, Jaeger E, Charkiewicz E, Pokorny R, Bolker M, Doehlemann G, Batschauer A (2016) White collar 1-induced photolyase expression contributes to UV-tolerance of *Ustilago maydis*. Microbiol Open 5(2):224–243
Buhler RMM, Muller BL, Moritz DE, Vendruscolo F, Oliveira DD, Ninow JL (2015) Influence of light intensity on growth and pigment production by *Monascus ruber* in submerged fermentation. Appl Biochem Biotechnol 176:1277–1289
Butler WL, Norris KH, Siefelman HW, Hendricks SB (1959) Detection, assay and preliminary purification of the pigment controlling photoresponsive development of plants. Biochemistry 45:1703–1709
Calvo AM, Wilson RA, Bok JW, Keller NP (2002) Relationship between secondary metabolism and fungal development. Microbiol Mol Biol Rev 66(3):447–459
Cercignani G, Frediani A, Lucia S, Petracchi D (2000) Competition-integration of blue and orange stimuli in Halobacterium salinarum cannot occur solely in SRI photoreceptor. Biophys J 79(3):1554–1560
Chambon C, Ladeveze V, Servouse M, Blanchard L, Javelot C, Viadescu B, Karst F (1991) Sterol pathway in yeast. Identification and properties of mutant strains defective in mevalonate diphosphate decarboxylase and farnesyl diphosphate synthetase. Lipids 26(8):633–636
Chandi GK, Gill BS (2011) Production and characterization of microbial carotenoids as an alternative to synthetic colors a review. Int J Food Prop 14(3):503–513
Chen HC, Loros JJ (2009) *Neurospora* sees the light. Comm Integ Biol 2(5):448–451
Chen CH, Ringelbery CS, Gross RH, Dunlap JC, Loross JJ (2009) Genome-wide analysis of light inducible responses reveals hierarchical light signalling in *Neurospora*. EMBO J 28:1029–1042
Cheong KK, Strub C, Montet D, Durand N, Alter P, Melle JC, Schorr GS, Fontana A (2016) Effect of different light wavelength on the growth and ochratoxin A production in *Aspergillus carbonarius* and *Aspergillus westerdijkiae*. Fungal Biol 120(5):745–751
Chopra I, Roberts M (2001) Tetracycline antibiotics: mode of action, application, molecular biology and epidemiology of bacterial resistance. Microbiol Mol Biol Rev 65(2):232–260
Coesel S, Mangogna M, Ishikawa T, Heijde M, Rogato A, Finazzi G, Todo T, Bowler C, Falciatore A (2009) Diatom PtCPF1 is a new cryptochrome/photolyase family member with DNA repair and transcription regulation activity. EMBO Rep 10(6):655–661
Colombo D, Ammirati E (2011) Cyclosporine in transplantation- a history of converging timelines. J Biolog Regul Homeostatic Agent 25(4):493–504
Cooney MK, Johnston WA, Pohl S, Bidigare RR (2006) Influence of photoperiod on pigmentation and metabolic efficiency of the marine aerobic anoxygenic photosynthetic bacterium *Erythrobacter longus* strain NK3Y. Aquat Microb Ecol 43:303–309
Cordell GA (2011) Sustainable medicines and global health care. Planta Med 77:1129–1138
Cragg GM, Newman DJ (2013) Natural products: a continuing source of novel drug leads. Biochim Biophys Acta 1830(6):3670–3695
Cragg GM, Schepartz SA, Suffness M, Grever MR (1993) The taxol supply crisis. New NCI policies for handling the large-scale production of novel natural product anticancer and anti-HIV agents.Journal of. Nat Prod 56(10):1657–1668
Das A, Khosla C (2009) Biosynthesis of aromatic polyketides in bacteria. Acc Chem Res 42(5):631–639
David B, Wolfender JL, Dias DA (2015) The pharmaceutical industry and natural products: historical status and new trends. Phytochem Rev 14(2):299–315
Davis SJ, Venar AV, Vierstra RD (1999) Bacteriophytochromes: phytochrome-like photoreceptors from non-photosynthetic eubacteria. Science 286(5449):2517–2520
Dehesa L, Abuchar A, Nuno-Gonzalez A, Vitiello M, Kerdel FA (2012) The use of cyclosporine in dermatology. J Drugs Dermatol 11(8):979–987
Dreyer J, Eichhorn H, Friedlin E, Kurnsteiner H, Kuck U (2007) A homologue of the *Aspergillus velvet* gene regulates both cephalosporin C biosynthesis and hyphal fragmentation in *Acremonium chrysogenum*. Appl Environ Microbiol 73(10):3412–3422
Dring MJ (1986) Pigment composition and photosynthesis action spectra of sporophytes of Laminaria (Phaeophyta) grown in different light qualities and irradiances. Br Phycol J 21(2):199–207

Egli TH, Quayle JR (1986) Influence of the carbon: nitrogen ratio of the growth medium on cellular composition and the ability of methylotrophic yeast *Hansenula polymorpha* to utilize mixed carbon sources. J Gen Microbiol 132:1779–1788

Esbelin J, Mallea S, Ram AFJ, Carlin F (2012) Role of pigmentation in protecting *Aspergillus niger* conidiospores against pulsed light radiation. Photochem Photobiol 89(3). https://doi.org/10.1111/php.12037

Estiarte N, Lawrence CB, Sanchis V, Ramos AJ, Sempere AC (2016) LaeA and VeA are involved in growth morphology, asexual development, and mycotoxin production in *Alternaria alternata*. Int J Food Microbiol 238:153–164

Estrada AF, Avalos J (2008) The whilte collar protein WcoA of *Fusarium fujikuroi* is not essential for photocarotenogenesis, but is involved in the regulation of secondary metabolisms and conidiation. Fungal Genet Biol 45:705–718

Feng YL, Shao YC, Chen FS (2012) Monascus pigments. Appl Microbiol Biotechnol 96(6):1421–1440

Fraikin GY, Strakhovskaya MG, Belenikina NS, Rubin AB (2015) Bacterial photosensory proteins: regulatory function and optogenetic application. Microbiology 84(4):461–472

Froehlich AC, Liu Y, Lorros JJ, Dunlap JC (2002) White collar-1, a circadian blue light receptor, binding to the frequency promoter. Science 297(5582):815–819

Froehlich AC, Noh B, Vierstra RD, Loros J, Dunlap JC (2005) Genetic and molecular analysis of phytochromes from the filamentous fungus Neurospora crassa. Eukaryot Cell 4(12):2140–2152

Fuller KK, Loros JJ, Dunlap JC (2015) Fungal photobiology: visible light as a signal for stress, space and time. Curr Genet 61:275–288

Gaynes R (2017) The discovery of penicillin-new insights after more than 75 years of clinical use. Emerg Infect Dis 23(5):849–853

Gharaie S, Vaas LAI, Rosberg AK, Windstam ST, Karlsson ME, Bergstrand KJ, Khalil S, Wohanka W, Alsanius BW (2017) Light spectrum modifies the utilization pattern of energy sources in *Pseudomonas* sp. DR 5-09. PLoS One 12(12):e0189862. https://doi.org/10.1371/journal.pone.0189862

Glukhova LB, Sokolyanskaya LO, Plotnikov EV, Gerasimchuk AL, Karnachuk OV, Solioz M, Karnachuk RA (2014) Increased mycelial biomass production by Lentinula edodes intermittently illuminated by green light emitting diodes. Biotechnol Lett 36(11):2283–2289

Gmoser R, Ferreira JA, Lennartsson PR, Taherzadeh MJ (2017) Filamentous ascomycetes fungi as a source of natural pigments. Fung Biol Biotechnol 4(4). https://doi.org/10.1186/s40694-017-0033-2

Gomelsky M, Hoff WD (2011) Light helps bacteria make important lifestyle decisions. Trends Microbiol 19(9):441–448

Gonzalez CR, Simo R, Sommaruga R, Gasol JM (2013) Away from darkness: a review on the effects of solar radiation on heterotrophic bacterioplankton activity. Front Microbiol 4. https://doi.org/10.3389/fmicb.2013.00131

Halbwirth H (2010) The creation and physiological relevance of divergent hydroxylation patterns in the flavonoid pathway. Int J Mol Sci 11(2):595–621

Haleem DAE, Amara A, Zaki S, Abulhamd A, Abulreesh G (2007) Biosynthesis of biodegradable polyhydroxyalkanotes biopolymers in genetically modified yeasts. Int J Environ Sci Technol 4(4):513–520

Harashima K, Kawazoe K, Yoshida I, Kamata H (1987) Light stimulated aerobic growth of *Erythrobacter* spp. OCh 114. Plant Cell Physiol 28(2):365–374

He Q, Cheng P, Yang Y, Wang L, Gardner KH, Liu Y (2002) White collar-1, a DNA binding transcription factor and a light sensor. Science 297:840–843

Heintzen C (2012) Plant and fungal photopigments. Adv Rev 1:411–432

Heravi MM, Zadsirjan V, Malmi M (2018) Application of asymmetric pictet-spengler reaction in the total synthesis of natural products and relevant biologically active compounds. Molecules 23(4):943

Herndll GJ, Niklas GM, Frick J (1993) Major role of ultraviolet-B in controlling bacterioplankton growth in the surface layer of the ocean. Lett Nat 361:717–719

Herrou J, Crosson S (2012) Function, structure and mechanisms in bacterial photosensory LOV proteins. Nat Rev Microbiol 9(10):713–723

Hoff WS, Jung KH, Spudich JL (1997) Molecualr mechanism of photosignaling by archaeal sensory rhodopsins. Annu Rev Biophys 26:223–258

Hortnagl P, Perez MT, Sommaruga R (2011) Contrasting effects of ultraviolet radiation on the growth efficicnecy of freshwater bacteria. Aquat Ecol 45(1):125–136

Huode A, Kademi A, Lebianc D (2004) Lipases and their industrial applications: an overview. Appl Biochem Biotechnol 118(1-3):155–170

Idnurm A, Heitman J (2005) Photosensing fungi : phytochrome in the spotlight. Curr Biol 15(20):829–832

Imbert M, Blondeau R (1999) Effect of light on germinating spores of Streptomyces viridosporus. FEMS Microbiol Lett 18:159–163

Ji Y, Bi JN, Yan B, Zhu XU (2006) Taxol-producing fungi: a new approach to industrial production of taxol. Chin J Biotechnol 22(1):1–6

Jongedijk E, Cankar K, Buchhaupt M, Schrader J, Bouwmeester H, Beekwilder J (2016) Biotechnological production of limonene in microorganisms. Appl Microbiol Biotechnol 100:2927–2938

Kapoor V, Singh R, Kumar V (2011) Influence of carbon and nitrogen sources on the α-amylase production by a newly isolated thermophilic *Streptomyces* spp. MS702 (MTCC10772). Asian J Biotechnol 3(6):540–553

Kate KT, Laird SA (2000) The commercial use of biodiversity: access to genetic research and benefit sharing. Earthscan Publication Ltd, London, p 74

Kato N, Brooks W, Calvo AM (2003) The expression of sterigmatocystin and penicillin genes in *Aspergillus nidulans* is controlled by VeA, a gene required for sexual development. Eukaryot Cell 2(6):1178–1186

Kawagoshi Y, Hino N, Fujimoto A, Nakao M, Fujita Y, Sugimura S, Furukawa K (2005) Effect of inoculum conditioning on hydrogen fermentation and pH effect on bacterial community relevant to hydrogen production. J Biosci Bioeng 100(5):524–530

Kim MS (2017) Phototaxis of Cyanobacteria under complex light environment. Am Soc Microbiol 8. doi: e00498-17

Kim H, Son H, Lee Y (2014) Effects of light on secondary metabolism and fungal development of *Fusarium graminearum*. J Appl Microbiol 116:380–389

Kirti K, Amita S, Priti S, Kumar AM, Jyoti X (2014) Colorful world of microbes: carotenoids and their application. Adv Biol. https://doi.org/10.1155/2014/837891

Klement T, Buchs J (2013) Itaconic acid- a biotechnological process in change. Bioresour Technol 135:422–431

Kolber ZS, Plumley FG, Lang AS, Beatty JT, Blankenship RE, Vndover CL (2001) Contribution of aerobic photoheterotrophic bacteria to the carbon cycle in the ocean. Science 292:2492–2495

Kula M, Rys M, Mozdzen K, Skoczowski A (2014) Metabolic activity, the chemical composition of biomass and photosynthetic activity of Chlorella vulgaris under different light spectra in photobioreactors. Eng Life Sci 14(1):57–67

Kumar RR, Prasad S (2011) Metabolic engineering of bacteria. Indian J Microbiol 51(3):403–409

Kuo FS, Chien YH, Chen CJ (2012) Effects of light sources on growth and carotenoid content of photosynthetic bacteria Rhodopseudomonas palustris. Bioresour Technol 113:315–318

Kutvonen H, Rajalaa P, Carpen L, Bomberg M (2015) Nitrate and ammonia as nitrogen sources for deep subsurface microorganisms. Front Microbiol 6. https://doi.org/10.3389/fmicb.2015.01079

Li SY, Wang P, Yuan W, Su ZS, Bullard SH (2016) Endocidal regulation of secondary metabolites in the producing organisms. Sci Rep 6:29–35

Lobanovska M, Pilla G (2017) Penicillin's discovery and antibiotic resistance: lessons for the future. Yale J Biol Med 90(1):135–145

Lopez-Figueroa F, Condex-Alvarez R, Gomez I (2003) Relations between electron transport rates determined by pulse amplitude modulated chlorophyll fluorescence and oxygen evolution in macroalgae under different light conditions. Photosynth Res 75(3):259–275

Losi A (2007) Flavin-based blue light photosensors: a photobiophysics update. Photochem Photobiol 83:1283–1300

Luo YZ, Li BZ, Liu D, Zhang L, Chen Y, Jia B, Zeng BX, Zhao HM, Yuan YJ (2016) Engineered biosynthesis of natural products in heterologous hosts. Chem Soc Rev Home 44(15):5265–5290

Mathan S, Subramanian V, Nagamony S (2013) Optimization and antimicrobial metabolites production from endophytic fungi Aspergillus terreus KC 582297. Eur J Exp Biol 3(4):138–144

Mei QM, Dvornyk V (2015) Evolutionary history of photolyases/cryptochrome superfamily in eukaryotes. PLoS One 10(9). https://doi.org/10.1371/journal.pone.0135940

Mikami Y (1988) Microbial conversion of terpenoids. Biotechnol Genet Eng Rev 6(1):271–320

Mikell JR, Herath W, Khan IA (2015) Eleven microbial metabolites of 6-hydroxyflavanone. Chem Pharm Bull 63:579–583

Mishra BB, Tiwari VK (2011) Natural products: an evolving role in future drug discovery. Eur J Med Chem 46:4769–4807

Miyake T, Mori A, Kii T, Okuno T, Usui Y, Sato F, Sammoto H, Watanbe A, Kariyama M (2005) Light effects on cell development and secondary metabolism in *Monascus*. J Ind Microbiol Biotechnol 32:103–108

Mouget JL, Rosa P, Vachoux C, Tremblin G (2005) Enhancement of marennine production by blue light in the diatom *Haslea ostrearia*. J Appl Phycol 17(5):437–445

Nagendran R, Lee YH (2014) Green and red light reduces the disease severity by *Pseudomonas cichorii* JBC1 in tomato plants via upregulation of defense related gene expression. Bacteriology 105(4):412–417

Nakashimada Y, Rachman MA, Kakizono T, Nishio N (2002) Hydrogen production of *Enterobacter aerogenes* altered by extracellular and intracellular redox states. Int J Hydrog Energy 27(11):1399–1405

Nedwell DB (1999) Effect of low temperature on microbial growth: lowered affinity for substrates limits growth at low temperature. FEMS Microbiol Ecol 30(2):101–111

Ng FM, Dawes EA (1973) Chemostat studies on the regulation of glucose metabolism in *Pseudomonas aeruginosa* by citrate. Biochem J 132:129–140

Panche AN, Diwan AD, Chandra SR (2016) Flavonoids: an overview. J Nutri Sci 5. https://doi.org/10.1017/ins.2016.41

Parkin TB, Brock TD (1980) The effects of light quality on the growth of phototrophic bacteria in lakes. Arch Microbiol 125(1):19–27

Pattanaik B, Lindberg P (2015) Terpenoids and their biosynthesis in cyanobacteria. Lifestyles 5(1):269–293

Penesyan A, Gillings M, Paulsenm IT (2015) Antibiotic discovery: combatting bacterial resistance in cells and in biofilm communities. Molecules 20(4):5268–5298

Prub S, Fetzner R, Seither K, Herr A, Pfeiffer E, Metzler M, Lawrence CB, Fisher R (2014) Role of *Alternaria alternata* blue light receptor (white collar 1) in spore formation and secondary metabolism. Appl Environ Microbiol 80(8):2582–2591

Pudi N, Varikuti GD, Badana AK, Gavara MM, Kumari S, Malla R (2016) Studies on optimization of growth parameters for enhanced production of antibiotic alkaloids by isolated marine actinomycetes. J Appl Pharm Sci 6(10):181–188

Raghavulu SV, Mohan SV, Goud RK, Sarma PN (2009) Effect of anodic pH microenvironment on microbial fuel cell (MFC) performance in concurrence with aerated and ferricyanide catholytes. Electrochem Commun 11(2):371–375

Raja A, Prabakarana P (2011) Actinomycetes and drug-an overview. Am J Drug Dis Dev 1(2):75–83

Ramirez DAT, Munoz SVB, Michel FC (2010) Effects of different wavelengths of light on lignin peroxidase production by the white-rot fungi Phanerochaete chrysosporium grown in submerged cultures. Bioresour Technol 101(23):9213–9220

Rao MPN, Xiao M, Li WJ (2017) Fungal and bacterial pigments: secondary metabolites with wide application. Front Microbiol 8:1113–1119

Ricciuti CP, Lubin LB (1976) Light induced inhibition of sporulation in *Bacillus licheniformis*. J Bacteriol 128(1):506–509

Romero JR, Hedtke M, Kastner C, Mullaer S, Fischer R (2010) Fungi, hidden in soil or up in the air: light makes a difference. Annu Rev Microbiol 64:585–610

Rubio JJL, Arnnz ME, Padmanabhan S, Murillo FJ (2001) A repressor-antirepressor pair links two loci controlling light induced carotenogenesis in *Myxococcus xanthus*. J Biol Chem 277:7262–7270
Ruiz B, Chavez A, Forero A, Garcia HY, Romero A, Sanchez M, Rocha D, Sanchez B, Rodriguez SR, Sanchez S, Langley E (2010) Production of microbial secondary metabolites: regulation by the carbon sources. Crit Rev Microbiol 36(2):146–167
Saez A, Gasol JM, Lefort T, Hofer J, Sommaruga R (2006) Effect of natural sunlight on bacterial activity and differential sensitivity of natural bacterioplankton groups in Northwestern Mediterranean coastal waters. Appl Environ Microbiol 72(9):5806–5813
Salas ML, Mounier J, Valence F, Coton M, Thierry A, Coton E (2017) Antifungal microbial agents for food biopreservation-a review. Microorganisms 5(3). https://doi.org/10.3390/microorganisms5030037
Sanchez S, Demain AL (2008) Metabolic regulation and overproduction of primary metabolites. Microb Biotechnol 1(3):283–319
Sanchez-Arreguin A, Perez-Martinez AS, Herrera-Estrella A (2012) Proteomic analysis of Trichoderma atroviride reveals independent roles for transcription factos BLR-1 and BLR-2 in light and darkness. Eukaryot Cell 11(1):30–41
Schmidt-Heydt M, Rufer C, Raupp F, Bruchmann A, Perrone G, Geisen R (2011) Influence of light on food relevant fungi with emphasis on ochratoxin producing species. Int J Food Microbiol 145:229–237
Seca AML, Pinto DCGA (2018) Plant secondary metabolites as anticancer agents: successes in clinical trials and therapeutic application. Int J Mol Sci 19(1). https://doi.org/10.3390/ijms19010263
Shanthakumar SP, Duraisamy P, Vishwanath G, Selvanesan BC, Ramaraj V, David BV (2015) Broad spectrum antimicrobial compounds from the bacterium *Exiguobacteriummexicanum* MSSRF9. Microbiol Res 178:59–65
Sharrock RA (2008) The phytochrome red/far-red photoreceptor superfamily. Genome Biol 9(8). https://doi.org/10.1186/gb-2008-9-8-230
Shi SB, Si T, Liu ZH, Zhang HF, Ang EL, Zhao HM (2016) Metabolic engineering of a synergistic pathway for n-butanol production in *Saccharomyces cerevisiae*. Sci Rep 6:2–10
Siddiqui MS, Thodey K, Trenchard I, Smolke C (2011) Advancing secondary metabolite biosynthesis in yeast with synthetic biology tools. FEMS Yeast Res 12(2). https://doi.org/10.1111/j.1567-1364.2011.00774.x
Sieracki ME, Sierburth JM (1986) Sunlight induced growth delay of plankton marine bacteria in filtered seawater. Mar Ecol Prog Ser 33:19–27
Silva RS, Tavares M, Trindade E, Dias JA (2014) Congenital sucrase-isomaltase deficiency: A case report Deficecongenito de sacarase-isomaltase. GE Port J Gastroenterol 21(6):250–253
Sineshchekov OA, Jung KH, Spudich JL (2002) Two rhodopsins mediate phototaxis to low and high intensity light in *Chlamydomonas reinhardtii*. Proc Natl Acad Sci 99(13):8689–8694
Singh R, Kumar M, Mittal A, Mehta PK (2017) Microbial metabolites in nutrition, healthcare and agriculture. 3 Biotech 7(1). https://doi.org/10.1007/s13205-016-0586-4
Singh SP, Singh P (2015) Effect of temperature and light on the growth of algae species: a review. Renew Sust Energ Rev 50:431–444
Solano F (2017) Melanin and melanin-related polymers as materials with biomedical and biotechnological application-cuttlefish ink and mussel foot proteins as inspired biomolecules. Int J Mol Sci 18(7). https://doi.org/10.3390/ijms18071561
Sousa AI, Martins I, Lillebo AI, Flindt M, Pardal MA (2007) Influence of salinity, nutrients and light on the germination and growth of Enteromorpha sp. Spores. J Exp Mar Biol Ecol 341(1):142–150
Spudich JL, Luecke H (2002) Sensory rhodopsin II : functional insights from structure. Curr Opsin Struct Biol 12(4):540–546
Sreerag RS, Jacob J, Nisha GV, Asha A, Sasidharan NK (2014) Influence of six nitrogen sources with fructose on antimicrobial metabolite production by bacterium associated with entomopathogenic nematode. Int J Pharm Pharm Sci 6(5):299–304

Srinubabu G, Lokeswari N, Jayaraju K (2006) Screening of nutritional parameter for the production of protease from Aspergillus oryzae. J Chem 4(2):208–215

Steenbergen JN, Alder J, Thorne GM, Tally FP (2005) Daptomycin: a lipopeptide antibiotic for the treatment of serious Gram-positive infection. J Antimicrob Chemother 55(31):283–288

Struvay C, Feller G (2012) Optimization to low temperature activity in psychrophilic enzymes. Int J Mol Sci 13(9):11643–11665

Swartz TE, Tseng TS, Frederickson MA, Paris G, Comerci DJ, Rajashekara G, Kim JG, Mudgett MB, Splitter GA, Ugalde RA, Goldbaum FA, Briggs WR, Bogomolni RA (2007) Blue light activated histidine kinases: two component sensors in bacteria. Science 317(5841):1090–1093

Takano H, Obitsu S, Beppu T, Ueda K (2005) Light-induced carotenogenesis in Streptomyces coelicolor A3(2): identification of an extracytoplasmic function sigma factor that directs photodependent transcription of the carotenoid biosynthesis gene cluster. J Bacteriol 187(5):1825–1832

Tudzynski B (2014) Nitrogen regulation of fungal secondary metabolism in fungi. Front Microbiol 5. https://doi.org/10.3389/fmicb2014.00656

Tudzynski B, Kawaide H, Kamiya Y (1998) Gibberellin biosynthesis in *Gibberella fujikuroi*: cloning and characterization of the copalyl diphosphate synthase gene. Curr Genet 34(3):234–240

Velmurugan P, Lee YH, Venil CK, Lakshmanaperumalsamy P, Chae JC, Oh BT (2010) 'Effect of light on growth, intracellular and extracellular pigment production by five pigment producing filamentous fungi in synthetic medium. J Biosci Bioeng 109(4):346–350

Wang J, Wen B, Xu QY, Xie XX, Chen N (2015) Optimization of carbon source and glucose feeding strategy for improvement of L-isoleucine production by *Escherichia coli*. Biotechnol Biotechnol Equip 29(2):374–380

Wang J, Yan D, Dixon R, Wang YP (2016a) Deciphering the principles of bacterial nitrogen dietary preferences: a strategy for nutrient containment. Am Soc Microbiol 7(4):e00792–e00716. https://doi.org/10.1128/mBio00792-16

Wang LL, Dai Y, Chen WP, Shao YC, Chen FS (2016b) Effects of light intensity and color on the biomass, extracellular red pigment and citrinin production of *Monascus ruber*. J Agric Food Chem 64:9506–9514

Wawrik B, Kerkhof L, Kukor J, Zylstra G (2005) Effect of different carbon sources on community composition of bacterial enrichments from soil. Appl Environ Microbiol 71(11):6776–6783

Wright M, Grant T, Delumeau O, Van-Duinen G, Firbank SJ, Lewis PJ, Murray JW, Newman JA, Quin MB, Race PR, Rohou A, Tichelaar W, Van-heel M, Lewis RJ (2008) Molecular architecture of the 'stressosome', a signal integration and transduction hub. Science 322(5898):92–96

Wu HY (2016) Effect of different light qualities on growth, pigment content, chlorophyll fluorescence and antioxidant enzyme activity in red alga *Pyropia haitanensis* (Bangiales, Rhodophyta). Biomed Res Int. https://doi.org/10.1155/2016/7383918

Wu JQ, Zhang QL, Deng W, Qian JC, Zhang SL, Liu W (2011) Toward improvement of Erhthromycin A production in an industrial *Saccharopolyspora erythraea* strain via facilitation of genetic manipulation with an artificial attB site for specific recombination. Appl Environ Microbiol 77(21):7508–7516

Yu HQ, Fang HHP (2003) Acidogenesis of gelatin rich wastewater in an upflow anaerobic reactor: influence of pH and temperature. Water Res 37(1):55–66

Yu XH, Liu HT, Klejnot J, Lin CT (2010) The cryptochrome blue light receptors. Arabidopsis Book 8:e0135. https://doi.org/10.1199/tab.0135

Yuan J, Ma Q, Ye L, Piao G (2016) The traditional medicine and modern medicine from natural products. Molecules 21(5). https://doi.org/10.3390/molecules21050559

Yurkov VV, Gemerden HV (1993) Impact of light/dark regime on growth rate, biomass formation and bacteriochlorophyll synthesis in *Erythromicrobium hydrolyticum*. Biochemistry 159:84–89

Ziemons D, Koutsantas K, Becker K, Dahlmann T, Kuck U (2017) Penicillin production in industrial strain *Penicillium chrysogenum*P2niaD18 is not dependent on the copy number of biosynthesis genes. Bio Med Cent Biotechnol 17. https://doi.org/10.1186/s12896-017-0335-8

Zoltowshi BD, Scwerdtfeger C, Widom J, Loros JJ, Bilwes AM, Dunlap JC, Crane BR (2007) Conformational switching in the fungal light sensor vivid. Science 316:1054–1057

Microbe-Based Biopesticide Formulation: A Tool for Crop Protection and Sustainable Agriculture Development

6

Akash Mishra, Anfal Arshi, Shraddha P. Mishra, and Madhu Bala

Abstract

Biopesticides have a pivotal role in agriculture and are the most environment-friendly approach in pest control management. Biopesticides are mainly based upon plants or microorganisms and their product to control pest without affecting non-target organisms as they do not contain any chemical which has a hazardous secondary effect on surrounding environment. Further, no destructive residue is left after a period of application of such product. Nowadays, research based on advanced technology such as genetic engineering and nanotechnology has improved the modern biopesticides in terms of their effectiveness and applicability.

In this chapter, we have tried to explain the overall process of biopesticide formulation technique mainly consisting of microbes. The delivery of a formulated product is the crucial step in the efficacy and success of biopesticides. Therefore, this chapter also provides lucrative information regarding various methods adopted for the application of biopesticides in the crop field.

Keywords

Microbe-based biopesticides · Biopesticide formulation · Delivery system · Nanoparticle

A. Mishra · A. Arshi (✉) · S. P. Mishra · M. Bala
Defence Institute of Bio-Energy Research (DIBER), Defence Research and Development Organization (DRDO), Ministry of Defence, Government of India, Haldwani, Uttarakhand, India
e-mail: madhubala@diber.drdo.in

P. K. Arora (ed.), *Microbial Technology for the Welfare of Society*, Microorganisms for Sustainability 17,
https://doi.org/10.1007/978-981-13-8844-6_6

6.1 Introduction

Human population is increasing day by day, and it imposes more food consumption worldwide. The rise in human population results in direct dependency on the use of pesticides for production of food materials at a very large scale. It is clear that the dependency on the use of pesticide will remain the same or get an increase in the future. Presently, to overcome the problems of food scarcity, innovative approaches for crop cultivation and increase in agro-based products are taking place. Advancement in technology and the use of machinery as well as the introduction of new pesticides in agricultural practices allow a farmer to manage crops and get good yield without extensive use of labor (Bolognesi 2003; SaeediSaravi and Shokrzadeh 2011). These pesticides reduce the problem of agricultural loss caused by pests and make a food material available in any season at a very normal price (Cooper and Dobson 2007).

Usually, a pesticide is made as selective as it should not act upon non-targeted pest, but it is very difficult to achieve the appropriate selectivity, and the adverse effects can be seen in non-target organisms including humans. Sometimes the use of pesticides can be considered as self-poisoning. Maksymiv (2015) reported that every year the fatal rate is 220,000 and 3 million cases are due to pesticide poisoning. The proper handling and use of pesticides make it beneficial for agriculture, but if utmost care is not taken in its use, the result can be in the form of contamination of soil, water, and air.

However, an alternative to these toxic pesticides can solve such problems. Recent researches based on the control of pests by living organisms and their products have gathered the attention towards a novel form of pesticides which can be called as the biopesticide. Biopesticides are non-toxic to the user and other organisms and therefore can be referred to as an environmental-friendly approach for the control of the pest. They act specifically upon a targeted pest only (Xu et al. 2011). In comparison to chemical pesticides, the adverse effect on the environment due to the use of biopesticides is negligible, and the repeated use of it decreases the dependency on the chemical pesticides without a reduction in crop yield (Alam 2000; Rao et al. 2007). Therefore, biopesticides are the best alternative to chemical pesticides in modern agricultural practices and can be implemented in the sustainable development of environment and agribusiness.

On the other hand, there is a lack of information among farmers regarding the use and effectiveness of biopesticide (Alam 2000). A mistaken belief about the effectiveness and product cost are the main reasons for its rejection. Therefore, in this chapter, we have summarized the types of chemical pesticide and adverse impact associated with them. Further, biopesticide as an alternative to chemical pesticide, its types, and formulation, as well as the delivery system of microbe-based biopesticide, are also being discussed.

6.2 Chemical Pesticides and Their Adverse Effect on Living Organisms and Environment

Nowadays, agricultural practices mainly depend upon the use of several kinds of chemical pesticides. According to Maksymiv (2015), "pesticide" can be referred to as a substance used to control/repel or kill a "pest" such as rodents, snails, weeds, insects, bacteria, and fungi. FAO in association with UNEP (1990) has also defined the term "pesticide" as "chemical intended to defend the attack of a pest on crop, humans and their animals." These chemicals can be growth regulators, defoliants, or substances used to prevent the deterioration of crop product during storage. However, the chemicals used for nutrient enhancement in a plant as well as animals do not come under the term pesticide.

6.2.1 Classification of Pesticides

Pesticides are of different physicochemical properties, identities, and uses. Therefore, they are classified according to the need. Here, we are presenting three main categories of pesticides: classification on the basis of mode of entry and action, classification on the basis of targeted pest species, and classification on the basis of chemical composition of the pesticide.

6.2.1.1 Classification on the Basis of Mode of Entry and Action

Each pesticide has a different route of entry and mode of action into the pest body. Hence, they are classified into different categories which may include non-systemic, systemic, stomach poisons, repellent, and fumigant.

Non-systemic (Contact) Pesticides The non-systemic pesticides can also be called as contact pesticides because it requires a physical contact to kill the targeted pest where it penetrates into the epidermis and causes poisoning, resulting into death of the pest. Diquat and paraquat dibromide are the examples of this type of pesticide.

Systemic Pesticides In this category, pesticides may get absorbed by the plant or animal body and reach up to the untreated tissue. For example, a systemic herbicide can transfer inside the plant body to reach untreated areas of root, stem, or leaves. The route of movement of pesticide may be multidirectional or unidirectional. Therefore, these types of pesticides can transfer through plant tissue and kill a targeted pest very effectively. Similarly, systemic insecticides are also effective against fleas, lice, and warble grubs and kill such pest if applied upon an animal body. Glyphosate and 2,4-dichlorophenoxyacetic acid (2,4-D) are examples of systemic pesticides (Buchel 1983).

Stomach Poisons Stomach poisons are those pesticides which enter inside the pest body through the mouth and digestive system, causing death due to poisoning. Such pesticides are very effective in vector control such as mosquito and blackfly larvae. If these poisons are applied in the water, it destroys the midgut or stomach of the vector. A good example of this type of pesticide is malathion.

Repellents These pesticides do not kill a pest but keep them away from a treated zone. Additionally, repellent makes a pest unable to trace the crop.

Fumigants Some pesticides form vapors and poisonous gases to kill a targeted pest when applied. Such pesticides come under the category of fumigants. The poisonous gases enter the body by the respiratory system of the pest and cause poisoning. The active ingredients of such pesticides may be liquids and get converted into gas during application. Fumigants can be best used for removal of pests from stored products like grains, vegetable, and fruits. Soil pest can also be controlled by the use of fumigants.

6.2.1.2 Classification on the Basis of Targeted Pest Species

Generally, chemical pesticides are specific to control a particular kind of pest. Therefore, they are arranged in a classification where a specific name is given to each type of pesticide which reflects their activity on the targeted pest. Under this method of classification, a Latin word "*cide*" (which means killer or to kill) is used to give a name to the pesticide after the targeted pest name (Table 6.1). However, the name of some pesticide does not end with the word "*cide*." These pesticide are classified according to their function. Examples of such pesticide are growth regulator that controls the growth of pest, defoliant which causes leaf shedding from the plant, an attractant that attracts pest to trap them, and chemosterilant which makes a pest sterile.

Table 6.1 Types of pesticide and their targeted pest

Type of pesticide	Targeted pests	Example
Insecticide	Insects	*Aldicarb, allethrin*
Fungicide	Fungi	*Azoxystrobin, pentachlorophenol*
Bactericide	Bacteria	*Copper complexes*
Herbicide	Weeds and other unwanted plants	*Atrazine, alachlor*
Acaricide	Mites feeding on plant and animal	*Bifenazate, chlorpyrifos, aldicarb*
Rodenticide	Mice and some other rodents	*Warfarin, pindone*
Algaecide	Algae	*Copper sulfate, diuron*
Larvicide	Larvae	*Methoprene*
Virucide	Viruses	*Scytovirin, ribavirin*
Molluscicide	Snail and other members of Mollusca	*Metaldehyde, thiacloprid*
Nematicide	Nematodes	*Aldicarb, carbofuran*
Avicide	Birds	*Avitrol, diazinon*
Piscicide	Fishes	*Rotenone*
Termiticide	Termites	*Fipronil*

Modified after Yadav and Devi (2017)

Some pesticides can be considered in more than one class because they control a number of different pests. Aldicarb is the best example for such type of pesticide as it acts upon insects, mites, and nematodes and hence can be considered as the insecticide, acaricide, and nematicide, respectively.

6.2.1.3 Classification on the Basis of Chemical Composition of Pesticide

It is the most common but important method to classify different pesticides which are based on the composition of chemicals and other ingredients used. Such pesticides are self-informer about their physical and chemical property, mode of action, and efficacy. On the basis of chemical composition, these types of pesticides can be classified into four main categories such as organophosphorus, organochlorines, synthetic pyrethroid, and carbamates (Buchel 1983).

Organophosphorus Organophosphorus pesticides possess multiple functions and control a broad range of targeted pests and, hence, are one of the broad-spectrum pesticides. The pest resistance rate is very slow for such pesticides, and also they are biodegradable causing a minimum adverse impact on the environment (Martin 1968). These pesticides are best used for invertebrates and vertebrates, as they possess cholinesterase inhibitors causing failure in nervous impulses which results in paralysis and death. The best examples of organophosphorus pesticides are malathion, parathion, diazinon, and glyphosate.

Organochlorine Organochlorine pesticides are basically chlorinated hydrocarbons that possess an organic compound attached with more than five chlorine atoms. This class includes those pesticides which were the first to be synthesized and applied in agricultural practices. The mode of action of such pesticide is the disruption in the nervous system of insects leading to paralysis and death. The common examples of organochlorine pesticides are DDT, endosulfan, lindane, aldrin, dieldrin, and chlordane. The main drawback of such pesticide is a long-term residual impact on the environment.

Synthetic Pyrethroids Synthetic pyrethroids are produced by the method of structure duplication of natural pyrethrins which makes it more stable and effective than a natural one. The main active components of synthetic pyrethrins are pyrethrins I and II and some other compounds like jasmolins and cinerins. These pesticides are more effective for the control of insect pest and fishes. However, these are one of the safe insecticides as they are non-persistent and photodegradable. Examples are permethrin and cypermethrin.

Carbamates The mode of action and structure of carbamates are moreover similar to organophosphate pesticides. However, they originate from carbamic acid, whereas organophosphates are derived from phosphoric acid. These pesticides are also easily degradable. Carbofuran, carbaryl, propoxur, and aminocarb are some of the examples of such pesticides.

6.2.2 Adverse Effects of Chemical Pesticides

Chemical pesticides have transformed the agricultural practices, especially in terms of better production and disease management. However, the adverse impact of them on the environment suppresses their benefits of use. The main disadvantages of chemical pesticides are the secondary effect on non-target species and disturbance in plant and animal biodiversity and terrestrial and aquatic ecosystem. Majewski and Capel (1995) have reported that about 80–90% of sprayed pesticide may get volatilize in the environment after application, even in a few days. These chemical pesticides (e.g., herbicide) get volatilize in the air from the treated plant and causes a remarkable damage to other non-target plants (Straathoff 1986). Due to the uncontrolled use of different chemical pesticides, the number of various species from land and aquatic biodiversity has reduced because not only targeted pest is killed by the effect of pesticides but other non-targeted organisms like birds and several other beneficial insects are also killed (Maksymiv 2015). Some rare species of Peregrine falcon, osprey, and bald eagle are the main examples of a threatened organism by the use of such pesticides (Helfrich et al. 2009). In the list of all pesticides, insecticides possess the first rank in toxicity, whereas fungicides and herbicides are at the second and third, respectively (Mahmood et al. 2015). Contamination due to pesticide is associated with not only surface water but also groundwater. These pesticides can percolate through soil layers and reach up to the water belt present below the Earth's surface. The contaminated water consumption affects all the organisms including humans (Cerejeira et al. 2003).

Food is the basic requirement in human and other living organism's life. Food toxicity caused by pesticide contamination may also impose a serious effect on human health. There are several diseases associated with pesticide toxicity in humans and are the result of long-term exposure to these chemical pesticides. Some of the diseases are psychological, neurological, behavioral, and immune system dysfunctions; infertility due to hormonal imbalances; genotoxicity; cancers; and blood disorders (Maksymiv 2015). Several toxicological studies on animals also confirm that pesticides are potent neurotoxins, carcinogens, and immunotoxins if exposed for a long period of time (Baker and Wilkenson 1990; Bolognesi and Merlo 2011).

6.3 Biopesticide as an Alternative Tool for Sustainable Agriculture Development

Overuse of chemical control methods to kill a pest has weakened the modern agricultural practices. According to Jeyaratnam (1990), around 25 million people (per year) working for agriculture in developing countries were in danger from the use of pesticides. Due to several undesirable effects of chemical pesticides (as discussed in Sect. 6.2.2.) on humans as well as the environment, approaches are being made to find an alternative as a solution to this problem. Acadmic and industrial researches have been coming in the forefront to make pesticides which are of biological origin, although this approach is not new, as biocontrol of pest has been practiced from ancient times. Saxena et al. (1989) have reported that farmers from many countries in Africa approached to bio-control the insect pests from their crops. They were aware to use the predatory bird to control insect pest.

Biopesticides can be derivative of natural materials like minerals or any other living organisms such as bacteria, plants, and animals (US EPA 2008). Generally, living organisms or chemicals produced by them can be used as biopesticides because they are the natural enemies of those pests which are harmful to the plant or any other host. In the use of biopesticides, the risk is negligible for human health and the environment. The most common living organisms used as biopesticides are *Bacillus thuringiensis* (as bioinsecticides), *Trichoderma* (as biofungicides), and *Phytophthora* (as bioherbicides) (Kandpal 2015).

6.3.1 Advantages in the Use of Biopesticides

Biopesticides in agricultural practices are getting more attention as they possess the following advantages in their use than chemical-based pest control methods:

1. Biopesticides are generally safe to be used as they are not harmful to non-targeted pests.
2. They are very specific to the target.
3. They are required in less quantity.
4. They are completely biodegradable.
5. They can be used as an alternative to chemical methods.
6. Chances of pest resistance development are very less.

Generally, biopesticides have a narrow range of targeted pests. However, the mode of action to kill a pest is very specific (Clemson HGIC 2007). They are designed in such a way to control the population of pest up to a significant level only rather than complete eradication of targeted pest (Lewis et al. 1997).

With the above-discussed advantages, biopesticides can be used as a modern tool for crop protection. They are the best alternative to chemical pesticides; their use and effects on the surrounding environment including human health can play a crucial role in the evolution of present methods available for sustainable development of agriculture.

6.3.2 Types of Biopesticides

According to Semeniuc et al. (2017), depending upon the nature of origin and main active ingredients, biopesticides can be of different types such as antagonists, botanicals, growth promoters, predators and pheromones, compost teas, etc. Here we are discussing two main types of biopesticides (plant extract-based and microorganism-based biopesticides).

6.3.2.1 Plant Extract-Based Biopesticides

Plants are the ultimate source of naturally occurring active ingredients which do have antimicrobial property and other pest control mechanisms. These active ingredients can be extracted (depending upon method of extraction) in the form of either plant extracts or essential oils (Vidyasagar and Tabassum 2013) from different parts of a plant such as seed, rhizome, root, stem, bark, leaves, flowers, and fruit (Lengai and Muthomi 2018). Dried mass of plant parts gives a good yield of active ingredient (Chougule and Andoji 2016). Some important examples of plants which are used for the preparation of biopesticides for the control of various targeted pests are shown in Table 6.2. In an experiment conducted by Muthomi et al. (2017), it was reported that several ethanolic extracts of plants such as turmeric (*Curcuma longa*), lemon (*Citrus limon*), garlic (*Allium sativum*), pepper (*Capsicum frutescens*), and ginger (*Zingiber officinale*) are very effective to control *Alternaria solani*, *Pythium ultimum*, *Rhizoctonia solani, and Fusarium oxysporum* f.sp. *lycopersici*. *Pseudomonas syringae* pv. tomato can also be controlled by *Rhus coriaria*-, *Eucalyptus globulus*-, and *Rosmarinus officinalis*-based biopesticides (Baştaş 2015). Derivatives of pyrethrum (*Chrysanthemum cinerariifolium*), neem (*Azadirachta indica)*, tobacco (*Nicotiana tabacum*), and sabadilla (*Schoenocaulon officinale*) are some of the most common examples of commercialized biopesticides of botanical origin (Dar et al. 2014).

6.3.2.2 Microorganisms-Based Biopesticides

Out of all commercialized biopesticides, microorganism-based biocontrol products possess a larger portion and may include bacteria, fungi, viruses, protozoans, and nematodes (Koul 2011). Singh (2014) reported that there are about 175 biopesticides based on microbes as an active agent and can be used for the management of insect, pathogens, weed, and nematodes. Table 6.3 shows some commercially used microbes as biocontrol agents. Important bacterial species which can be used as biopesticides are *Pseudomonas*, *Bacillus*, *Enterobacter*, *Burkholderia*, *Serratia*, *Xanthomonas, and Streptomyces*, whereas *Beauveria*, *Paecilomyces*, *Metarhizium*, *Trichoderma*, *Pythium*, *Penicillium*, *Verticillium*, and *Fusarium* are some examples of fungal species having a potential to be used as biopesticides (Kachhawa 2017). To kill plant pathogens, these biocontrol agents may exhibit the mode of action such as antibiosis and parasitism, competition, secretion of volatile natural compounds, and hyperparasitism (Suprapta 2012). Agricultural field specifically rhizosphere provides a favorable habitat to the biocontrol agents where they compete with other pathogens and develop their biocontrol potential so that they can be used as biopesticides (Song et al. 2012).

6.4 Formulation of Microbial Biopesticides and Its Delivery System

For efficacy and success, an appropriate formulation strategy is a must for biological products which make it target specific. According to Woods (2003), a developed biological product should contain a functional biological agent along with good

Table 6.2 List of some important plants used as biopesticides and their targeted pests

Source plant	Targeted pest	Host plant
Azadirachta indica (neem)	*Aphis craccivora*	*Vigna unguiculata*;
	Amrasca devastans	*Gossypium hirsutum*;
	Myzus persicae	*Solanum tuberosum*;
	Sitobion avenae	*Triticum* sp.
	Lipaphis erysimi	*Brassica* sp.;
	Bemisia tabaci	*Prunus salicina*;
	Sciothrips cardamomi	*Lycopersicon esculentum*;
	Rhizopus and Aspergillus sp.	*Capsicum chinense*;
	Monilinia fructicola	*Cardamomum* sp.
	Trichothecium roseum	
	Meloidogyne javanica	
Allium sativum (garlic)	*Curvularia lunata*	*Triticum* sp.;
	Aspergillus niger	*Lycopersicon esculentum*;
	Candida albicans	*Abelmoschus esculentus*;
	Trichophyton rubrum	*Panax* sp.
	Drechslera tritici-repentis	*Capsicum* sp.
	Bipolaris sorokiniana	Human and Animal sp.
	Rhizoctonia solani	*Oryza* sp.
	Colletotrichum sp.	*Gossypium hirsutum*;
	Bacillus subtilis	Stored grain products;
	Salmonella senflenberg	*Vigna unguiculata*;
	Staphylococcus aureus	*Brassica oleracea*;
	Staphylococcus epidermidis	
	Sitotroga cerealella	
	Brevicoryne brassicae	
Euphorbia sp.	*Pseudomonas aeruginosa*	Human and Animal sp.;
	Enterobacter aerogenes	*Arachis hypogaea*
	Salmonella typhi	
	Aspergillus flavus	
Curcuma longa (turmeric)	*Tribolium castaneum*	*Triticum aestivum*;
	Bactrocera zonata	*Prunus persica*;
	Trichoplusia ni	*Brassica oleracea*;
	Alternaria solani	*Solanum lycopersicum*;
	Streptococcus pyogenes	Human sp.;
	Ralstonia solanacearum	Animal sp.
	Escherichia coli	
	Listeria monocytogenes	
	Bacillus subtilis	
Tagetes spp.	*Fusarium oxysporum*	*Gladiolus grandifloras*;
	Klebsiella pneumoniae	*Leucadendron*;
	Brevicoryne brassicae	Human and Animal sp.;
	Plutella xylostella	*Brassica oleracea*
	Mamestra brassicae	
	Meloidogyne incognita	

(continued)

Table 6.2 (continued)

Source plant	Targeted pest	Host plant
Cinnamomum zeylanicum	*Botrytis cinerea*	*Zea mays*;
	Penicillium expansum	*Pinus densiflora*
	Aspergillus oryzae	
	Fusarium solani	
	Escherichia coli	
	Staphylococcus aureus	
	Meloidogyne sp.	
Thymus vulgaris (*garden thyme*)	*Penicillium* spp.	*Gallus gallus domesticus*;
	Saccharomyces spp.	*Triticum aestivum*;
	Aspergillus niger	*Solanum lycopersicum*;
	Tilletia tritici	*Citrus aurantium*;
	Xanthomonas vesicatoria,	*Cajanus cajan*;
	Escherichia coli	*Gossypium hirsutum*
	Salmonella typhimurium	
	Diaphorina citri	
	Pratylenchus brachyurus	
Jatropha spp.	*Aphis fabae*	*Vigna unguiculata*;
	Sitophilus zeamais	*Zea mays*;
	Tribolium castaneum	*Triticum aestivum*;
	Oryzaephilus surimanensis	*Solanum melongena*
	Aspergillus flavus	
	Alternaria alternate	
	Meloidogyne incognita	
Zingiber officinale (ginger)	*Fusarium oxysporum,*	*Solanum lycopersicum*;
	Aspergillus spp.	*Arachis hypogaea*;
	Escherichia coli	*Coffea* spp.;
	Salmonella typhi	*Mangifer aindica*;
	Tribolium castaneum	*Oryza sativa*;
	Necrobia rufipes	*Brassica oleracea*;
	Dermestes maculatus	*Clarias gariepinus*

Adapted from Lengai and Muthomi (2018)

physical property and ease of use during storage and application. There are certainly important criteria associated with the formulation of microbial biopesticides, which may include the following: a biopesticide should be cost-effective in production to increase its commercial value, stability with respect to storage and viability of biocontrol agent, and ease in handling and application, and it should always be effective in the control of desired pests. These criteria can be fulfilled by formulating biocontrol agents in different ways (Seaman 1990; Mollet and Grubenmann 2001).

During the formulation process, the microbial component is mixed with a carrier depending upon the type of product. Some other adjuvants are also added not only to combat environmental factors but also for the better survival of the microbial agent and improved efficacy of the final product.

Based on the physical property, microbial biopesticide is mainly of two types: dry and liquid-based formulations. Liquid formulations can be polymer-based,

Table 6.3 List of some important microorganisms used as biocontrol agent and their targeted pest

Biocontrol agent (microbe)		Targeted pest	Host plant
Fungus	*Trichoderma* spp.	*Fusarium* spp.	*Phoenix dactylifera*;
		Pythium spp.	*Zea mays*;
		Alternaria spp.	*Solanum* spp.;
		Rhizoctonia spp.	*Capsicum annum*;
		Verticillium spp.	*Lycopersicon esculentum*
		Penicillium stekii	
		Botrytis cinerea	
		Phytophthora spp.	
		Meloidogyne spp.	
	Beauveria spp.	*Spodoptera litura*	*Pinus tabulaeformis*;
		Uvarovistia zebra	*Phaseolus vulgaris*;
		Cyclocephala lurida	*Brassica oleracea*;
		Frankliniella occidentalis	*Avena sativa*;
		Hylobius abietis	*Triticum aestivum*
		Oryzaephilus surinamensis	
		Tuta absoluta	
		Polyphylla fullo	
	Paecilomyces spp.	*Meloidogyne javanica*	*Solanum* sp.;
		Phytophthora palmivora	*Vigna mungo;*
			Theobroma cacao
	Metarhizium spp.	*Musca domestica*	*Citrus sinensis*;
		Spodoptera litura	*Polyphylla fullo*;
		Ceratitis capitate	*Phaseolus vulgaris*;
		Polyphylla fullo	*Solanum lycopersicum*
		Tetranychus urticae	
		Maruca vitrata	
		Tuta absoluta	
		Callosobruchus maculatus	
Bacteria	*Bacillus* spp.	*Macrophomina phaseolina*	*Glycine max*;
		Fusarium graminearum	*Triticum aestivum*;
		Colletotrichum spp.	*Capsicum annuum*;
		Phaeomoniella chlamydospora	*Vitis vinifera*;
		Fusarium solani	*Vicia faba*;
		Meloidogyne incognita	*Vigna mungo*;
		Spodoptera frugiperda	*Solanum lycopersicum*
		Xanthomonas campestris pv. *campestris*	
	Pseudomonas spp.	*Vibrio harveyi*	*Mangifera indica*;
		Fusarium oxysporum	*Solanum* spp.;
		Alternaria alternata	*Vitis vinifera*;
		Sclerotium rolfsii	*Oryza sativa*;
		Botryodiplodia theobromae	*Triticum aetivum*
		Rhizoctonia solani	
		Aspergillus aculeatus	
	Verticillium spp.	*Trialeurodes vaporariorum*	*Phaseolus vulgaris*;
		Bemisia tabaci	*Euphorbia pulcherrima*

Modified after Lengai and Muthomi (2018)

oil-based, water-based, or in combination of all of these. In water-based formulations like suspo-emulsions, suspension concentrate, capsule suspension, etc., an inert material is required such as stickers, stabilizers, coloring agents, surfactants, additional nutrients, and antifreeze compounds (Gasic and Tanovic 2013). For dry formulations, different techniques such as freeze-drying, spray drying, and air-drying are used. These dry formulations are developed by the addition of dispersant, binder and wetting agents, etc. (Tadros 2005; Brar et al. 2006; Knowles 2008).

Once microbial biopesticide is formed, it can be applied to the crop field. There are several methods for the application and delivery of such products depending upon its type. An illustrative table (Table 6.4) is given below for some of the major delivery systems used in microbial biopesticide formulation techniques.

6.4.1 Development and Stages in Bioformulation

Developmental stages in formulation of microbe-based biocontrol product contain the interrelated multistep process. To achieve the aim of biocontrol, search/selection and screening of a potent microbial organism are a must. It is really a tough task which depends upon the characteristics of the pathogen, crop, and agricultural practices (Junaid et al. 2013). An illustrative chart is given below (Fig. 6.1) to understand the possible steps in microbe-based biocontrol product formulation.

6.5 Advancement in the Development of Formulation of Biopesticide

The science involved in the formulation of biopesticide is still young and developing. Therefore, an extensive research is required in some of its steps such as formulation/production, delivery, and commercialization of the finalized product (Kumar and Singh 2015). Nowadays, biopesticides are being formulated by natural products produced by beneficial microorganisms and locally grown plants (garlic, neem, tulsi, etc.) which provide an alternative to dependency on chemical pesticides and are easily available, resulting in more consumption of biopesticides. The enhanced production of these natural compounds in beneficial organism or search of new biometabolites can be achieved by advanced technologies like recombinant DNA technology. Fitches et al. (2004) have reported that novel fusion proteins are the result of advanced technology and maybe next-generation biopesticides, as these are toxic to a targeted pest but not for other organisms. These toxins are combined with a carrier protein for oral entry into an insect pest, while earlier, the injection of toxin was required to kill the targeted pest. In the microbial system, the recombinant protein can be developed by new technologies and further obtained in bulk by scaling up during industrial production and formulated for commercial proposes. There are a number of innovative approaches through which the efficiency of a biopesticide can be enhanced. The use of nanotechnology and genetic engineering is one of them. These advanced technologies can develop an improved and acceptable version of pest control measures.

Table 6.4 Types of biopesticide formulation techniques and their delivery system

Type of formulation	Delivery system	Example	Formulation process	Use	References
Dry formulations	Direct application	Dusts (DP)	Microbes are loaded onto a fine solid mineral particle (talc, fly ash, clay, etc.) ranging from 50 to 100 μm	Directly applied to the target by mechanical or manual methods	Knowles (2001)
	Seed dressing formulations	Powders for seed dressing (DS)	Active ingredient (organism) is mixed with a powdered carrier material and other adjuvants for seed coating	Seed can be coated by tumbling it with powder for adherence of microbes used	Woods (2003)
		Granules (GR) and microgranules (MG)	Similar but larger in size to dust formulation (100–1000 microns for granules and 100–600 microns for microgranules)	Specifically used for the application of desired products to the soil to control nematodes, weeds, and insects living in soil	Tadros (2005), Knowles (2005) and Lyn et al. (2010)
			Made up of mineral materials like silica, polymers, kaolin, starch, and ground plant residues		
	Dilution in water	Water-dispersible granules (WG)	Granules developed for making a uniform suspension in water and overcome the problem of dustiness by powder formulation	Good storage stability with safety and greater convenience in application	Knowles (2008)

(continued)

Table 6.4 (continued)

Type of formulation	Delivery system	Example	Formulation process	Use	References
			Granules are formulated by several processing techniques like fluid bed granulation, extrusion granulation, spray drying, etc.		
		Wettable powders (WP)	A dry, finely ground formulation having particle size about 5 microns and requiring mixing it with water before application	Primitive technique for the application of organisms	Brar et al. (2006)
Liquid formulations	Dilution in water	Emulsions	Consist of oil droplets dispersed in water or vice versa (dispersed droplet size ranges from 0.1 to 10 μm)	More stable and contamination-free	Gasic and Tanovic 2013
		Suspension concentrates (SC)	A finely ground solid particle having a mixture of thickening agents, wetting/ dispersing agents, antifoaming agents, etc. dispersed in water. Particle size ranges from 1 to10 μm	Easy and safe in the application, environmentally safe	Woods (2003) and Knowles (2005)

(continued)

Table 6.4 (continued)

Type of formulation	Delivery system	Example	Formulation process	Use	References
		Oil dispersions (OD)	Active ingredients in solid form are dispersed in non-aqueous liquid usually oil. It generally requires dilution prior to use	Improved retention, spreading, and penetration; results in broader pest control	Vernner and Bauer (2007)
		Suspo-emulsions (SE)	A form of the mixture of suspension concentrate and emulsion techniques	Provides greater stability to the final product	Gasic and Tanovic (2013)
		Capsule suspensions (CS)	A stable technique for encapsulation of active ingredients (organism) in materials like gelatine, starch, etc. which require dilution before use	A highly effective form of biopesticide formulation as the controlled release of active ingredients makes it more stable in use	Brar et al. (2006)
		Ultra-low volume formulations	A formulation technique where the active ingredient is in high concentration	Extremely soluble in crop-compatible liquid (ultra-low volume liquid	Woods (2003)

6.5.1 Nanoparticle-Based Biopesticide for Protection of Plant from Pathogens

With the increase of multidrug resistance (MDR) in pathogens for antibiotics and other pesticides, scientists have moved to produce biometabolites by advanced technologies. In this context, nanotechnology aids in the production of nanoparticles, which are more effective and resistance-free antimicrobial compounds. Many nanoparticles are produced to date, but silver nanoparticles (AgNPs) are the most studied one (Mishra et al. 2015). Silver nanoparticle have been reported of being potential growth inhibitors and antimicrobial for types of microorganisms such as viruses, fungi, and bacteria. If compared to other metals, these are more lethal for

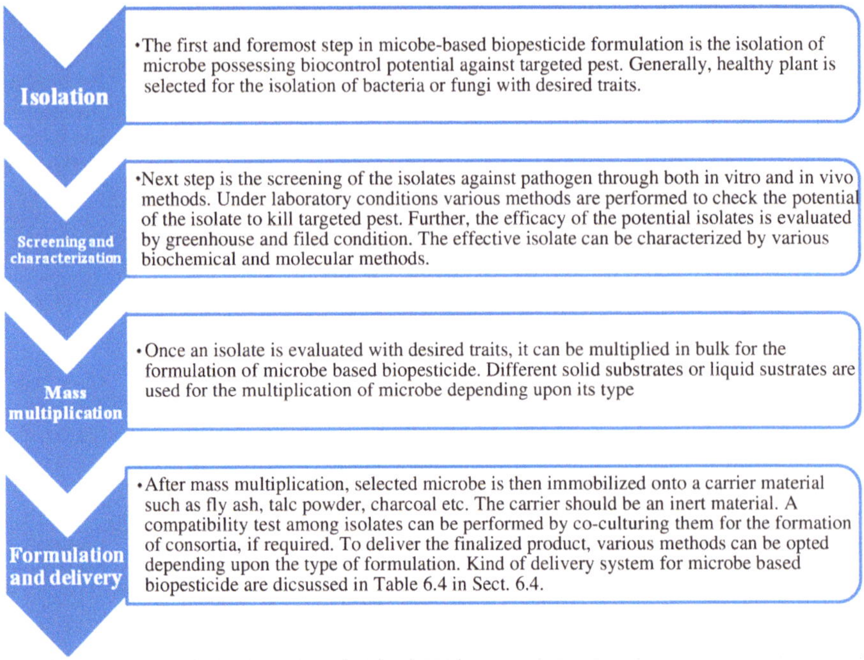

Fig. 6.1 Steps in microbe-based biopesticide formulation

microorganisms than the higher organism. The exact mechanism of the mode of action on a microbe is clearly not known and is a matter of debate. However, several theories regarding the possible mode of actions are given by various scientists for nanoparticles.

As per a theory proposed by Sondi and Salopek-Sondi (2004), silver nanoparticles get attached to the cell wall of a bacteria and break it, resulting in structural changes and alteration in cell permeability and ultimately death of the bacteria. On the accumulation of a large number of AgNPs, "pits" are formed on the surface of the bacterial cell. Another theory is given by Kim et al. (2007), according to which, free radicals formed by silver nanoparticles are the major cause of cell death. Matsumura et al. (2003) have reported that the interaction between silver ions produced by the nanoparticle and thiol group of different enzymes essential for the life of a bacterial cell results in inactivation of them. According to an important theory suggested by Hatchett and Henry (1996), an interaction between nanoparticles and sulfur/phosphorus may affect DNA replication, causing the death of the microbe.

Mostly, these theories are based on the study on the effects of nanoparticles on human pathogens. Since the cellular machinery is almost the same for both human and plant pathogens, therefore, these theories can also be applied for plant pathogens as well.

6.5.2 Use of Genetically Improved Strain of Microbes in Biocontrol Formulation

The use of improved biocontrol agents is the upcoming future of plant disease management. The microbial agents such as bacteria and fungi have antagonism potential or plant growth promotion ability and therefore are used as biocontrol agents. Besides the use of nanoparticles, modification at the genetic level in wild strains of a microbe may improve its biocontrol potential. Various molecular techniques allow us to modify a wild strain and enhance its disease suppression activity. Modification in a gene can also result in the formation of an entirely new strain with resistance for abiotic and biotic factors, enhanced production of antimicrobial compounds, and better competitiveness for nutrients and extended host range.

6.6 Current Status of Biopesticide Utilization

At present, the estimated production of biopesticides at global context is approximately 3000 tons per year, but compared to other control methods, its role in plant protection is only 2%. However, the demand has been increased from the last two decades as these biopesticides are having some advantages over chemical pesticides such as easy registration of the biocontrol product, organic food production, and aid in waste-free agricultural practices. With the above-discussed properties, the use of pesticide is gradually increasing at the rate of 10% per year (Kumar and Singh 2015).

Several bacterial key species play a vital role in the success of biopesticide market. *Bacillus thuringiensis* is one of them. Out of the total biopesticides produced, 90% of them are derived from this entomopathogenic bacterium. According to Hubbard et al. (2014), 30 OECD countries manufacture 225 biopesticides based on the microbes. With respect to similarity, more than 200 biopesticides gained market in the USA, and out of which 60 are present in the EU market. On the basis of use, NAFTA countries (USA, Mexico, and Canada) have a higher rate of use (45%), while Asia accounts for only 5% of the use of biopesticides (Bailey et al. 2010).

The policy of a country to utilize biocontrol method is a critical point in the establishment and success of biopesticide market. Due to several merits of biopesticide, many countries are promoting the use of it and have started doing amendments in the policies implemented for chemical pesticide production and utilization. Moreover, the system designed for control methods is based on the use of chemical pesticides and is the reason of barrier in the market for entry of biopesticide products (Kumar and Singh 2014), although the barrier gaps are also present in the form of demerits associated with the chemical pesticide which provide an alternative in the use of it.

6.7 Future Challenges

For the overall success in commercialization of biopesticides, it requires an advanced research in the formulation process and its stages such as selection of biocontrol agent, production of formulation, and its delivery system. There is a need for extensive research in the area of the efficacy of biocontrol agents used. Production of biopesticide at a very affordable price is also a matter of concern in developing countries. Maintaining quality of biopesticide is the crucial feature on which work has to be done. Thus, these are some main features and challenges of the near future for the research associated with advancement in biopesticides.

6.8 Conclusion

Chemical-based pesticides have transformed the conventional practices implemented in agriculture and control various kinds of pests, as well as they produce a huge amount of food material which is being used for providing food security worldwide. But the negative impact on humans and the environment is unavoidable and makes them inapplicable. Meanwhile, advanced research in the field of biopesticide has provided an alternative to it and delivers an eco-friendly approach for pest control management. The microbe-based biopesticides are the most effective and, therefore, appreciated form of biopesticide as they contain naturally occurring living organism which makes them an environmentally accepted method to control pests. The technological progress in the formulation and production of these microbe-based biopesticides has now made the process so easy and cost-effective as well. Further, advanced technologies such as nanotechnology have boosted the efficacy of microbe-based biopesticides. Therefore, the use of biopesticide containing microbes is an environment-friendly tool of protection of crop and sustainable development of the environment.

Acknowledgment The authors are very thankful to the director of Defence Institute of Bio-Energy Research (DIBER), DRDO, Ministry of Defence, Government of India (GOI), India, for providing financial and infrastructural support.

References

Alam G (2000) A study of biopesticides and biofertilizers in Haryana, India Gatekeeper Series No. 93. IIED, UK

Bailey KL, Boyetchko SM, Längle T (2010) Social and economic drivers shaping the future of biological control: a Canadian perspective on the factors affecting the development and use of microbial biopesticides. Biol Control 52:221–229

Baker SR, Wilkenson CF (1990) The effects of pesticides on human health. Princeton Sci Pub, Princeton

Baştaş KK (2015) Determination of Antibacterial Efficacies of Plant Extracts on Tomato Bacterial Speck Disease. J Turkish Phytopathol 1(3):1–10

Bolognesi C (2003) Genotoxicity of pesticides: a review of human biomonitoring studies. Mutat Res 543:251–272

Bolognesi C, Merlo FD (2011) Pesticides: human health effects. In: Nriagu JO (ed) Encyclopedia of environmental health. Elsevier, Burlington

Brar SK, Tyagi VRD, Valero JR (2006) Recent advances in downstream processes and formulations of *Bacillus thuringiensis* based biopesticide. Process Biochem 41:323–342

Buchel KH (1983) Chemistry of Pesticides. Wiley, New York

Cerejeira MJ, Viana P, Batista S, Pereira T, Silva E, Valerio MJ (2003) Pesticides in Portuguese surface and ground waters. Water Res 37(5):1055–1063

Chougule PM, Andoji YS (2016) Antifungal Activity of Some Common Medicinal Plant Extracts against Soil Borne Phytopathogenic Fungi Fusarium oxysporum Causing Wilt of Tomato. Int J Dev Res 3:7030–7033

Clemson Extension, Home and Garden Information Center (2007) Organic pesticides and biopesticides. http://hgic.clemson.edu/factsheets/HGIC2756.htm. Accessed 20 Aug 2008

Cooper J, Dobson H (2007) The benefits of pesticides to mankind and the environment. Crop Prot 26(9):1337–1348

Dar AS, Nisar AD, Mudasir AB, Mudasir HB (2014) Prospects, utilization and challenges of botanical pesticides in sustainable agriculture. Int J Mol Biol Biochem 1:1–14

Fitches E, Edwards MG, Mee C, Grishin E, Gatehouse AMR (2004) Fusion proteins containing insect-specific toxins as pest control agents: snowdrop lectin delivers fused insecticidal spider venom toxin to insect haemolymph following oral ingestion. J Insect Physiol 50:61–71

Gasic S, Tanovic B (2013) Biopesticide formulations, possibility of application and future trends. J Pestic Phytomedicine Belgrade 2:97–102

Hatchett DW, Henry S (1996) Electrochemistry of sulfurad layers on low-index faces of silver. The. J Phys Chem 100:9854–9859

Helfrich LA, Weigmann DL, Hipkins P, Stinson ER (2009) Pesticides and aquatic animals: a guide to reducing impacts on aquatic systems. Virginia Polytechnic Institute and State University. Available from https://pubs.ext.vt.edu/420/420-013/420-013.html. Accessed 17 Jan 2015

Hubbard M, Hynes RK, Erlandson M, Bailey KL (2014) The biochemistry behind biopesticide efficacy. Sust Chem Process 2:18

Jeyaratnam J (1990) Acute pesticide poisoning: a major global health problem. World Health Stat Q 43(3):139–144

Junaid JM, Dar NA, Bhat TA, Bhat AH, Bhat M (2013) A (2013) Commercial Biocontrol Agents and Their Mechanism of Action in the Management of Plant Pathogens. Int J Modern Plant Animal Sci 1(2):39–57

Kachhawa D (2017) Microorganisms as a Biopesticides. J Entomol Zool Stud 3:468–473

Kandpal V (2015) Biopesticides. Int J Environ Res Dev 4(2):191–196

Kim JS, Kuk E, Yu K, Kim JH, Park SJ, Lee HJ, Kim SH, Park YK, Park YH, Hwang CY, Kim YK, Lee YS, Jeong DH, Cho MH (2007) Antimicrobial effects of silver nanoparticles. Nanomedicine 3:95–101

Knowles A (2001) Trends in pesticide formulations. PJB Publications Ltd. D215, Agrow Reports, UK, pp 89–92

Knowles A (2005) New developments in crop protection product formulation. T and F Informa UK Ltd., Agrow Reports UK, pp 153–156

Knowles A (2008) Recent developments of safer formulations of agrochemicals. Environmentalist 28(1):35–44. https://doi.org/10.1007/s10669-007-9045-4

Koul O (2011) Microbial biopesticides: opportunities and challenges. CAB Rev Perspect Agric Vet Sci Nutr Nat Res 6:56

Kumar S, Singh A (2014) Biopesticides for integrated crop management: environmental and regulatory aspects. J Fertil Pestic 5:e121

Kumar S, Singh A (2015) Biopesticides: present status and the future prospects. J Fertil Pestic 6:2. https://doi.org/10.4172/2471-2728.1000e129

Lengai GMW, Muthomi JW (2018) Biopesticides and their role in sustainable agricultural production. J Biosci Med 6:7–41

Lewis W, van Lenteren C, Phatak S, Tumlinsen J (1997) A total system approach to sustainable pest management. Proc Nat Acad Sci USA 94:12243–12248

Lyn ME, Burnett D, Garcia AR, Gray R (2010) Interaction of water with three granular biopesticide formulations. J Agric Food Chem 58(1):1804–1814

Mahmood I, Imadi SR, Shazadi K, Gul A, Hakeem KR (2015) Effects of pesticides on environment. In: Hakeem KR et al (eds) Plant, soil and microbes. Springer, Cham, p 253. https://doi.org/10.1007/978-3-319-27455-3_13

Majewski M, Capel P (1995) Pesticides in the atmosphere: distribution, trends, and governing factors. In: Pesticides in the hydrologic system, vol 1. Ann Arbor Press Inc., Boca Raton, FL, p 118

Maksymiv I (2015) Pesticides: benefits and hazards. J Vasyl Stefanyk Precarpathian Nat Univ 1:70–76. https://doi.org/10.15330/jpnu.2.1.70-76

Martin H (1968) Pesticides manual. British Crop Protection Council, London

Matsumura Y, Yoshikata K, Kunisaki S, Tsuchido T (2003) Mode of bacterial action of silver zeolite and its comparison with that of silver nitrate. Appl Environ Microbiol 69:4278–4281

Mishra S, Keswani C, Singh A, Singh BR, Singh SP, Singh HB (2015) Microbial nanoformulation: exploring potential for coherent nano-farming. In: Gupta VK et al (eds) The handbook of microbial bioresources. CAB International, Wallingford

Mollet H, Grubenmann A (2001) Formulation technology. Wiley-VCH Verlag, Weinheim, Germany, pp 389–397

Muthomi JW, Lengai GMW, Wagacha JM, Narla RD (2017) In vitro activity of plant extracts against some important plant pathogenic fungi of tomato. Aust J Crop Sci 6:683–689. https://doi.org/10.21475/ajcs.17.11.06.p399

Rao GVR, Rupela OP, Rao VR, Reddy YVR (2007) Role of biopesticides in crop protection: present status and future prospects. Indian J Plant Protect 35(1):1–9

SaeediSaravi SS, Shokrzadeh M (2011) Role of pesticides in human life in the modern age: a review. In: Stoytcheva M (ed) Pesticides in the modern world – risks and benefits. InTech

Saxena KN, Pala OKeyo A, Seshu Reddy K, Emolo E, Ngode L (1989) Insect pest management and socio-economic circumstances of small-scale farmers for food crop production in western Kenya: a case study. Int J Trop Insect Sci 10(4):443–462

Seaman D (1990) Trends in the formulation of pesticides: an overview. Pestic Sci 29(4):437. https://doi.org/10.1002/ps.2780290408

Semeniuc CA, Pop CR, Rotar AM (2017) Antibacterial activity and interactions of plant essential oil combinations against gram-positive and gram-negative bacteria. J Food Drug Anal 25:403–408. https://doi.org/10.1016/j.jfda.2016.06.002

Singh HB (2014) Management of plant pathogens with microorganisms. Proc Indian Natl Sci Acad 2:443–454

Sondi I, Salopek-Sondi B (2004) Silver nanoparticles as antimicrobial agent: a case study of E. coli as a model for Gram-negative bacteria. J Colloid Interface Sci 275:177–182

Song CH, Islam RMD, Jeong YT, Lee YS (2012) Isolation and identification of antifungal compounds from bacillus subtilis c9 inhibiting the growth of plant pathogenic fungi. Mycobiology 1:59–66

Straathoff H (1986) Investigations on the phytotoxic relevance of volatilization of herbicides. Mededelingen 51(2A):433–438

Suprapta DN (2012) Potential of microbial antagonists as biocontrol agents against plant fungal pathogens. Int Soc Southeast Asian Agric Sci J 2:1–8

Tadros F (2005) Applied surfactants, principles, and applications. Wiley-VCH Verlag GmbH and Co. KGaA, Weinheim, pp 187–256

US Environmental Protection Agency, Biopesticide Fact Sheets (2008) Biopesticide fact sheets. http://www.epa.gov/opp00001/biopesticides/ingredients/. Accessed 20 Oct 2008

Vernner R, Bauer P (2007) Q-TEO, a formulation concept that overcomes the incompatibility between water and oil. Pfalzenschutz-Nachrichten Bayer 60(1):7–26

Vidyasagar GM, Tabassum N (2013) Antifungal investigations on plant essential oils: a review. Int J Pharm Pharm Sci 2:19–28

WHO/UNEP (1990) Public health impact of pesticides use in agriculture. WHO, Geneva, Switzerland

Woods TS (2003) Pesticide formulations. In: AGR 185 in encyclopedia of agrochemicals. Wiley, New York, pp 1–11

Xu XM, Jeffries P, Pautasso M, Jeger MJ (2011) Combined use of biocontrol agents to manage plant diseases in theory and practice. Phytopathology 101(9):1024–1031

Yadav IC, Devi NL (2017) Pesticides classification and its impact on human and environment. Environ Sci Engg 6: Toxicology

Techno-Economic Assessment of Microbe-Assisted Wastewater Treatment Strategies for Energy and Value-Added Product Recovery

7

Bikash Kumar, Komal Agrawal, Nisha Bhardwaj, Venkatesh Chaturvedi, and Pradeep Verma

Abstract

In the twentieth century, wastewater has emerged as one of the most appalling problems facing mankind. In recent times, numerous steps have been taken to conserve the water bodies, and a variety of wastewater treatment strategies have been developed to treat wastewater in order to make it reusable. The high operational cost associated with these strategies makes the process economically unfeasible. Therefore, looking into the high nutrient content of wastewaters from domestic and industrial establishments, it has been proposed that these treatment plants may be integrated with energy generation (bioenergy) and resource recovery (N, P, K fertilizers and molecular intermediates as value-added products) for making the overall process self-sustainable. Overall, the man-made problem caused due to wastewater can be used as an opportunity for economic benefits through technological advancements. The present chapter evaluates technical and economic aspects of various wastewater treatment strategies with special emphasis on energy and value-added product recovery. It will not only highlight crucial features of each process but also suggest probable areas of improvements keeping in mind the future prospects for establishing self-sustainable wastewater treatment plants.

Keywords

Microbe assisted · Wastewater treatment · Resource recovery · Value-added products · Techno-economic

B. Kumar · K. Agrawal · N. Bhardwaj · P. Verma (✉)
Bioprocess and Bioenergy Laboratory, Department of Microbiology, Central University of Rajasthan, Ajmer, Rajasthan, India
e-mail: pradeepverma@curaj.ac.in; vermaprad@yahoo.com

V. Chaturvedi
SMW College, MG Kashi Vidyapith, Varanasi, Uttar Pradesh, India

P. K. Arora (ed.), *Microbial Technology for the Welfare of Society*, Microorganisms for Sustainability 17,
https://doi.org/10.1007/978-981-13-8844-6_7

7.1 Introduction

Air, water, and soil are some of the most important factors for the survival of life on earth. Water is vital for existence of life and also acts as universal solvent. It is a vital component of all living organisms, and without it, life is impossible. Leonardo da Vinci had rightly described water as "the vehicle of nature" ("vetturale di natura"). The entire human civilization has evolved around water like Nile River was lifeline for Egyptian civilization and Indus River for the Indus Valley civilization (Pradeep and Anshup 2009). It is truly said that accessibility to clean water is a clear sign of wealth, health, serenity, beauty, and originality. Water, which is free of hazardous chemicals and microorganisms, is considered as pure and also necessary for human health (Pradeep and Anshup 2009). However, due to different anthropogenic activities, water is being polluted and unfit for human use, leading to water crisis. Further, climate change causing irregular rainfall has added more woes to this escalating problem. Many countries in Africa and Asia are facing severe water crisis. This water shortage is also accompanied by the depletion of the resources (energy and important chemicals) available to growing human society.

Researchers are working on development of various technologies to utilize the wastewater. The conventional effluent and sewage treatment plants for reclamation of water resources are chemical and energy intensive and also require various post-treatment approaches because of the unwanted by-products formed. These methods require high capital investment, operational and maintenance cost, including larger areas, larger infrastructures, or centralized systems (Capodaglio 2017). In order to tackle these problems, several integrated approaches for simultaneous wastewater treatment and resource recovery had been developed. It is an already established fact that microbes (bacteria, fungi, algae, cyanobacteria) play a crucial role in wastewater treatment or water purification and can be pivotal in resource recovery. The sewage sludge, dairy wastewater, industrial and domestic effluents are some of the sources of wastewater. The huge amount of wastewater is available for application, and there is no lack of raw materials for development of integrated approach. The recovery of different resources from wastewater can minimize the environmental footprint of wastewater treatment (Yan et al. 2018b) and simultaneously result in recovery of resources such as energy, N, P, K fertilizers, different organic compounds, and essential chemicals and nutrients.

The microbe-mediated wastewater treatment can help in generation of electricity and bio-methane; the wastewater can act as substrate for the growth of micro-algae and cyanobacteria for generation of bioethanol and biodiesel. In the last two decades, a concept of microbial nutrient recovery cell (MNRC) was derived which is used by the metallurgy scientist and microbiologist for generation of costly metals and industrially important chemicals, respectively. This book chapter gives an insight into types of wastewater available for resource recovery, different microbe-assisted techniques available, treatment and resource recovery, and cost analysis of the process, thus giving a complete techno-economic perspective of microbe-mediated wastewater treatment strategy for resource recovery. There are different types of wastewater available for microbial assisted treatment for the generation of bioenergy and extraction of value-added products. The wastewater can be broadly classified into three units as described below.

7.1.1 Domestic Wastewater

Rapid economic/industrial development and population growth have led to the increased migration of people to urban areas in search of jobs and financial stability. This has put severe pressure on fulfilling the need of food, water, energy, and other resources. Due to large population, the amount of domestic wastewater released to the environment has greatly increased. Human households generate enormous amounts of wastewater on a daily basis. The reports from several government agencies suggest that an average household generates approximately 300–500 gallons of wastewater daily. The wastewaters are generated from washing, bathing, kitchen, and toilets, that are released to sewage systems. This wastewater is then treated, and then it is either reused or released to rivers, etc. The domestic wastewater consists of mostly kitchen waste, human fecal waste, etc., which are mostly biodegradable (Liu et al. 2018). Along with these biodegradable components are some non-degradable plastic waste which increases pollution by choking the sewer and drainage system.

The established wastewater and sewer treatment plants are exposed to increased pressure, and their running cost is very high. Therefore, a large amount of domestic wastewater is discharged to water bodies such as rivers and lakes. This has led to deterioration of groundwater and surface water quality, which triggers several potential health and environmental hazards to both animals and mankind. Thus, in the last few decades, attempts are made to turn this crisis into opportunity as domestic wastewater contains organic waste (from kitchen and toilet) and essential elements such as nitrogen, phosphorus, potassium, etc. Various research groups are working on different microbial treatment strategies for simultaneously treating this waste, generation of electricity, and recovery of different resources such as volatile fatty acids (Li and Li 2017), nitrogen, phosphorus, and potassium (Shin and Bae 2018) and several other essential organic chemicals. The microbe-based systems are considered self-sufficient for energy requirement with involvement of less amount of chemical for recycling water and resources. To combat this problem, an efficient, economical, and feasible process needs to be developed. A number of techniques have been tested so far by several scientists and stakeholders; however, the best suitable process is yet to be established. The domestic wastewater is basically rich in carbon (C), nitrogen (N), and phosphorus (P) which can be recovered through different recovery mechanisms and techniques involved. After treatment of wastewater, a huge amount of sewage sludge is generated; these are a potential substrate of different organic matter with high nutritional value (proteins, lipids, and carbohydrates) and essential organic chemicals such as PHA for microbes which could be used as a raw material (Balasubramanian and Tyagi 2016; Cole et al. 2016) in different resource recovery techniques.

7.1.2 Industrial Wastewater

Industrialization is considered as one of the major parameters to evaluate the progress of a nation. In the last century, rapid urbanization and industrialization have led to great socio-economic changes in several nations. Till the beginning of the eighteenth century, rate of industrialization was very slow. With the introduction of industrial revolution in the late eighteenth century came the fossil fuel-based engines and an increase in number of chemical and textile industries. These industries use freshwater as one of the important raw materials from washing and cooling (heat absorption), thus leading to the entry of a large number of chemicals leading to water pollution (Han-chang 2002). This wastewater needs to be treated prior to its release to the environment. But the exponential growth of industries is not accompanied by the similar intensity of wastewater treatment setup. This is because the conventional processes are cost intensive and require big infrastructure, and the industrialists in order to make higher profits have neglected to follow the effluent treatment strategies which need to be implemented. Different combination of pollutants is generated by different industries with different chemical and physical properties. On the basis of type of industries, the wastewater is divided into the following categories:

7.1.2.1 Agro-industries

A number of agro-industries, e.g., food processing units, use larger quantity of freshwater, resulting in the huge amount of wastewater after processing of the food items. Some of the important agro-industries are as follows:

(i) *Canning Wastewater*

Canning is used to preserve processed food materials in an airtight sealed jar or can. The canning industries generates huge amount of cane processing water basically rich in phytochemical compounds released during processing of plants parts used as food materials. For example, citrus canning industries involve sequential acidic and alkaline treatment of the citrus membrane, which involves intermediate washing with water that results in huge amount of processing water with very high chemical oxygen demand (COD~10,000 mg/L). It is estimated that for the production of 1 ton of peeled segment used in canning, it will result in generation of 3.6 tons of effluent wastewater with high COD. These effluents contain beneficial phytochemicals such as pectin, flavonoids, and oligosaccharides. These phytochemicals are of great commercial importance such as food, feed, and medicine. Therefore, wastewater treatment strategies may be developed to avoid environmental problems along with the recovery of phytochemicals as valuable organic food compounds (Yan et al. 2018a).

(ii) *Molasses Wastewater*

One of the major by-products of the beet sugar and cane sugar refining industries is molasses wastewater with high chemical oxygen demand (COD) (80,000–130,000 mgL^{-1}), thereby making it one of the most polluted wastewaters released by any food industry (Onodera et al. 2013; Ren et al. 2018). However, molasses wastewater has potential to act as medium for microbial growth as it mainly contains various natural sugars, along with nitrogen, salts, and vitamins added during sugar processing from sugarcane and beetroot (Ren et al. 2018; Avci et al. 2014). The molasses act as an important raw material for production of alcohols, hydrogen, and several amino acids (Ren et al. 2018; Yan et al. 2012; Sirianuntapiboon and Prasertsong 2008).

(iii) *Dairy and Livestock Wastewater*

Dairy waste mainly consists of the cheese whey permeates, and livestock industries generate huge amount of wastewater containing cattle fecal waste (cow, buffalo, swine, etc.). Dairy waste has high BOD and COD, along with presence of antibiotics (i.e., tetracycline, sulfonamides, macrolides, and fluoroquinolones) which are used extensively in the dairy industries, thus making treatment or disposal a major obstruction. Dairy waste also consists of huge amount of casein which can be recovered as a value-added product during treatment. Similarly swine and cow fecal materials are rich in phosphorus and nitrogen; the manure which is a major type of agriculture waste is rich in ammonium and phosphorus (Chandra et al. 2018; Kim et al. 2008). The direct releases of these fecal matters and livestock waste/wastewater have several negative environmental consequences such as pollution of freshwater water bodies; hence, it should be properly treated prior to its discharge. Current treatment approaches for livestock waste/wastewater focus on removal of organics and nutrients via biological processes (Wu et al. 2018; Kim et al. 2008). A popular treatment method, i.e., anaerobic digestion (AD), can effectively reduce organic concentration and recover useful biogas as bioenergy.

(iv) *Brewery Wastewater*

Several grains such as barley, oats, rice, wheat, and millets are used extensively in brewery industries. The freshwater is used in washing and rinsing of these grains, machines and barrels, filters, bottles, etc. The brewery industry wastewater consists of the suspended solids, detergents, and high concentration of COD and BOD due to soluble and insoluble inorganics (Han-chang 2002). The brewery wastewater does not consist of toxic effluents and mostly consists of biodegradable substances; thus, it can be subjected to microbial digestions for effluent treatments (Lu et al. 2019; Han-chang 2002).

7.1.2.2 Paper and Pulp Wastewater

The demand of paper is very high throughout the world, which makes the paper and pulp industry one of the biggest industries. In 2016, as per the record of FAOSTAT, the world production of paper is 410.9 million ton (FAOSTAT). It is a common perspective that the larger the industry, the higher the amount of waste generated that eventually affects the environment. Paper and pulp industry is the largest consumer of freshwater where it uses 5–100 m^3 of water in different steps of pulping for 1 ton of paper produced. The amount of water utilization depends upon the characteristics of substrates, paper type/quality, and extent of water being reused (Doble and Kumar 2005). The overall process generates huge amount of wastewater during different stages of papermaking, which makes paper and pulp industry the third largest generator of wastewater after metal processing and chemical industries (Ashrafi et al. 2015; Savant et al. 2006). The standard industrial technologies and approaches for solutions of problems arising from the industries determine the condition of surrounding environment and the quality of life. Mostly in developing countries, the wastewater generated is more as they are less aware about the water reuse, wastewater treatment plants are poorly regulated, and they lack strict guidelines for water quality measures before its release to environment, whereas in developed countries, they reuse the water more readily along with different technological advancements in treatment process; thus, the amount of wastewater generated has low toxicity (Toczyłowska-Mamińska 2017). However, now the world is getting aware about the toxicity and negative impact of paper and paper mill effluents, and so the governments are tightening the regulations related to wastewater treatment measures. The paper and pulping industry wastewater characteristics at different stages of the paper making depend on the type of process, the type of wood materials, the process technology applied, the internal recirculation of effluent, and the amount of water reused. The paper mill effluents have high COD (1100–2000 mg/L), high total suspended solids (TSS) (300–510 mg), several organic compounds and inorganic compounds such as organic halides (12.5 mg/L), chlorinated compounds (chlorinated hydrocarbons, chlorate, catechols, dioxins, furans, guaiacols, phenols, syringols, and vanillins), volatile organic compounds (VOC), residual lignin, and resin acid which mostly originate from lignins, resins, tannins, and chlorine compounds (Vashi et al. 2018; Farooqi and Basheer 2017; Ashrafi et al. 2015). The paper and pulp wastewater has detrimental impacts on the environment. There are several conventional methods such as physicochemical and biological treatment (aerobic granulation) methods and some hybrid technologies such as MFC-BES and pilot-scale column-type sequencing batch reactor (Farooqi and Basheer 2017). These hybrid methods are based on the concept to develop self-sustainable, energy-efficient systems which make the wastewater reusable by removing pollutants and also help in recovery of energy (electricity) and value-added compounds.

7.1.2.3 Textile/Dyeing Industry Wastewater

Industrial revolution started with mechanization of the textile mills, and till date it contributes in large amount to the wastewater generated and released in the environment. Textile processing and dyeing involve use of several acids, alkalis, bleaching agents (peroxide), starch, surfactant, dyes, and metals (Ozturk and Cinperi 2018). As the process involves several washing and rinsing processes, some of these chemicals are washed away during each step, and thus, textile mill effluents consist of these components in less or higher quantity. Due to the presence of these chemicals, textile wastewater has relatively high toxicity, COD, BOD, intensity of color, and salts (Holkar et al. 2016). Various technologies have been developed in order to treat this wastewater for its reclamation and recovery of various industrially important compounds (Sahinkaya et al. 2018).

7.1.2.4 Tannery Industry Wastewater

Leather tannery industries contribute largely to wastewater generation as the processes involved in the tannery are water intensive. The quality of tannery wastewater depends on the different mechanical and chemical processes involved in the leather processing. The water-intensive process involved in these industries basically includes soaking and washing, liming, plumming, and batting followed by drumming and rinsing. The tannery wastewater has high COD (1500–2500 mg/L); high chloride content (5 g/L); highly alkaline, heavy metals such as chromium; and high quantity of settable substances (10–20 g/L), emulsified fat which causes foaming in tannery wastewater (Han-chang 2002). The tannery wastewater can be efficiently used for recovery of different biodegradable compounds and metals used during leather processing and generation of bioenergy (biogas).

7.1.2.5 Pharmaceutical Industry Wastewater

Several chemical manufacturing units such as pharmaceuticals, organic dyeing materials, glue, adhesives, soaps, synthetic detergents, insecticides, pesticides, and herbicides generate wastes based on the raw materials used and working process. The large chemical industries simultaneously produce several chemicals and pharmaceutical products; thus, wastewater includes extraction from natural and synthetic compounds, specific poisonous substances, nutrients, and several organic compounds (Stadlmair et al. 2018). Therefore, the BOD/COD is lower than 30% as COD is in the range of 5000–15000 mg/L and BOD is relatively low that results in poor biodegradability, varying range of pH, and bad color (Han-chang 2002). These wastewaters require intensive treatment strategies (Stadlmair et al. 2018; Shi et al. 2017) and as they contain a wide range of chemicals so the recovery potential of these wastewaters is really high.

7.1.2.6 Petrochemical Industry Wastewater

Petrochemicals are group of compounds/chemicals derived from petroleum and natural gas. These widespread applications of petrochemicals have led to the contamination of almost every natural resource, say, air, water, and soil. Areas near petroleum refineries have high rate of surface soil and water pollution (Shokrollahzadeh et al. 2008). The animals and plants are adversely affected by products or by-products generated from these refineries. The petrochemical refineries generate huge quantity of the wastewater which has high COD, BOD, oil, grease, metal salts, volatile compounds (Behnami et al. 2018), phenols, and mineral oil (Han-chang 2002). Several wastewater treatment strategies such as activated sludge treatment and membrane bioreactor (MBR) have been developed for reclamation of the depleting water resources.

It is very clear from the above discussion that water is very essential in almost all human activities such as household, industrial, and agricultural. But the clean water resources are depleting, and climate change has added more worries due to uneven rainfall. Therefore, there is urgent necessity to treat wastewater and extract all possible resources based on circular economy concept or best from waste. In the next part, we will discuss different microbe-based technologies developed for the achievement of the above-mentioned objective of clean water and resource recovery.

7.2 Microbe-Assisted Technologies for Wastewater Treatment: Techno-Economic Evaluation

Several new umbrella concepts such as circular bio-economy and NEWEL (Nutrients-Energy-Water-Environment-Land) (Mo and Zhang 2013) are established. Under these umbrella concepts, innovative research measures have resulted in development of broad spectrum of technologies for utilization of wastewater as resource rather than just another waste generated through anthropogenic activity. The development of these technologies has led to different comparative analysis and technical and economic evaluation in order to identify a self-sustainable technique, which is an economically feasible alternative to the physical and chemical processes. Under this heading, we will discuss the different microbe-assisted process developed for wastewater treatment based on the literature survey, the technical steps involved in developed processes, and economic or market feasibility of the technique.

7.2.1 Microbial Fuel Cell (MFC) Technology: Bio-electrochemical System (BES)

The novel approach of microbial fuel cell (bio-electrochemical system) can be exploited for wastewater treatment in order to meet the energy and water crisis. The MFC involves microbial conversion of chemical energy stored in biodegradable organic materials by bio-electrochemical catalytic activity. This electrochemical energy involves transfer of electron between cathode and anode, which leads to generation of electricity (Kumar et al. 2018); similarly reaction between electron and proton results in formation of methane, hydrogen, hydrogen peroxide, and several by-product recoveries such as redox chemicals, heavy metals, and different value-added compounds (Jadhav et al. 2017).

The MFC consists of anode (oxidative) and cathode (reductive) fuel cells to produce energy and other value-added products by integration of electrochemical and biochemical processes. Figure 7.1 represents the schematic design of MFC, where in anodic chamber, electrochemically active microbes catalyze the oxidation of organic electron donors and deliver the electrons to anode, where it is captured as electrical energy. In order to maintain the electro-neutrality, the catalytic conversion also results in generation of protons (H^+) in anodic chamber, and in order to maintain the electro-neutrality, protons travel through semi-permeable cation exchange membrane to cathodic chamber (Kumar et al. 2018). In bio-electrochemical system under the influence of external potential, protons transferred to cathodic chambers are utilized for generation of value-added chemicals/compounds (Jadhav et al. 2017). Bio-electrochemical system offers a flexible platform for oxidation of pollutants for energy generation and simultaneous reduction-oriented methods for product recovery. Thus, it has provided an integrated solution for wastewater treatment and resource recovery in the form of clean water, bioenergy, and chemicals. The

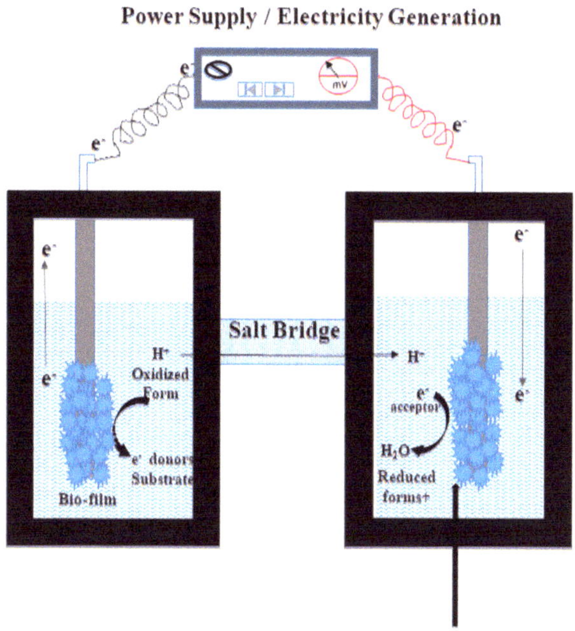

Fig. 7.1 Schematic diagram of MFC/bio-electrochemical system

recovery of different resources reduces the additional energy required in individual synthesis process of each resource. In the technological advancement of MFC, where an external voltage is applied in the MEC, this external voltage acts as the driving force in accelerating the microbial electro-catalysis for production of high-value chemicals such as methane, hydrogen gas, hydrogen peroxide, and caustic soda at the cathode at very low energy cost (Sharma et al. 2014; Foley et al. 2010; Rabaey and Rozendal 2010; Rozendal et al. 2008a). This electricity-driven method is now also applied in the area of bioremediation and exploited for inorganic resource recovery as well. These chemicals have higher market value as compared to the external electricity applied, and several studies associated with life cycle assessment of BES suggested that BES is not only economical but also can help in providing several significant environmental benefits (Foley et al. 2010). Table 7.1 contains details of different value-added products extracted during different wastewater treatments along with the microbes involved in the process.

7.2.2 Economics of the BES system

The techno-economic feasibility of the BES system at larger scale beyond laboratory scale (few milliliters to liters) and pilot scale (30–50 l capacity) needs to be assessed. In the early twentieth century, most of the feasibility studies demonstrated

Table 7.1 Recovery of value-added products from various wastes using bio-electrochemical cells and involved microbes

Resource		Techniques involved and recovery potential	Microbes used	Reference
Inorganic	*Heavy metals*	Different heavy metal ions such as Au^{3+} ($E^{0'}$=1.001 V), Ag^{+}($E^{0'}$=0.799 V), Cr^{4+}($E^{0'}$=1.330 V), Cu^{2+} ($E^{0'}$=0.337 V), Fe^{3+}($E^{0'}$=0.770 V), Hg^{2+}($E^{0'}$=0.911 V), and V^{5+}($E^{0'}$=0.991 V) were recovered from industrial effluents in the absence of any external voltage. BES also helped in recovering metals having lower redox potential such as Cd^{2+}($E^{0'}$=0.400 V), Ni^{2+}($E^{0'}$=0.250 V), Pb^{2+}($E^{0'}$=0.130 V), and Zn^{2+}($E^{0'}$=0.762 V) in the presence of external power with BES having biotic anode and abiotic cathode	*Trichococcus pasteurii*, *Pseudomonas aeruginosa*, *β-proteobacteria*, *Actinobacteria*, *Acinetobacter* sp., *Shewanella oneidensis* (Cr), *Rhodoferax ferrireducens* (V), *Bacillus selenitireducens* (Ar), *Shewanella putrefaciens* (AU)	Mathuriya and Yakhmi (2014)
		Cr^{4+}, Cu^{2+}, and Cd^{2+} were recovered completely in a self-driven and sustainable MFC-MEC with initial feed concentration along with simultaneous generation of bioelectricity	Electroplating industry wastewater was used as inoculums	Li et al. (2017)
		Silver recovery from wastewater treatment from silver(I) diammine complex N.A.D.	Anaerobic sludge from the digester of a brewery wastewater treatment plant as inoculum	Ho et al. (2018)
		Designed a hybrid, BES, and electrolysis reactor to recover heavy metals from fly ash leachate, where Cu (II) (98.5%) from Cu_2O was recovered in hybrid system during cathodic reduction; lead (98.1%) and zinc (95.4%) were recovered during electrolysis	Mixed culture	Tao et al. (2014)
		A dual chamber MFC-MEC was used in treatment of wastewater from acid mine drainage (with single or mixed metal) for recovery of H_2 and heavy metals such as Cu^{2+}, Fe^{2+}, and Ni^{2+} along with concurrent electricity generation	Mixed sludge	Jadhav et al. (2017)

	Selenium	Industrial units such as glass manufacturing and electronic industries have high concentration of dissolved selenium in selenate (VI) and selenite (IV) form. The dissolved selenate of selenite in wastewater is converted to elemental Se (0) and deposited as nano-sized bright red spherical particle over anode	*Shewanella* sp. *Shewanella oneidensis*	Jadhav et al. (2017);Lee et al. (2007)
	Uranium	The radioactive wastewater containing soluble uranium can be used in bio-electrochemical system, resulting in uranium recovery (87%) in a biotic cathodic chamber in the presence of *Geobacter sulfurreducens* by reduction soluble U (VI) to relatively insoluble U (IV) with organic components serving as electron donor	*Geobacter sulfurreducens*	Gregory and Lovley (2005)
Nutrient recover	*C, N, P, K*	The urine (rich in C, N, P, K) is an ideal source of ammonia and electricity. Ammonia recovery in MFC was performed by treating the urine in the anodic chamber, which results in diffusion of ammonium ions toward cathodic chamber where it is absorbed on electrode. This process resulted in high current density (as high as 3.6 A/m^2) and 61% or more NH^{4+} ion recovery	Microflora obtained from activated sludge and effluent of other MFC was used as inoculum	Ieropoulos et al. (2012); Kuntke et al. (2012)
		BES has been used effectively to recover phosphate from swine wastewater rich in phosphate. The precipitation of struvite ($NH_4MgPO_4 \cdot 6H_2O$) on cathode surface takes place and resulted in phosphate recovery up to 27%. Similarly, recovery of orthophosphate (600 mg/L) from digested sewage sludge along with concurrent electricity generation in the presence of *E. coli* was successfully performed	*E. coli*	Ichihashi and Hirooka (2012); Xiao et al. (2012); Fischer et al. (2011)
	Sulfur	Recovery of sulfur in elemental form is recovered by bio-electrochemical processes; in this developed process, cathode was enriched with autotrophic sulfate-reducing bacteria (SRB) and sulfur-oxidizing bacteria (SOB), and the bacterial metabolic process resulted in recovery of sulfur	Sulfur-reducing bacteria and sulfur-oxidizing bacteria	Blázquez et al. (2016)

(continued)

Table 7.1 (continued)

Resource		Techniques involved and recovery potential	Microbes used	Reference
Algal biomass	*Electricity, algal oil, and carbon sequestration*	In a phototrophic bioreactors/MFC, the algal biomass can be generated under the presence of photosynthetic light. The algae use light and CO_2 and produce several organic compounds and help in carbon sequestration. These phototrophic algae can also be used as a medium for electricity generation and biofuel production. In this process, the photosynthetic process takes place in cathodic chamber which results in growth of the algal biomass. The photosynthetic process generates oxygen that can be used for cathodic reduction and helps in generation of electricity and clean water. The recovered biomass can be used as substrate for photo-bio-electrochemical cells and source of algal bio-oils	*Chlorella vulgaris* *Chaetoceros* sp.	Jadhav et al. (2017); Rajesh et al. (2015); Wang et al. (2010)
Intermittent bio-chemical compounds (industrial chemicals/ gases)	*Methane*	MFC is used for electricity generation using different wastewater, i.e., sewage and industrial effluents. These effluents are rich in methanogens; thus, sometimes electrical energy generation is limited by the generation of methane and H_2 gas. High recovery of methane (up to 0.55 mol/gVSS day) by *Methanobacterium palustre* and electron capture efficiency (over 80%) by hydrogenophilic methanogenic culture were observed in the cathodic chamber of MEC	*Methanobacterium palustre*	Villano et al. (2010)
		Methane recovery at a rate of 0.006 m3/(m3 day) was observed using acetate-based synthetic wastewater	*Methanobacterium palustre*	Cheng et al. (2009)
	H_2, CO_2, CO gas	Several other gases such as H_2, CO_2, and CO were also recovered from wastewater treatment by MFC-MEC system. These gases have widespread application in different industrial process. A novel biological H_2 generation of 0.02 m^3/day in cathodic chamber fed with acetate and external applied voltage of 0.5 V	Mixed culture of bio-electrochemically active microbes containing inoculums obtained from the sulfate-rich paper mill waste	Jadhav et al. (2017); Lu and Ren (2016); Rozendal et al. (2006, 2008b)
	Ethanol recovery	Apart from the general lignocelluloses to ethanol generation, that bio-electrochemical conversion waste biomass to ethanol has been developed and demonstrated by many researchers	Mixed microbial consortia	Rosenbaum et al. (2011); Thygesen et al. (2010); Steinbusch et al. (2008)

	Hydrogen peroxide	Hydrogen peroxide is an essential industrial chemical which can be generated in bio-electrochemical system during treatment of gray and black water. The H_2O_2 recovery rate was 1.9 ± 0.2 kg H_2O_2/m^3 day and as accompanied production of H_2O_2 along with electricity generation. However, the power yield is low	Microbial consortium collected from MFC used for carbon and nitrogen removal	Mathuriya and Yakhmi (2014); Harnisch and Schröder (2010); Rozendal et al. (2009)
	Urea	The urea can be recovered from the human urine waste in a granulated activated carbon urease bioreactor where urea in the presence of the immobilized urease enzyme is converted to carbolic acid and ammonia. The ammonia is used in bio-electrochemical cells, where it gets oxidized to produce nitrogen and water and generates electricity	Urease enzyme from *Canavalia ensiformis* (jack bean) was immobilized on granular activated carbon	Nicolau et al. (2014)
	Medium chain triglycerides – the biological precursors of renewable fuels	Food materials are rich in medium chain fatty acids which can be recovered from the food wastewater using BES. The cathodic microorganism mediated the electron supply that helps in the production of ethanol using hydrogen. Similarly, acetate reduction in the presence of H_2 as electron donor helps in the production of medium chain fatty acids such as caprylate (C8) and caproate (C6). These fatty acids act as precursors for fuels and chemicals. Bio-electrochemical generation of caprylate (36 mg/L), caproate (739 mg/L), butyrate (263 mg/L), and ethanol (27 mg/L) was reported by Eerten-Jansen et al. (2013), where ethanol production was mediated by mixed anaerobic bacterium culture	Mixed culture *Clostridium kluyveri*, *Clostridium tyrobutyricum*, and *Eubacterium pyruvativorans*	Van Eerten-Jansen et al. (2013); Aoyama et al. (2007)

(continued)

Table 7.1 (continued)

Resource		Techniques involved and recovery potential	Microbes used	Reference
	Biopolymer synthesis	Polyhydroxyalkanoates (PHAs), biodegradable plastics from bacterial cell, are a suitable alternative to polypropylene, but their cost of production is very high as compared to low-cost petrochemical derived plastics. Thus, the application is limited. However, in recent times, some scientists have demonstrated the synthesis of these PHAs in cathodic chamber of a BES using the wastewater rich in carbon and nutrient sources. This PHA biosynthesis involves in BES biofilm on cathode consisting of micro-aerophilic microbes utilize CO_2 and generate organic compounds such as alcohols, diols, carboxylic acid, and polymer (poly-β-hydroxybutyrate) in the presence of external cathodic potential	Micro-aerophilic microbes' aerobic microbial consortia *Alcaligenes eutrophus* ATCC 17697T	Mohan et al. (2014); Chen et al. (2013b); Braunegg et al. (2004); Lee (1996); Ishizaki and Tanaka (1991)
	Bio-refinery intermediates	Agriculture-based industries and lignocellulosic bio-refineries generate wastes which can be potentially used as substrates for the BES. The substrates are rich in polyphenolic lignin, hemicelluloses, and celluloses that get reduced by hydrogen produced in the cathodic chamber yielding the high value polyphenolic group of metabolites such as equol and resveratrol which are widely used as antioxidants and in bioplastic synthesis	Mixed microbial culture	Jadhav et al. (2017); Chen et al. (2013a);Thygesen et al. (2010)

in the literature are based on the pilot scales with simpler pollutants such as synthetic wastewater which definitely does not consider the solid waste matters and environmental effect on the pollutants before its entry to the BES system. These studies are limited to a smaller span of operating time. As the BES technology was evolving in past 10–15 years, several attempts have been made to study the technical and economic feasibility in terms of key performance parameters such as electrode material, electrode connections, flow modes, and different range of actual wastewater with different pollutant removal (Hiegemann et al. 2016; Liang et al. 2013). The scale-up studies by Feng et al. (2014) and Cotterill et al. (2017) demonstrated the scale-up with 1000-L stackable horizontal MFC and 175-L microbial electrochemical cell (MEC), respectively. It is evident from the observation made during these studies that the laboratory scale reactor design may not work at large scale as evidenced by the lower pollutant removal and hydrogen yield as compared to laboratory scale. It is a well-established fact that development of any technology on larger scale is driven by capital involvement in operation of the system such as land required for the setup, the cost of electrodes and reactors, etc. However, as per studies available in the literature, BES technology is more studied technology for simultaneous waste treatment and resource recovery. The amount of research going around on the BES technology suggests that BES will have lower operating cost than the traditional aerobic treatments as it involves anaerobic process in the anode section. But the cost of land required, electrode separator, membrane, and reactor material and its construction may lead to enhancement in the capital requirement; however, the concept of resource recovery during the process may balance the enhanced cost and help in development of a sustainable system. As demonstrated by Ge and He (2016), they suggested that 200-L MFC could be cost competitive to 10000 gpd traditional wastewater treatment system in a decentralized system. The mass transfer balance in BES system is also very important. The mass movement must be limited to its own specific environment; any deviation from that may adversely affect the BES performance. Therefore, the reactors must be designed very carefully to prevent any leakage/overflow in substrate or liquid; the distance between the electrodes and separators must be precisely determined.

7.2.3 Microbial Enzyme-Based Wastewater Treatment

Microorganisms which are capable of generating different catalytic enzymes are used in bioremediation of the wastewater containments. However, direct application of microorganism is a slow process and an energy-intensive process as it requires ambient environment for the growth of microbes (Sharma et al. 2018; Ghosh et al. 2017). Thus, in the last few decades, focus has been shifted to the microbial enzymes separated from microbes (Thatoi et al. 2014). Enzymes are biologically derived macromolecules which act as catalyst for biochemical degradation of different pollutants (Kalogerakis et al. 2017). The major advantages of application of enzymes are as follows: (a) high selectivity to specific substrate, (b) nontoxic by-product formation by enzymatic biotransformation, (c) high mobility due to small size, and

(d) limits the application of high energy and harsh chemical employed in the physicochemical processes (Sharma et al. 2018; Maloney et al. 2015; Gianfreda and Bollag 2002).

Water Research Commission Project report (No:1170/1/04) by Rhodes University BioSURE Process® identified the active role of group of hydrolase enzymes such as endoglucanases, glucosidases, lipases, phosphatases, proteases, and sulfatases (Watson et al. 2004; Whiteley et al. 2002a, b, 2003). They demonstrated the presence of these hydrolase enzymes in biosulfidogenic reactors used for the industrial wastewater treatment. They also demonstrated that these enzymes could be used for treatment of the wastewater from acid mine drainage, abattoirs, textile dyeing, and tanneries (Agrawal et al. 2018; Mutambanengwe and Oyekola 2008). As per literature survey, it is much evident that very less work has been carried out on direct involvement of enzyme-mediated resource recovery. But the enzymes are highly selective and result in breakdown of several toxic organic and inorganic compounds into different nontoxic residues. These residues can be separated before their release from the treatment plants. The enzyme-mediated microbial fuel cell/biological fuel cell (Kumar et al. 2018) is one such another approach for the involvement of enzymes in electricity generation. Therefore, enzyme is not directly involved in resource recovery but can be used as a biocatalyst of electricity generation and resource recovery.

7.3 Methodology Used for the Enhancement of Enzyme Performance in Wastewater Treatment

7.3.1 Immobilization

The application of enzyme-mediated transformation and valorization of raw materials dates back 50 years. The enzymes were used in different batch reactors, which has major drawbacks such as high operating costs (enzyme production cost is more), loss of catalytic activity due to inactivation, etc. During the early twentieth century, the idea of enzyme mobilization has evolved as an interesting alternative to overcome above-mentioned limitations. During the immobilization, biocatalysts are ensured in a localized space which helps in the prevention of loss of the enzyme, enhances the shelf life of the enzyme, and above all increases reusability of the enzyme. Different immobilization techniques have been proposed such as adsorption or covalent binding of enzyme on solid support like nanoparticles, inclusion in a capsule or magnetic beads, embedding in matrix, fiber, or resins, etc. Bayramouglu and Arica (2008) demonstrated that covalently immobilized horseradish peroxidase on magnetic beads showed higher phenol conversion, high activity, and stability as compared to its free enzyme. Thus, it can be suggested that HRP can be successfully used in a large-scale continuous enzymatic degradation of phenolic pollutants. HRP has wide substrate specificity from an azo dye Remazol Blue (Bhunia et al. 2001) to chlorinated dibenzodioxins and dibenzofurans (Köller et al. 2000). Peroxidase-based bioreactors can be designed to treat wastewater by immobilization into such medium to allow efficient interactions with substrate/pollutants for its degradation. Several nano-based imbedding materials suggested are carbon nanotubes (Campbell et al. 2013, 2014b; Dinu et al. 2010), graphene oxide (Zhang et al. 2010), graphene oxide sheets (Campbell et al. 2014b), metal-oxide particles (Campbell et al. 2014a), nanotubular aluminosilicate (Zhai et al. 2013), and nanodiamonds (Krueger and Lang 2012) on which the peroxidases are embedded through encapsulation into a gel (De Lathouder et al. 2008) or membrane (Zhang et al. 2012a; Shen et al. 2011). The material should be chosen such that it has minimum interaction with nanosupport or encapsulator in order to preserve the specificity and catalytic behavior of enzyme. However, improvement in stability, selectivity, and efficiency of nano-based immobilized enzyme has provided ample opportunity for its application in wastewater treatment (Zhang et al. 2012b). Similarly, Jamie et al. (2016) demonstrated covalent immobilization of lipase enzyme from *Candida rugosa* that was embedded in modified multiwalled carbon nanotubes (MWCNTs). These lipase-based MWCNTs were used in oily wastewater treatment which showed enhanced resistance of enzyme to severe conditions under industrial applications. The CNT-immobilized enzyme showed 93 times higher catalytic activity as compared to those immobilized on other support material. It also helped in retention of about 98% of biological activity. Laccase from *Phoma* sp. UHH 5-1-03 was cross-linked by electron beam irradiation to polyvinylidene fluoride membrane. The immobilized laccase addressed high removal efficiency of >85% for the acetaminophen and mefenamic acid from a municipal wastewater containing pharmaceutically active compounds (PhACs; applied as a mixture of acetaminophen, bezafibrate, indometacin, ketoprofen, mefenamic acid, and naproxen). The immobilized laccase also displayed higher wastewater stability as compared to non-immobilized laccase (Jahangiri et al. 2018).

7.3.2 Biocatalytic Membrane Reactor/Enzymatic Membrane Reactor (BMR/EMR)

Enzyme-based membrane systems were introduced in order to overcome the limitations of enzyme immobilization technique, i.e., decrease in enzyme activity due to steric hindrance effects that arise due to distortion of enzyme structure during immobilization process and the interfacial limitations. The enzymatic/biocatalytic membrane reactor (EMR/BMR) is a bioreactor in which a biochemical transformation takes place in the presence of enzyme and a selective membrane is used to separate the enzymes and end product generated (Vladisavljević 2015; Rios et al. 2004). Different types of configurations of membrane catalytic bioreactors are based on relative positions of the catalyst such as the following: (a) the enzyme is separated with the help of a membrane; (b) the enzyme is incorporated within the membrane wall as a filter; (c) the enzyme is encapsulated in the core-shell microcapsule; and (d) a matrix is prepared where the enzyme is encapsulated (Vladisavljević 2015) (Fig. 7.2).

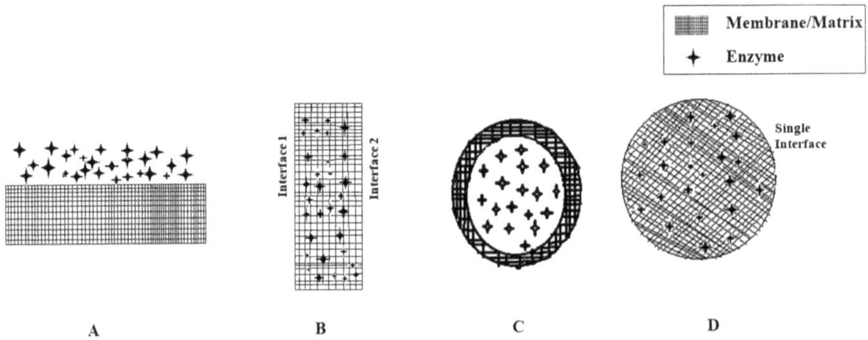

Fig. 7.2 Types of membranes used in EMR. (**A**). Enzyme not embedded in membrane. (**B**) Enzyme embedded in the membrane. (**C**) Enzyme is encapsulated within a core-shell capsule. (**D**) Enzyme is encapsulated in a matrix-type capsule (Adapted from: Vladisavljević 2015)

7.3.2.1 Types of Enzyme-Based Membrane Bioreactors and Their Application in Wastewater Treatment

(i) *Immobilized Enzyme Membrane Reactor:*

In this approach, enzymes are immobilized onto a membrane by covalent attachment, electrostatic deposition, gel formation, and physical and chemical mediated adsorption. Wastewater is transferred through the membrane where it interacts with the enzyme, and the products/treated water diffuses from the reaction side to the other side of the membrane, and there they are recovered as a permeate. A batch ultrafiltration cell-based bioreactor having flat polyacrylonitrile membrane with crude enzyme from *Pseudomonas* sp. having catechol 2,3-dioxygenase activity has been used in coke wastewater having phenolic effluent. The phenolic degradation of 40–80% was achieved with this system (Vladisavljević 2015; Bohdziewicz 1997; Bodzek et al. 1994). A capillary hollow fiber membrane bioreactor having polyphenol oxidase (EC 1.14.18.1) obtained from *Agaricus bisporus* immobilized in a polysulfone membrane was capable of removing up to 25% of phenolics in 8 h from coal-gas conversion plant effluents (Edwards et al. 1999). Lante et al. (2000) demonstrated application of laccase from *Pyricularia oryzae* imbedded in a polyethersulfone membrane of a SPIRA-CEL spiral wound module for treatment of synthetic wastewater with eighteen different phenolic substrates. The laccase immobilization resulted in good operational stability and shows potential physico-chemical properties for decreasing phenol substance concentration in a synthetic phenolic wastewater. Jolivalt et al. (2000) demonstrated immobilization of laccase obtained from a white rot fungus *Trametes versicolor* on a modified polyvinylidene difluoride (PVDF) microfiltration membrane of a frame plate reactor. This system resulted in efficient removal of phenylurea herbicide (N',N'-(dimethyl)-N-(2-hydroxyphenyl)urea) from wastewater.

(ii) *Extractive Membrane Bioreactor*

The extractive *membrane bioreactor* consists of a separate membrane system and a biological component (enzyme/microbes). The wastewater first enters into the

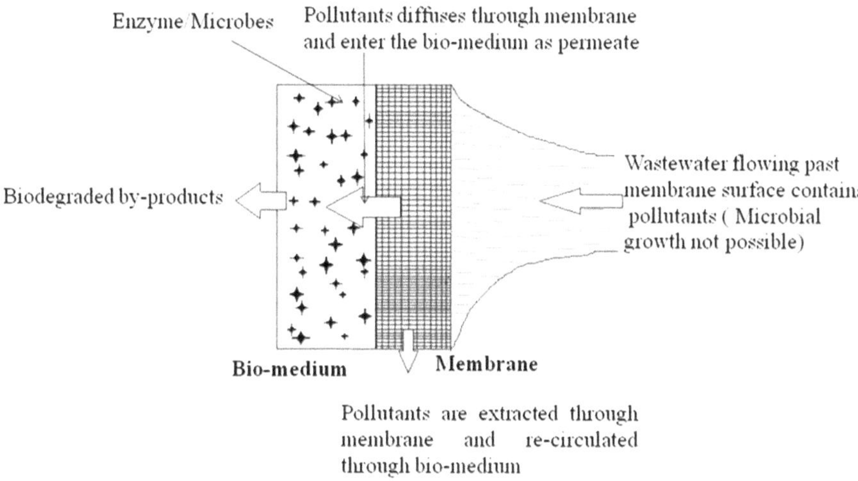

Fig. 7.3 Sketch of an extractive membrane bioreactor (Based on Livingston 1994; Vladisavljević 2015)

membrane where the pollutants which are hostile to the enzyme or microbes are separated out. The pollutants separated through the membrane are recycled into the bio-medium where the biodegradation occurs. In the recycling unit, the nutrient, pH, and temperature are regulated for proper activity of the biological system (Livingston 1994). If enzyme-producing microbes are directly used, then nutrients, oxygen, and pH are regulated. Livingston (1994) has also given an insight into the technology where he explained about an integrated system in which a series of membranes may be used for different pollutants and the biological priority pollutants can be subjected to biodegradation. Kojima et al. (1995) demonstrated an application of hollow fiber bioreactor fitted with polyethersulfone membrane; the bio-medium is supplied with glucose oxidase enzyme obtained from the *Aspergillus niger* that helped in the removal of organic components (glucose) from the synthetic wastewater (Fig. 7.3).

(iii) *Membrane Separation Reactor (MSR)*

It is a stirred tank reactor fitted with membrane module. The membrane is used to bound a dissolved or dispersed catalyst (may be enzyme molecules or cells) in the batch reactor vessel. The untreated permeate leaves the reactor through the permeable membrane, whereas the excess solvent and product are withdrawn regularly in feed and bleed fashion. MSR are used by Gallifuoco et al. (2001) for de-polymerization of polygalacturonic acid using pectolyase obtained from *Aspergillus japonicus*; the membrane material used was polyethersulfone. Similar reactor was used by Lopez et al. (2002) in dye decolorization by using the manganese peroxidase enzyme obtained from *Bjerkandera* sp. Soybean peroxidase obtained from ground soya bean hulls was used in reactors for degradation of phenols in an MSR; however, better results were obtained when soybean hulls were used directly that contributed to degradation of the phenols as well as adsorption of excess dye (Flock et al. 1999).

(iv) *Anoxic-Oxic Membrane Bioreactor (A/O MBR)*

An anoxic-oxic membrane bioreactor is designed which consists of the modified activated sludge process, anoxic system coupled with contact stabilization, and membrane system (Komala et al. 2011). This system helps in overcoming the fouling of membrane system. Fouling is loss of membrane performance which occurs due to deposition of the suspended particle on its pore (Wang et al. 2017). Fouling causes increase in pressure and decrease in the flux (Komala et al. 2011). Different methods such as aeration and backwashing with water or permeate pump are applied to reduce fouling. Among them, aeration is one of the best suited methods as it increases the flow circulation by inducing shear stress on the membrane. Aeration also helps in providing oxygen to the biomass, maintaining the activated sludge suspension apart from reducing fouling by scouring the membrane (Le-Clech et al. 2006). This AO-MBR system was employed for the dye biodegradation of Remazol Black V (azo dye) with a co-substrate tempe industrial wastewater as carbon source (Komala et al. 2011). A long-term performance of the system was tested for dye degradation. The effect of aeration on fouling of membrane was also evaluated, and it was reported that in long-term operation, for stable flux and membrane recovery, a filtration-backwash time of 61 minutes with aeration intensity of 0.7–1 bar was optimum. Xu et al. (2014) demonstrated that Fenton-Anoxic-Oxic/MBR system can be successfully employed for reclamation of water from the pharmaceutical waste, i.e., avermectin fermentation wastewater. The toxicity of the effluents was reduced significantly along with pollutant removal and improved biodegradability. They also demonstrated that HCl + NaClO system with aeration is used for removing fouling, increasing the flux of MBR and acceptable trans-membrane pressure (TMP). Xiang et al. (2003) demonstrated a pilot-scale ($10m^3/d$) plant of AP-MBR for treating the dyeing wastewater of woolen mill (without wasting sludge) for 125 days operation. The water obtained from the treatment plant is reusable in plant.

7.3.2.2 Economics of the Enzyme-Based Technologies

The enzyme-based reactors are dependent on biological catalyst which has certain advantages as well as disadvantages. The advantages are as follows: the processes used in the treatment of many pollutants are very specific in nature, so selective degradation can be done. However, cost of production, stability, and activity at specific temperature and pH require infrastructure or sophisticated setup and selection of the special membrane material so that stability is not hampered much by steric hindrance. Fouling is one of the major problems in the MBR, so A/O MBR is introduced. Several other concerns limiting the application of the MBR on large scale are membrane flux, membrane life (adversely affected by fouling), and the high price of membrane price. Liu et al. (2010) had performed a case study of an A/O MBR-based sewage treatment in Qingdao Liuting International Airport on the techno-economic evaluation of the operation and maintenance. He suggested that the relationship between the costs of MBR system is negative logarithm of membrane flux and membrane life while the relationship between membrane price and costs of running MBR system is in linear relationship. However, this is a highly efficient technique for the wastewater treatment, and resource can be recovered as byproducts of degraded organic pollutants. Different materials scientist and stakeholders are interested in developing this technology on larger scale and wider scale of pollutants. With the improvement of membrane technology, reduction of the membrane prices and overall running cost may be decreased.

7.4 Direct Application of Microbes

7.4.1 High Rate Algal Pond (HRAP) Systems

The introduction of high rate algal ponds (HRAPs) for treatment of wastewater was discovered around 50 years ago; they are widely used for growing microalgae using wastewater as substrate for their growth and in turn treat the wastewater, and the algal biomass generated is used for recovery of value-added products and bioenergy (Craggs et al. 2014; Oswald and Golueke 1960) (Fig. 7.4).

HRAP consists of a shallow open pond (depth of 0.2-1.0 m) divided into a channel by central wall fitted with a paddle wheel for proper circulation of the water throughout the channel, where average water velocity ranges from 0.15 to 0.30 m/s (Sutherland et al. 2015; Craggs et al. 2014). This shallow depth attributed with high nitrogen (N) and phosphorus (P) content of wastewater along with the turbulent eddies resulting in vertical mixing by paddle wheel helps in algal growth by enhancing exposure of cells to sunlight, preventing sedimentation, and enhancing the diffusion of nutrient across the cell boundary layer (Sutherland et al. 2015; Hadiyanto et al. 2013; Park et al. 2011). The appearance of dead zones is common in long channel of large-scale operations (Grobbelaar 2012). Carbon dioxide is added in the HRAP under carbon-limited situation to enhance algal growth (Craggs et al. 2012). The anaerobic ponds or gravity settlers help in removing and digesting the

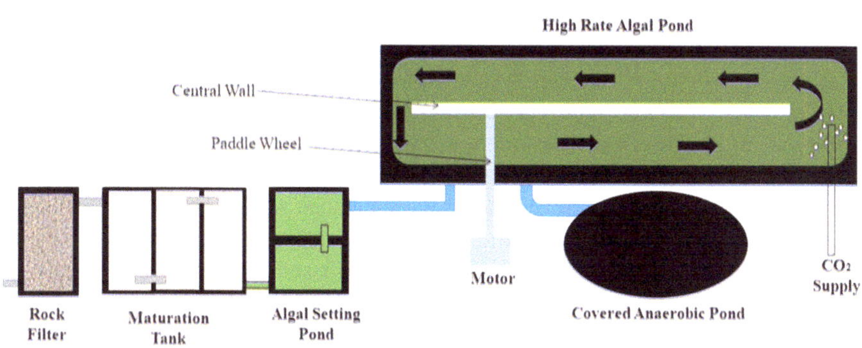

Fig. 7.4 Advanced pond system comprising of covered anaerobic pond, HRAP with CO_2 supply, algal settling pond, maturation tank, and rock filter (Adapted from Craggs et al. 2014)

wastewater solids (Craggs et al. 2013). HRAP systems require low energy (~1 kWh m^3 of water treated) (Arashiro et al. 2018; Garfí et al. 2017; Passos et al. 2017), are less expensive, and require less maintenance as compared to conventional techniques such as activated sludge system (Arashiro et al. 2018; Garfí et al. 2017; Molinos-Senante et al. 2014; Craggs et al. 2014;). HRAP is used for simultaneous wastewater treatment and the resource recovery; this process helps in generation of economically viable feedstock which can be subjected to different treatment strategies for generation of biofuels such as biodiesel by trans-etherification of lipid fraction (Rodolfi et al. 2009), bio-methane by anaerobic digestion of the whole biomass (Yen and Brune 2007), bio-oil by pyrolysis of dry biomass (Miao and Wu 2004) or hydrothermal liquefaction of wet biomass (Biller and Ross 2011), and bio-ethanol by fermentation of the polysaccharide/carbohydrate part (Harun et al. 2010). The HRAP also helps in recovery of NPK fertilizers; the protein extracted from the algal biomass is rich in different essential amino acids and phyto-hormones (Coppens et al. 2016; Garcia-gonzalez and Sommerfeld 2016; Uysal et al. 2015; Jäger et al. 2010) which can be extracted as value-added product and help in making the entire process economically sustainable.

7.4.2 Anammox Process

Autotrophic anaerobic ammonium oxidation (*anammox*) bacteria have unique capability to metabolize ammonium and nitrate or nitrite to produce nitrogen gas. Anammox process was first observed by Mulder et al. (1995) in denitrifying fluidized bed reactor designed to treat effluent of a methanogenic reactor, and they patented the process. Most of the genera of anammox bacteria have been discovered in wastewater treatment plants, and some are also identified from laboratory-scale bioreactors. The six different genera of the anammox bacteria are *Anammoxoglobus*, *Anammoximicrobium*, *Brocadia*, *Jettenia*, *Kuenenia*, and *Scalindua*, and they all belong to order *Planctomycetales* (Li et al. 2018). The discovery of anammox process has led to understanding that half of the total nitrogen turnover in marine environment was mainly mediated by these bacteria (Li et al. 2018; Kuenen 2008). Thus, it was believed that anammox process has great potential in removal of ammonium from waste gas or wastewater (Kuenen 2008).

Different type of wastewater systems utilizes anammox process for wastewater treatment and resource recovery. Since the discovery of anammox system was observed in the wastewater treatment system, it has been proven that wastewater with highly contaminated nitrogen and low organic content can be treated with the help of anammox bacteria. Several already established wastewater techniques have employed the application of anammox for recovery of nitrogen. Table 7.2 describes different treatment strategies involving anammox-based treatment and its applications in nutrient recovery.

Table 7.2 Different wastewater treatment strategies using anammox bacterial system

Treatment strategies	Wastewater types	Applications	References
Sequencing batch reactor (SBR)	Ammonium-rich wastewater treatment	Nitrogen removal	Yang et al. (2016)
Fluidized bed reactor	Effluent from methanogenic reactors	Denitrification of effluents from methanogenic reactor	Mulder et al. (1995)
Fixed-bed reactor	Concentrated wastewater streams produced in food and agro-industry/synthetic wastewater	Ammonium removal from sludge digestion effluent and nitrogen recovery	Strous et al. (1997)
Upflow anaerobic sludge bed (UASB)	Pulp and paper wastewater	Enhancement in nitrogen removal efficiency	Tang et al. (2009)
	Colistin sulfate and kitasamycin pharmaceutical manufacturing wastewater	High value for nitrogen removal from pharmaceutical wastewater and met the discharge standard of water	Tang et al. (2011a)
	Synthetic wastewater rich in $(NH_4)_2SO_4$ and $NaNO_2$	High rate of nitrogen removal	Tang et al. (2011b)
	Domestic wastewater, landfill leachate	High rate of nitrogen removal by using granules in up flow reactors	Tang et al. (2017)
Membrane bioreactor (MBR)	Synthetic wastewater with glucose substrate	High anammox activity was observed on MBR as compared to SBR	Tao et al. (2012)
Rotating biological contactor (RBC)	High-salinity synthetic wastewater	Nitrogen removal	Windey et al. (2005)
Moving bed biofilm reactor (MBBR)	Municipal wastewater	Nitrogen removal and analysis of effect of seasonal temperature variation on the anammox bacteria	Gilbert et al. (2014)
Integrated fixed-biofilm activated sludge (IFAS) reactor	Synthetic wastewater and sludge dewatering liquors	High nitrogen removal efficiency and characterization of microbial community in the biofilm	Zhang et al. (2015b)
	High ammonium wastewater	Nitrogen removal, characterization of the role of granular sludge in anammox-based techniques	Zhang et al. (2015a)
Anammox bio-cathode in MFC	Autotrophic anammox bacteria, as a sustainable biocatalyst/bio-cathode in microbial desalination cells (MDCs) for energy-positive wastewater treatment	Increased coulombic efficiency in MDCs and nitrite and ammonium removal (90%) from wastewater. Increased carbon and nitrogen removal from anode and cathode chambers	Kokabian et al. (2018)

7.4.3 Photosynthetic Bacteria

The photosynthetic bacterium uses light as source of energy and different organic materials as carbon substrate and proton donor in autotrophic and heterotrophic growth. The different photosynthetic bacteria are grouped under different family like groups of microbes such as *Chromatiaceae*, *Chlorobiaceae*, *Ectothiorhodospiraceae*, *Heliobacteriaceae*, and *Rhodospirillaceae* (Li et al. 2011; Dong and Cai 2001). These photosynthetic bacteria play a vital role in nutrient cycle and different biological processes, i.e., carbon sequestration, dehydrogenation, denitrification, sulfide oxidation, etc. (Han et al. 2008). Photosynthetic bacteria have shown great prospective in area of simultaneous wastewater treatment and resource recovery. PSB helps in COD and ammonia nitrogen removal up to 85–99% (Meng et al. 2018; Yang et al. 2017; Saejung and Thammaratana 2016). PSB biomass after cleaning wastewater can be used as by-products for feeding fish and livestock as feed or feed additive as they are rich in single cell protein, thus helping in reducing production cost of fish (Li et al. 2011). SCP have several health benefits for livestock and fish such as the following: it promotes growth, it enhances resistance against diseases, etc. PSB biomass can be used as source for extraction of value-added products such as coenzyme Q10 (CoQ10) and carotenoids (Meng et al. 2018; Hao et al. 2017; Jeong et al. 2008). Therefore, the PSB has evolved as an attractive tool for treating nontoxic wastewater for simultaneous wastewater treatment and resource recovery. It has an added advantage over algal based technology as it can be used for wastewater with high COD (Meng et al. 2018). As compared to other conventional technology, PSB has several advantages during its application process such as the following: it can work with high organic loading, the space requirement for bacterial growth is less, thus leading to low investment involved as the nutrients are mostly available in wastewater, and it requires less power consumption (Meng et al. 2018; Li et al. 2011). PSB can also be used in biological hydrogen production (Meng et al. 2018). Different wastewater types have been reported to be treated by the action of PSB tabulated in Table 7.3. Photosynthetic microbes are used in different bioreactors such as membrane sequencing batch reactor (Kaewsuk et al. 2010) (MSBR), photo-bioreactor, and photo anaerobic membrane bioreactor (Hülsen et al. 2016).

7.5 Economics of the Direct Involvement of Microorganisms in Treatment of Wastewater

Microorganism-based biological treatment has been preferred over chemical-based traditional methods. Most of the reactors designed for wastewater treatment nowadays involve the microbes or the bio-molecules generated by the microbes. Some of the systems such as algal based system and photosynthetic-based system involve microbes directly. The applications of microbes in MFC have been explained in a separate section. Considering the economics of the HRAP and PSB systems, Harun et al. (2011) demonstrated the techno-economic evaluation of the microalga

Table 7.3 Different wastewater treatment strategies using photosynthetic bacterial system

Wastewater type	Photosynthetic bacteria	Wastewater treatment efficiencies	Resource recovery	References
Domestic wastewater	*Rhodocyclus*, *Rhodopseudomonas*, and *Rhodobacter*	Total COD < 50 mgL^{-1}	The phototrophic bacteria act as vertical integrated producers for microbial protein	Hülsen et al. (2016)
		Total nitrogen < 5.0 mg L^{-1}		
		Total phosphorus < 1 mg L^{1}		
		Wastewater in discharge limits		
Fishing wastewater	*Rhodovulum sulfidophilum*	COD removal up to 85% from sardine processing wastewater	Increased biomass production rich in single cell protein of *R. sulfidophilum*. This biomass can be used as feed additive for aquaculture such as shrimp culture	Azad et al. (2003)
Livestock wastewater	*Rubrivivax gelatinosus*	COD removal up to 91% from poultry slaughter house wastewater	Increased biomass production up to 0.085 g biomass (d.w.) L^{1} day^{1}	Ponsano et al. (2008)
Starch wastewater	*Rhodopseudomonas palustris*	BOD removal – 95%	Biomass yield – 0.51 g dried solid/g BOD removed with crude protein – 0.58 g/g dried solid	Prachanurak et al. (2014)
		COD removal – 88%		
Brewery wastewater	*Rhodopseudomonas*	Effluent COD was reduced to below 80 mg/L, meeting the national discharge standard	The bacterial biomass was rich in bacterio-chlorophyll, carotenoids, coenzyme Q10, polysaccharides, and protein	Meng et al. (2018)

[a]Different other wastewaters such as monosodium glutamate wastewater (Wang and Liu 1999), citric acid wastewater (Han et al. 2008), and pharmaceutical wastewater (Mingxing and Yanling 2010) are also reported to be treated in the presence of phototrophic bacteria

photo-bioreactor in biogas and biodiesel generation. They demonstrated that the biodiesel production was integrated with bio-methane production. The methane generated was able to surrogate for the energy required during different steps of biodiesel production from algal biomass such as micro-algal cultivation, extraction, dewatering, and trans-esterification. They theoretically calculated that the energy requirement for the overall process is reduced by 33% and the carbon emission is also reduced by approximately 75%. As evident that microalgae growth can be nurtured in wastewater where the wastewater will provide the essential nutrients such as carbon, nitrogen, etc., in turn the algal biomass results in pollutant treatment. Meng et al. (2018) suggested that HRAP design require one-fifth of the cost required by any other lagoon system for cleaning wastewater to acceptable water quality. The land requirement is smaller for construction of HRAP as compared to other lagoon system. The best advantage of the microbe-based system is that the microbe produces different metabolites, bio-molecules during the process which are of great economical value. Therefore, strategies may be designed/implemented to recover these components apart from wastewater units for making the process carbon neutral and cost neutral and even generating revenues/profits in some cases.

7.6 Conclusion and Future Perspective

This chapter has elucidated in detail the role of microbes and how different microbe-assisted products/systems can be used in wastewater treatment directly or in combination. Currently, the bio-electrochemical system is considered as one of the leading technologies due its potential for direct electricity generation apart from wastewater treatment with simultaneous recovery of economically important chemical intermediates. The scaling up of the system is a major hindrance and limited by loss in electricity yield, cost of electrodes, issues associated with the continuous running of the process, and sludge formation and its separation. The application of photosynthetic microbes has also shown great potential but is limited due to cost involved in designing larger specialized photo bioreactors and the requirement of large land areas for HRAP. The application of microbial enzyme in different bioreactors for wastewater treatment has shown great potential; however, cost of enzyme production, stability, and reusability are a major concern. Although various techniques such as immobilization and membrane-based technology have enhanced the reusability of enzyme, these techniques are faced with decreased specificity and reactivity due to steric hindrance caused by immobilizing material. Thus, scientists have designed several nanoparticle-based immobilization systems, but the overall process at larger scale is still costly.

It is evident that the microbial based technologies are in the naïve stage and need lots of scientific, economic, as well as social impact studies before their application as a large-scale technology. The incorporation of microbes with different treatment strategies has opened a new venture for scientist from different streams such as materialist, physicist, microbiologist, chemist, economist, and sociologist in order to take the technology from the laboratory to land, overcoming the above-described limitations associated with different developed techniques.

Acknowledgment Authors acknowledge Department of Biotechnology, Government of India (Grant No. BT/304/ NE/TBP/2012 and BT/PR7333/PBD/26/373/2012), for financially supporting the work. B.K. acknowledges Jawaharlal Nehru Memorial Fund, New Delhi, for providing Doctoral Studies Scholarship. K.A. acknowledges the financial support provided by Central University of Rajasthan, Ajmer, India. N.B. gratefully acknowledges University Grants Commission, Government of India, for providing RGNF fellowship.

References

Agrawal K, Chaturvedi V, Verma P (2018) Fungal laccase discovered but yet undiscovered. Bioresour Bioprocess 5(4):1–12. https://doi.org/10.1186/s40643-018-0190-z

Aoyama T, Nosaka N, Kasai M (2007) Research on the nutritional characteristics of medium-chain fatty acids. J Med Investig 54:385–388

Arashiro LT, Montero N, Ferrer I et al (2018) Life cycle assessment of high rate algal ponds for wastewater treatment and resource recovery. Sci Total Environ 622–623:1118–1130. https://doi.org/10.1016/j.scitotenv.2017.12.051

Ashrafi O, Yerushalmi L, Haghighat F (2015) Wastewater treatment in the pulp-and-paper industry: a review of treatment processes and the associated greenhouse gas emission. J Environ Manag 158:146–157. https://doi.org/10.1016/j.jenvman.2015.05.010

Avci A, Kiliç NK, Dönmez G, Dönmez S (2014) Evaluation of hydrogen production by *Clostridium* strains on beet molasses. Environ Technol 35:278–285

Azad SA, Vikineswary S, Chong VC, Ramachandran KB (2003) *Rhodovulum sulfidophilum* in the treatment and utilization of sardine processing wastewater. Lett Appl Microbiol 38(1):13–18. https://doi.org/10.1046/j.1472-765X.2003.01435.x

Balasubramanian S, Tyagi RD (2016) Value-added bio-products from sewage sludge. In: Christian Larroche C, Sanroman M, Du G, Pandey A (eds) Current developments in biotechnology and bioengineering bioprocesses, bioreactors and controls. Elsevier, Amsterdam, pp 27–42

Bayramouglu G, Arica MY (2008) Enzymatic removal of phenol and p-chlorophenol in enzyme reactor: horseradish peroxidase immobilized on magnetic beads. J Hazard Mater 156:148–155

Behnami A, Farajzadeh D, Isazadeh S et al (2018) Diversity of bacteria in a full-scale petrochemical wastewater treatment plant experiencing stable hydrocarbon removal. J Water Process Eng 23:285–291. https://doi.org/10.1016/j.jwpe.2018.04.015

Bhunia A, Durani S, Wangikar PP (2001) Horseradish peroxidase catalyzed degradation of industrially important dyes. Biotechnol Bioeng 72:562–567

Biller P, Ross AB (2011) Potential yields and properties of oil from the hydrothermal liquefaction of microalgae with different biochemical content. Bioresour Technol 102:215–225

Blázquez E, Gabriel D, Baeza JA, Guisasola A (2016) Treatment of high-strength sulfate wastewater using an autotrophic biocathode in view of elemental sulfur recovery. Water Res 105:395–405

Bodzek M, Bohdziewicz J, Kowalska M (1994) Preparation of membrane-immobilised enzymes for phenol decomposition. J Chem Technol Biotechnol 61(3):231–239

Bohdziewicz J (1997) Biodegradation of phenol by enzymes from *Pseudomonas* sp. immobilized onto ultrafiltration membrane. Process Biochem 33(8):811–818

Braunegg G, Bona R, Koller M (2004) Sustainable polymer production. Polym-Plast Technol Eng 43:1779–1793

Campbell AS, Dong C, Dordick JS, Dinu CZ (2013) BioNano engineered hybrids for hypochlorous acid generation. Process Biochem 48:1355–1360

Campbell AS, Dong C, Maloney A et al (2014a) A systematic study of the catalytic behavior at enzyme–metal-oxide nanointerfaces. Nano Life 4(2):1450005

Campbell AS, Dong C, Meng F et al (2014b) Enzyme catalytic efficiency: a function of bio-nano interface reactions. ACS Appl Mater Interfaces 6:5393–5403

Capodaglio A (2017) Integrated, decentralized wastewater management for resource recovery in rural and peri-urban areas. Resources 6(2):22. https://doi.org/10.3390/resources6020022
Chandra R, Castillo-Zacarias C, Delgado P, Parra-Saldívar R (2018) A biorefinery approach for dairy wastewater treatment and product recovery towards establishing a biorefinery complexity index. J Clean Prod 183:1184–1196. https://doi.org/10.1016/j.jclepro.2018.02.124
Chen B-Y, Hsueh C-C, Liu S-Q et al (2013a) Deciphering mediating characteristics of decolorized intermediates for reductive decolorization and bioelectricity generation. Bioresour Technol 145:321–325
Chen BY, Liu S-Q, Hung JY et al (2013b) Reduction of carbon dioxide emission by using microbial fuel cells during wastewater treatment. Aerosol Air Qual Res 13:266–274
Cheng S, Xing D, Call DF, Logan BE (2009) Direct biological conversion of electrical current into methane by electromethanogenesis. Environ Sci Technol 43:3953–3958. https://doi.org/10.1021/es803531g
Cole AJ, Neveux N, Whelan A et al (2016) Adding value to the treatment of municipal wastewater through the intensive production of freshwater macroalgae. Algal Res 20:100–109. https://doi.org/10.1016/j.algal.2016.09.026
Coppens J, Grunert O, Hende S, Den V et al (2016) The use of microalgae as a high-value organic slow-release fertilizer results in tomatoes with increased carotenoid and sugar levels. J Appl Phycol 28:2367–2377. https://doi.org/10.1007/s10811-015-0775-2
Cotterill SE, Dolfing J, Jones C et al (2017) Low temperature domestic wastewater treatment in a microbial electrolysis cell with 1 m^2 anodes: towards system scale-up. Fuel Cells 17:584–592
Craggs R, Sutherland D, Campbell H (2012) Hectare-scale demonstration of high rate algal ponds for enhanced wastewater treatment and biofuel production. J Appl Phycol 24:329–337. https://doi.org/10.1007/s10811-012-9810-8
Craggs RJ, Lundquist TJ, Benemann JR (2013) Wastewater treatment and algal biofuel production. In: Borowitzka MA, Moheimani NR (eds) Algae for biofuels and energy. Springer, Dordrecht, pp 153–163
Craggs R, Park J, Heubeck S, Sutherland D (2014) High rate algal pond systems for low-energy wastewater treatment, nutrient recovery and energy production. N Z J Bot 52:60–73. https://doi.org/10.1080/0028825X.2013.861855
De Lathouder KM, Smeltink MW, Straathof AJJ et al (2008) Hydrogel coated monoliths for enzymatic hydrolysis of penicillin G. J Ind Microbiol Biotechnol 35:815–824
Dinu CZ, Zhu G, Bale SS et al (2010) Enzyme-based nanoscale composites for use as active decontamination surfaces. Adv Funct Mater 20:392–398
Doble M, Kumar A (2005) Biotreatment of industrial effluents. Elsevier, Amsterdam
Dong XZ, Cai MY (2001) Determinative manual for routine bacteriology. Beijing Scientific Press, Beijing, pp 370–390
Edwards W, Bownes R, Leukes WD et al (1999) A capillary membrane bioreactor using immobilized polyphenol oxidase for the removal of phenols from industrial effluents. Enzym Microb Technol 24(3–4):209–217
Farooqi IH, Basheer F (2017) Treatment of adsorbable organic halide (AOX) from pulp and paper industry wastewater using aerobic granules in pilot scale SBR. J Water Process Eng 19:60–66. https://doi.org/10.1016/j.jwpe.2017.07.005
Feng Y, He W, Liu J et al (2014) A horizontal plug flow and stackable pilot microbial fuel cell for municipal wastewater treatment. Bioresour Technol 156:132–138
Fischer F, Bastian C, Happe M et al (2011) Microbial fuel cell enables phosphate recovery from digested sewage sludge as struvite. Bioresour Technol 102:5824–5830
Flock C, Bassi A, Gijzen M (1999) Removal of aqueous phenol and 2-chlorophenol with purified soybean peroxidase and raw soybean hulls. J Chem Technol Biotechnol 74:303–309
Foley J, De Haas D, Hartley K, Lant P (2010) Comprehensive life cycle inventories of alternative wastewater treatment systems. Water Res 44:1654–1666
Gallifuoco A, Alfani F, Cantarella M, Viparelli P (2001) Studying enzyme-catalyzed depolymerizations in continuous reactors. Ind Eng Chem Res 40:5184–5190

Garcia-gonzalez J, Sommerfeld M (2016) Biofertilizer and biostimulant properties of the microalga *Acutodesmus dimorphus*. J Appl Phycol 28(2):1051–1061. https://doi.org/10.1007/s10811-015-0625-2

Garfí M, Flores L, Ferrer I (2017) Life cycle assessment of wastewater treatment systems for small communities: activated sludge, constructed wetlands and high rate algal ponds. J Clean Prod 161:211–219. https://doi.org/10.1016/j.jclepro.2017.05.116

Ge Z, He Z (2016) Long-term performance of a 200 liter modularized microbial fuel cell system treating municipal wastewater: treatment, energy, and cost. Environ Sci Water Res Technol 2:274–281

Ghosh A, Dastidar MG, Sreekrishnan TR (2017) Bioremediation of chromium complex dyes and treatment of sludge generated during the process. Int Biodeterior Biodegradation 119:448–460. https://doi.org/10.1016/j.ibiod.2016.08.013

Gianfreda L, Bollag J-M (2002) Isolated enzymes for the transformation and detoxification of organic pollutants. In: Burns RG, Dick RP (eds) Enzymes in the environment: activity, ecology, and applications. CRC Press, New York, pp 499–542

Gilbert EM, Agrawal S, Karst SM et al (2014) Low temperature partial nitritation/anammox in a moving bed bio film reactor treating low strength wastewater. Environ Sci Technol 48(15):8784–8792

Gregory KB, Lovley DR (2005) Remediation and recovery of uranium from contaminated subsurface environments with electrodes. Environ Sci Technol 39:8943–8947

Grobbelaar JU (2012) Microalgae mass culture: the constraints of scaling-up. J Appl Phycol 24:315–318. https://doi.org/10.1007/s10811-011-9728-6

Hadiyanto H, Elmore S, Van Gerven T, Stankiewicz A (2013) Hydrodynamic evaluations in high rate algae pond (HRAP) design. Chem Eng J 217:231–239. https://doi.org/10.1016/j.cej.2012.12.015

Han L, Huang C, Wang C (2008) Application of photosynthetic bacteria to the wastewater treatment. Weifang High Vocat Educ 1:18

Han-chang S (2002) Industrial wastewater-types, amounts and effects. Encycl Life Support Syst I:1–6

Hao J, Wang X, Wang H (2017) Investigation of polyhydroxyalkanoates (PHAs) biosynthesis from mixed culture enriched by valerate-dominant hydrolysate. Front Environ Sci Eng 11:5

Harnisch F, Schröder U (2010) From MFC to MXC: chemical and biological cathodes and their potential for microbial bioelectrochemical systems. Chem Soc Rev 39:4433–4448. https://doi.org/10.1039/c003068f

Harun R, Danquah MK, Forde GM (2010) Microalgal biomass as a fermentation feedstock for bioethanol production. J Chem Technol Biotechnol 85:199–203

Harun R, Davidson M, Doyle M et al (2011) Technoeconomic analysis of an integrated microalgae photobioreactor, biodiesel and biogas production facility. Biomass Bioenergy 35:741–747. https://doi.org/10.1016/j.biombioe.2010.10.007

Hiegemann H, Herzer D, Nettmann E et al (2016) An integrated 45 L pilot microbial fuel cell system at a full-scale wastewater treatment plant. Bioresour Technol 218:115–122

Ho NAD, Babel S, Sombatmankhong K (2018) Bio-electrochemical system for recovery of silver coupled with power generation and wastewater treatment from silver(I) diammine complex. J Water Process Eng 23:186–194. https://doi.org/10.1016/j.jwpe.2018.04.001

Holkar CR, Jadhav AJ, Pinjari DV et al (2016) A critical review on textile wastewater treatments: possible approaches. J Environ Manag 182:351–366

Hülsen T, Barry EM, Lu Y et al (2016) Domestic wastewater treatment with purple phototrophic bacteria using a novel continuous photo anaerobic membrane bioreactor. Water Res 100:486–495. https://doi.org/10.1016/j.watres.2016.04.061

Ichihashi O, Hirooka K (2012) Removal and recovery of phosphorus as struvite from swine wastewater using microbial fuel cell. Bioresour Technol 114:303–307

Ieropoulos I, Greenman J, Melhuish C (2012) Urine utilisation by microbial fuel cells; energy fuel for the future. Phys Chem Chem Phys 14:94–98

Ishizaki A, Tanaka K (1991) Production of poly-β-hydroxybutyric acid from carbon dioxide by Alcaligenes eutrophus ATCC 17697T. J Ferment Bioeng 71:254–257
Jadhav DA, Ghosh Ray S, Ghangrekar MM (2017) Third generation in bio-electrochemical system research – a systematic review on mechanisms for recovery of valuable by-products from wastewater. Renew Sust Energ Rev 76:1022–1031. https://doi.org/10.1016/j.rser.2017.03.096
Jäger K, Bartók T, Ördög V, Barnabás B (2010) Improvement of maize (Zea mays L.) anther culture responses by algae-derived natural substances. S Afr J Bot 76:511–516. https://doi.org/10.1016/j.sajb.2010.03.009
Jahangiri E, Thomas I, Schulze A et al (2018) Characterisation of electron beam irradiation-immobilised laccase for application in wastewater treatment. Sci Total Environ 624:309–322. https://doi.org/10.1016/j.scitotenv.2017.12.127
Jamie A, Alshami AS, Maliabari ZO et al (2016) Immobilization and enhanced catalytic activity of lipase on modified MWCNT for oily wastewater treatment. Environ Prog Sustain Energy 35:1441–1449
Jeong S-K, Ahn S-C, Kong I-S, Kim J-K (2008) Isolation and identification of a photosynthetic bacterium containing a high content of coenzyme Q 10. Fish Aquat Sci 11:172–176
Jolivalt C, Brenon S, Caminade E et al (2000) Immobilization of laccase from *Trametes versicolor* on a modified PVDF microfiltration membrane: characterization of the grafted support and application in removing a phenylurea pesticide in wastewater. J Membr Sci 180:103–113
Kaewsuk J, Thorasampan W, Thanuttamavong M, Tae G (2010) Kinetic development and evaluation of membrane sequencing batch reactor (MSBR) with mixed cultures photosynthetic bacteria for dairy wastewater treatment. J Environ Manag 91:1161–1168. https://doi.org/10.1016/j.jenvman.2010.01.012
Kalogerakis N, Fava F, Corvini PF-X (2017) Bioremediation advances. N Biotechnol 38:41–42. https://doi.org/10.1016/j.nbt.2017.07.004
Kim JR, Zuo Y, Regan JM, Logan BE (2008) Analysis of ammonia loss mechanisms in microbial fuel cells treating animal wastewater. Biotechnol Bioeng 99:1120–1127
Kojima H, Ogawa M, Tamal S et al (1995) Oxidation ofglucose by glucose oxidase entrapped in hollow fibre membrane. J Chem Eng Jpn 28:623–626
Kokabian B, Gude VG, Smith R, Brooks JP (2018) Evaluation of anammox biocathode in microbial desalination and wastewater treatment. Chem Eng J 342:410–419. https://doi.org/10.1016/j.cej.2018.02.088
Köller G, Möder M, Czihal K (2000) Peroxidative degradation of selected PCB: a mechanistic study. Chemosphere 41:1827–1834
Komala PS, Effendi AJ, Wneten IG, Wisjnuprapto (2011) Performance of Anoxic-oxic Membrane bioreactor for azo dye biodegradation. Int J Civil Environ Eng 11:28–34
Krueger A, Lang D (2012) Functionality is key: recent progress in the surface modification of nanodiamond. Adv Funct Mater 22:890–906
Kuenen JG (2008) Anammox bacteria : from discovery to application. Nat Rev Microbiol 6:320–326
Kumar B, Agrawal K, Bhardwaj N et al (2018) Advances in concurrent bioelectricity generation and bioremediation through microbial fuel cells. In: Microbial fuel cell technology for bioelectricity. Springer, pp 211–239
Kuntke P, Śmiech K, Bruning H et al (2012) Ammonium recovery and energy production from urine by a microbial fuel cell. Water Res 46:2627–2636
Lante A, Crapisi A, Krastanov A, Spettoli P (2000) Biodegradation of phenols by laccase immobilised in a membrane reactor. Process Biochem 36:51–58
Le-Clech P, Chen V, Fane TAG (2006) Fouling in membrane bioreactors used in wastewater treatment. J Membr Sci 284:17–53
Lee SY (1996) Bacterial polyhydroxyalkanoates. Biotechnol Bioeng 49:1–14
Lee J-H, Han J, Choi H, Hur H-G (2007) Effects of temperature and dissolved oxygen on Se (IV) removal and Se (0) precipitation by *Shewanella* sp. HN-41. Chemosphere 68:1898–1905

Li R h, Li X y (2017) Recovery of phosphorus and volatile fatty acids from wastewater and food waste with an iron-flocculation sequencing batch reactor and acidogenic co-fermentation. Bioresour Technol 245:615–624. https://doi.org/10.1016/j.biortech.2017.08.199

Li L, Yang X, Li A et al (2011) Research progress of photosynthetic bacteria in wastewater treatment. Appl Mech Mater 71–78:2831–2835. https://doi.org/10.4028/www.scientific.net/AMM.71-78.2831

Li M, Pan Y, Huang L et al (2017) Continuous flow operation with appropriately adjusting composites in influent for recovery of Cr (VI), Cu (II) and Cd (II) in self-driven MFC--MEC system. Environ Technol 38:615–628

Li J, Li J, Gao R et al (2018) A critical review of one-stage anammox processes for treating industrial wastewater: optimization strategies based on key functional microorganisms. Bioresour Technol 265:498–505. https://doi.org/10.1016/j.biortech.2018.07.013

Liang P, Wei J, Li M, Huang X (2013) Scaling up a novel denitrifying microbial fuel cell with an oxic-anoxic two stage biocathode. Front Environ Sci Eng 7:913–919

Liu Z, Wang J, Wang C (2010) Tech-economic analysis on anoxic/Oxic membrane bioreactor (A/O-MBR) for domestic wastewater treatment. International conference challenges environment science computer engineering. Tech-economic 56–59. https://doi.org/10.1109/CESCE.2010.209

Liu Y, Huang L, Dong G et al (2018) Enhanced granulation and methane recovery at low load by downflow sludge circulation in anaerobic treatment of domestic wastewater. Bioresour Technol 249:851–857. https://doi.org/10.1016/j.biortech.2017.10.091

Livingston AG (1994) Extractive membrane bioreactors: a new process technology for detoxifying chemical industry wastewaters. J Chem Technol Biotechnol 60(2):117–124

Lopez C, Mielgo I, Moreira MT et al (2002) Enzymatic membrane reactors for biodegradation of recalcitrant compounds: application to dye decolourisation. J Biotechnol 99:249–257

Lu L, Ren ZJ (2016) Microbial electrolysis cells for waste biorefinery: a state of the art review. Bioresour Technol 215:254–264

Lu H, Peng M, Zhang G et al (2019) Brewery wastewater treatment and resource recovery through long term continuous-mode operation in pilot photosynthetic bacteria-membrane bioreactor. Sci Total Environ 646:196–205. https://doi.org/10.1016/j.scitotenv.2018.07.268

Maloney AJ, Dong C, Campbell AS, Dinu CZ (2015) Emerging enzyme-based technologies for wastewater treatment. ACS Symp Ser 1192:69–85. https://doi.org/10.1021/bk-2015-1192.ch005

Mathuriya AS, Yakhmi JV (2014) Microbial fuel cells to recover heavy metals. Environ Chem Lett 12:483–494. https://doi.org/10.1007/s10311-014-0474-2

Meng F, Yang A, Wang H et al (2018) One-step treatment and resource recovery of high-concentration non-toxic organic wastewater by photosynthetic bacteria. Bioresour Technol 251:121–127. https://doi.org/10.1016/j.biortech.2017.12.002

Miao X, Wu Q (2004) High yield bio-oil production from fast pyrolysis by metabolic controlling of Chlorella protothecoides. J Biotechnol 110:85–93. https://doi.org/10.1016/j.jbiotec.2004.01.013

Mingxing L, Yanling Z (2010) Research on the technological parameters of the treatment of wastewater from traditional Chinese medicine production by photosynthetic bacteria+ coagulation. Ind Water Treat 30:33–35

Mo W, Zhang Q (2013) Energy-nutrients-water nexus: integrated resource recovery in municipal wastewater treatment plants. J Environ Manag 127:255–267. https://doi.org/10.1016/j.jenvman.2013.05.007

Mohan SV, Velvizhi G, Krishna KV, Babu ML (2014) Microbial catalyzed electrochemical systems: a bio-factory with multi-facet applications. Bioresour Technol 165:355–364

Molinos-Senante M, Gómez T, Garrido-Baserba M et al (2014) Assessing the sustainability of small wastewater treatment systems: a composite indicator approach. Sci Total Environ 497–498:607–617. https://doi.org/10.1016/j.scitotenv.2014.08.026

Mulder A, Van De Graaf AA, Robertson LA, Kuenen JG (1995) Anaerobic ammonium oxidation discovered in a denitrifying fluidized bed reactor. FEMS Microbiol Ecol 16(3):177–183

Mutambanengwe C, Oyekola O (2008) Production of enzymes for industrial wastewater treatment: proof of concept and application. WRC Report No 1541/1/08. 1–58

Nicolau E, Fonseca JJ, Rodríguez-Martínez JA et al (2014) Evaluation of a urea bioelectrochemical system for wastewater treatment processes. ACS Sustain Chem Eng 2:749–754. https://doi.org/10.1021/sc400342x

Onodera T, Sase S, Choeisai P et al (2013) Development of a treatment system for molasses wastewater: the effects of cation inhibition on the anaerobic degradation process. Bioresour Technol 131:295–302

Oswald WJ, Golueke CG (1960) Biological transformation of solar energy. Adv Appl Microbiol 2:223–262

Ozturk E, Cinperi NC (2018) Water efficiency and wastewater reduction in an integrated woolen textile mill. J Clean Prod 201:686–696

Park JBK, Craggs RJ, Shilton AN (2011) Wastewater treatment high rate algal ponds for biofuel production. Bioresour Technol 102:35–42. https://doi.org/10.1016/j.biortech.2010.06.158

Passos F, Gutiérrez R, Uggetti E et al (2017) Towards energy neutral microalgae-based wastewater treatment plants. Algal Res 28:235–243. https://doi.org/10.1016/j.algal.2017.11.006

Ponsano EHG, Paulino CZ, Pinto MF (2008) Phototrophic growth of *Rubrivivax gelatinosus* in poultry slaughterhouse wastewater. Bioresour Technol 99:3836–3842. https://doi.org/10.1016/j.biortech.2007.06.063

Prachanurak P, Chiemchaisri C, Chiemchaisri W, Yamamotob K (2014) Biomass production from fermented starch wastewater in photo-bioreactor with internal overflow recirculation. Bioresour Technol 165:129–136. https://doi.org/10.1016/j.biortech.2014.03.119

Pradeep T, Anshup (2009) Noble metal nanoparticles for water purification: a critical review. Thin Solid Films 517:6441–6478. https://doi.org/10.1016/j.tsf.2009.03.195

Rabaey K, Rozendal RA (2010) Microbial electrosynthesis—revisiting the electrical route for microbial production. Nat Rev Microbiol 8:706–716

Rajesh PP, Jadhav DA, Ghangrekar MM (2015) Improving performance of microbial fuel cell while controlling methanogenesis by *Chaetoceros* pretreatment of anodic inoculum. Bioresour Technol 180:66–71

Ren HY, Kong F, Ma J et al (2018) Continuous energy recovery and nutrients removal from molasses wastewater by synergistic system of dark fermentation and algal culture under various fermentation types. Bioresour Technol 252:110–117. https://doi.org/10.1016/j.biortech.2017.12.092

Rios GM, Belleville MP, Paolucci D, Sanchez J (2004) Progress in enzymatic membrane reactors – a review. J Membr Sci 242:189–196. https://doi.org/10.1016/j.memsci.2003.06.004

Rodolfi L, Chini Zittelli G, Bassi N et al (2009) Microalgae for oil: strain selection, induction of lipid synthesis and outdoor mass cultivation in a low-cost photobioreactor. Biotechnol Bioeng 102:100–112

Rosenbaum M, Aulenta F, Villano M, Angenent LT (2011) Cathodes as electron donors for microbial metabolism: which extracellular electron transfer mechanisms are involved? Bioresour Technol 102:324–333

Rozendal RA, Hamelers HVM, Euverink GJW et al (2006) Principle and perspectives of hydrogen production through biocatalyzed electrolysis. Int J Hydrog Energy 31:1632–1640. https://doi.org/10.1016/j.ijhydene.2005.12.006

Rozendal RA, Hamelers HVM, Rabaey K et al (2008a) Towards practical implementation of bioelectrochemical wastewater treatment. Trends Biotechnol 26:450–459

Rozendal RA, Jeremiasse AW, Hamelers HVM, Buisman CJN (2008b) Hydrogen production with a microbial biocathode. Environ Sci Technol 42:629–634. https://doi.org/10.1021/es071720+

Rozendal RA, Leone E, Keller J, Rabaey K (2009) Efficient hydrogen peroxide generation from organic matter in a bioelectrochemical system. Electrochem Commun 11:1752–1755. https://doi.org/10.1016/j.elecom.2009.07.008

Saejung C, Thammaratana T (2016) Biomass recovery during municipal wastewater treatment using photosynthetic bacteria and prospect of production of single cell protein for feedstuff. Environ Technol 37:3055–3061

Sahinkaya E, Sahin A, Yurtsever A, Kitis M (2018) Concentrate minimization and water recovery enhancement using pellet precipitator in a reverse osmosis process treating textile wastewater. J Environ Manag 222:420–427. https://doi.org/10.1016/j.jenvman.2018.05.057

Savant DV, Abdul-Rahman R, Ranade DR (2006) Anaerobic degradation of adsorbable organic halides (AOX) from pulp and paper industry wastewater. Bioresour Technol 97:1092–1104. https://doi.org/10.1016/j.biortech.2004.12.013

Sharma M, Bajracharya S, Gildemyn S et al (2014) A critical revisit of the key parameters used to describe microbial electrochemical systems. Electrochim Acta 140:191–208

Sharma B, Dangi AK, Shukla P (2018) Contemporary enzyme based technologies for bioremediation: a review. J Environ Manag 210:10–22. https://doi.org/10.1016/j.jenvman.2017.12.075

Shen Q, Yang R, Hua X et al (2011) Gelatin-templated biomimetic calcification for β-galactosidase immobilization. Process Biochem 46:1565–1571

Shi X, Leong KY, Ng HY (2017) Anaerobic treatment of pharmaceutical wastewater: a critical review. Bioresour Technol 245:1238–1244. https://doi.org/10.1016/j.biortech.2017.08.150

Shin C, Bae J (2018) Current status of the pilot-scale anaerobic membrane bioreactor treatments of domestic wastewaters: a critical review. Bioresour Technol 247:1038–1046. https://doi.org/10.1016/j.biortech.2017.09.002

Shokrollahzadeh S, Azizmohseni F, Golmohammad F et al (2008) Biodegradation potential and bacterial diversity of a petrochemical wastewater treatment plant in Iran. Bioresour Technol 99:6127–6133. https://doi.org/10.1016/j.biortech.2007.12.034

Sirianuntapiboon S, Prasertsong K (2008) Treatment of molasses wastewater by acetogenic bacteria BP103 in sequencing batch reactor (SBR) system. Bioresour Technol 99:1806–1815

Stadlmair LF, Letzel T, Drewes JE, Grassmann J (2018) Enzymes in removal of pharmaceuticals from wastewater: a critical review of challenges, applications and screening methods for their selection. Chemosphere 205:649–661. https://doi.org/10.1016/j.chemosphere.2018.04.142

Steinbusch KJJ, Hamelers HVM, Buisman CJN (2008) Alcohol production through volatile fatty acids reduction with hydrogen as electron donor by mixed cultures. Water Res 42:4059–4066

Strous M, Gerven VE, Zheng P et al (1997) Ammonium removal from concentrated waste streams with the anaerobic ammonium oxidation (Anammox) process in different reactor configurations. Water Res 31:1955–1962

Sutherland DL, Howard-Williams C, Turnbull MH et al (2015) Enhancing microalgal photosynthesis and productivity in wastewater treatment high rate algal ponds for biofuel production. Bioresour Technol 184:222–229. https://doi.org/10.1016/j.biortech.2014.10.074

Tang C, Zheng P, Mahmood Q (2009) The shear force amendments on the slugging behavior of upflow Anammox granular sludge bed reactor. Sep Purif Technol 69:262–268. https://doi.org/10.1016/j.seppur.2009.07.029

Tang C, Zheng P, Chen T et al (2011a) Enhanced nitrogen removal from pharmaceutical wastewater using SBA-ANAMMOX process. Water Res 45:201–210. https://doi.org/10.1016/j.watres.2010.08.036

Tang C, Zheng P, Wang C et al (2011b) Performance of high-loaded ANAMMOX UASB reactors containing granular sludge. Water Res 45:135–144. https://doi.org/10.1016/j.watres.2010.08.018

Tang C, Duan C, Yu C, Song Y (2017) Removal of nitrogen from wastewaters by anaerobic ammonium oxidation (ANAMMOX) using granules in upflow reactors. Environ Chem Lett 15:311–328. https://doi.org/10.1007/s10311-017-0607-5

Tao Y, Gao D, Fu Y et al (2012) Bioresource technology impact of reactor configuration on anammox process start-up: MBR versus SBR. Bioresour Technol 104:73–80. https://doi.org/10.1016/j.biortech.2011.10.052

Tao H-C, Lei T, Shi G et al (2014) Removal of heavy metals from fly ash leachate using combined bioelectrochemical systems and electrolysis. J Hazard Mater 264:1–7

Thatoi H, Das S, Mishra J et al (2014) Bacterial chromate reductase, a potential enzyme for bioremediation of hexavalent chromium: a review. J Environ Manag 146:383–399. https://doi.org/10.1016/j.jenvman.2014.07.014

Thygesen A, Thomsen AB, Possemiers S, Verstraete W (2010) Integration of microbial electrolysis cells (MECs) in the biorefinery for production of ethanol, H_2 and phenolics. Waste Biomass Valoriz 1:9–20

Toczyłowska-Mamińska R (2017) Limits and perspectives of pulp and paper industry wastewater treatment – A review. Renew Sust Energ Rev 78:764–772. https://doi.org/10.1016/j.rser.2017.05.021

Uysal O, Uysal FO, Ekinci K (2015) Evaluation of microalgae as microbial fertilizer. Eur J Sustain Dev 4:77–82

Van Eerten-Jansen MCAA, Ter Heijne A, Grootscholten TIM et al (2013) Bioelectrochemical production of caproate and caprylate from acetate by mixed cultures. ACS Sustain Chem Eng 1:513–518

Vashi H, Iorhemen OT, Tay JH (2018) Aerobic granulation: a recent development on the biological treatment of pulp and paper wastewater. Environ Technol Innov 9:265–274. https://doi.org/10.1016/j.eti.2017.12.006

Villano M, Aulenta F, Ciucci C et al (2010) Bioelectrochemical reduction of CO_2 to CH_4 via direct and indirect extracellular electron transfer by a hydrogenophilic methanogenic culture. Bioresour Technol 101:3085–3090

Vladisavljević GT (2015) Biocatalytic membrane reactors. In: Saha B (ed) Catalytic reactors. Berlin/Boston, Walter de Gruyter GmbH, pp 51–102

Wang FF, Liu WW (1999) Application of photosynthetic microbe to the treatment of aginomoto sewage. Fujian Environ 16:17–20

Wang X, Feng Y, Liu J et al (2010) Sequestration of CO_2 discharged from anode by algal cathode in microbial carbon capture cells (MCCs). Biosens Bioelectron 25:2639–2643

Wang K, Abdala AA, Hilal N et al (2017) Mechanical characterization of membranes. In: Hilal N, Ismail AF, Matsuura T, Oatley-Radcliffe D (eds) Membrane characterization, pp 259–306

Watson SD, Akhurst T, Whiteley CG et al (2004) Primary sludge floc degradation is accelerated under biosulphidogenic conditions: enzymological aspects. Enzyme Microb Technol 34:595–602. https://doi.org/10.1016/j.enzmictec.2004.01.004

Whiteley C, Pletschke B, Rose P, Ngesi N (2002a) Specific sulphur metabolites stimulate $β$-glucosidase activity in an anaerobic sulphidogenic bioreactor. Biotechnol Lett 24:1509–1513

Whiteley CG, Heron P, Pletschke B et al (2002b) The enzymology of sludge solubilisation utilising sulphate reducing systems: properties of proteases and phosphatases. Enzyme Microb Technol 31:419–424. https://doi.org/10.1016/S0141-0229(02)00100-X

Whiteley CG, Enongene G, Pletschke BI et al (2003) Co-digestion of primary sewage sludge and industrial wastewater under anaerobic sulphate reducing conditions: enzymatic profiles in a recycling sludge bed reactor. Water Sci Technol 48:129–138

Windey K, Bo I, De ÃWV (2005) Oxygen-limited autotrophic nitrification – denitrification (OLAND) in a rotating biological contactor treating high-salinity wastewater. Water Res 39:4512–4520. https://doi.org/10.1016/j.watres.2005.09.002

Wu Z, Zou S, Zhang B et al (2018) Forward osmosis promoted in-situ formation of struvite with simultaneous water recovery from digested swine wastewater. Chem Eng J 342:274–280. https://doi.org/10.1016/j.cej.2018.02.082

Xiang Z, Yao Ho F, Yuan-song W (2003) A pilot scale anoxic/oxic membrane bioreactor (A/O MBR) for woolen mill dyeing wastewater treatment. J Environ Sci (China) 15:449–455

Xiao L, Young EB, Berges JA, He Z (2012) Integrated photo-bioelectrochemical system for contaminants removal and bioenergy production. Environ Sci Technol 46:11459–11466

Xu Q, Hamid A, Wen X et al (2014) Fenton-Anoxic–Oxic/MBR process as a promising process for avermectin fermentation wastewater reclamation. Sep Purif Technol 134:82–89. https://doi.org/10.1016/j.seppur.2014.07.029

Yan X, Bilad MR, Gerards R et al (2012) Comparison of MBR performance and membrane cleaning in a single-stage activated sludge system and a two-stage anaerobic/aerobic (A/A) system for treating synthetic molasses wastewater. J Membr Sci 394:49–56

Yan L, Ye X, Linhardt RJ et al (2018a) Full recovery of value-added compounds from citrus canning processing water. J Clean Prod 176:959–965. https://doi.org/10.1016/j.jclepro.2017.12.018
Yan T, Ye Y, Ma H et al (2018b) A critical review on membrane hybrid system for nutrient recovery from wastewater. Chem Eng J 348:143–156. https://doi.org/10.1016/j.cej.2018.04.166
Yang Y, Zhang L, Han X et al (2016) Determine the operational boundary of a pilot-scale single-stage partial nitritation/anammox system with granular sludge. Water Sci Technol 73:2085–2092. https://doi.org/10.2166/wst.2016.052
Yang A, Zhang G, Yang G et al (2017) Denitrification of aging biogas slurry from livestock farm by photosynthetic bacteria. Bioresour Technol 232:408–411
Yen H-W, Brune DE (2007) Anaerobic co-digestion of algal sludge and waste paper to produce methane. Bioresour Technol 98:130–134
Zhai R, Zhang B, Wan Y et al (2013) Chitosan--halloysite hybrid-nanotubes: Horseradish peroxidase immobilization and applications in phenol removal. Chem Eng J 214:304–309
Zhang F, Zheng B, Zhang J et al (2010) Horseradish peroxidase immobilized on graphene oxide: physical properties and applications in phenolic compound removal. J Phys Chem C 114:8469–8473
Zhang Y, Liu T, Wang Q et al (2012a) Synthesis of novel poly (N, N-diethylacrylamide-co-acrylic acid)(P (DEA-co-AA)) microgels as carrier of horseradish peroxidase immobilization for pollution treatment. Macromol Res 20:484–489
Zhang Z, Donaldson AA, Ma X (2012b) Advancements and future directions in enzyme technology for biomass conversion. Biotechnol Adv 30:913–919
Zhang L, Liu M, Zhang S et al (2015a) Integrated fixed-biofilm activated sludge reactor as a powerful tool to enrich anammox biofilm and granular sludge. Chemosphere 140:114–118. https://doi.org/10.1016/j.chemosphere.2015.02.001
Zhang L, Zhang S, Peng Y et al (2015b) Nitrogen removal performance and microbial distribution in pilot- and full-scale integrated fixed-biofilm activated sludge reactors based on nitritation-anammox process. Bioresour Technol 196:448–453. https://doi.org/10.1016/j.biortech.2015.07.090

Type III Polyketide Synthases: Current State and Perspectives

8

Thangamani Rajesh, Manish K. Tiwari, Sairam Thiagarajan, Pranav S. Nair, and Marimuthu Jeya

Abstract

Microorganisms produce an array of secondary metabolites primarily for their growth, self-defense, communication, and survival in natural environments. Among several other microbial secondary metabolites, antibiotic (antimicrobial, antiviral, and antifungal activities) compound effects are vital for humans and other higher living organisms to combat pathogens. Even though several such antibiotic compounds are identified in the past decades, very few have been commercialized and made available for practical applications. Polyketides are secondary metabolites that are produced as a result of enzyme-catalyzed condensation/polymerization reactions of simple fatty acids resulting in products with diverse structural and functional properties. The enzymes involved in the production of polyketides are polyketide synthases (PKSs), which are mainly involved in condensation of acyl-thioester units into functional secondary metabolites. Based on the origin and structural diversity, PKSs are classified as type I PKSs, type II PKSs, and type III PKSs. Type I PKSs are large multifunctional proteins with multi-domain architecture and are mostly of fungal origin, whereas

T. Rajesh · P. S. Nair
Biotechnology Division, CSIR- National Engineering and Environmental Research Institute, Chennai Zonal Center, Chennai, India

M. K. Tiwari
Department of Chemistry, University of Copenhagen, Copenhagen, Denmark

S. Thiagarajan
PSG Centre for Molecular Medicine and Therapeutics, PSG Institute of Medical Sciences and Research, Coimbatore, Tamil Nadu, India

M. Jeya (✉)
Marine Biotechnology Division, National Institute of Ocean Technology, Chennai, Tamil Nadu, India
e-mail: jeyambt@niot.res.in

P. K. Arora (ed.), *Microbial Technology for the Welfare of Society*, Microorganisms for Sustainability 17,
https://doi.org/10.1007/978-981-13-8844-6_8

type II PKSs are mono-functional proteins predominantly present in bacteria and actinomycetes. Type III PKSs are primarily of plant origin, and very few have been reported from microbial sources. In this chapter, the occurrence and distribution of type III PKSs, the genetic architecture of the genes involved in type III polyketide biosynthesis, the protein structure of type III PKSs, and the commercial importance of type III PKSs are discussed in detail.

Keywords
Type III polyketide synthases · Fungi · Product specificity · Protein engineering · Commercial implications

8.1 Introduction

Polyketides are a structurally and functionally distinct group of secondary metabolites produced by plants, bacteria, fungi, and other eukaryotes. These polyketide natural products are synthesized by a specific class of enzymes (PKSs) that are classified based on their sequence information, structural organization, and product specificity (Rimando and Baerson 2007). Type I PKSs are polyproteins with multidomain architecture that ultimately form bulky multi-functional complexes involved in polyketide biosynthesis in a modular or iterative fashion. Classical examples of type I PKS include erythromycin biosynthesis gene cluster from bacterium *Saccharopolyspora erythraea* and 6-methylsalicylic acid (6-MSA) biosynthesis in fungus *Penicillium patulum*. In *S. erythraea*, type I PKS enzymatic process is mediated via three~350 kDa polypeptides, linearly arranged catalyzing a single round of elongation and complete or partial reduction. Individual modules in the biosynthetic proteins of type I PKS include acyltransferase domain (AT domain) and acyl carrier protein domain (ACP domain) and a thioesterase domain (TE domain). In *P. patulum* biosynthesis of 6-MSA is performed in an iterative fashion wherein homotetrameric type I PKS consists of ketosynthase domain (KS domain), malonyl-CoA:ACP transacylase domain (MAT domain), dehydratase domain (DH domain), β-ketoreductase domain (KR domain), and an ACP domain (Fig. 8.1). The PKS biosynthesis system ultimately catalyzes the reaction to link four acetate units that ultimately cyclize through intramolecular aldol condensation and aromatization reactions (Pfeifer and Chaitan 2001).

Type II PKSs are multienzyme complexes consisting of distinct, separable proteins similar to type II fatty acid synthase system in plants and bacteria. Despite being dissociable, type II class of enzymes forms in vivo complexes similar to those of the type I PKS systems. Interestingly, most of the type II PKS complexes lack AT domains (Fig. 8.1) similar to those seen in FAS and type I PKS enzymes but exhibit acyltransferase activity (Hertweck et al. 2007).

Classical examples of type II PKSs include (i) actinorhodin biosynthesis in *Streptomyces coelicolor*, (ii) tracenomycins (tetracenomycin and elloramycin) biosynthesis in *Streptomyces glaucescens* and *Streptomyces olivaceus*, and (iii)

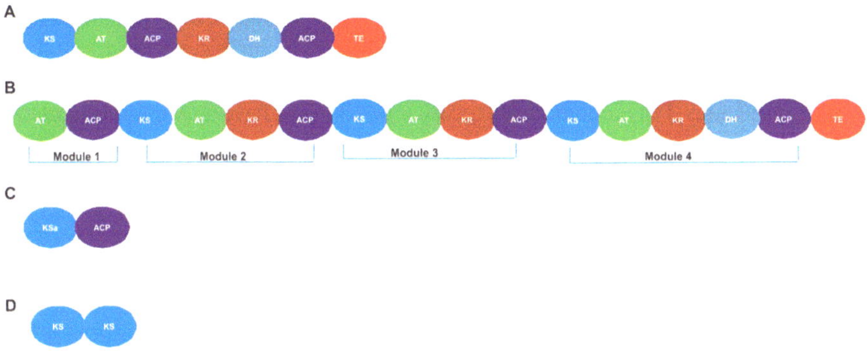

Fig. 8.1 Representative structure of PKS enzymes. (**a**) Iterative type I PKS, (**b**) non-iterative, modular type I PKS, (**c**) minimal type II PKS, and (**d**) type III PKS. Abbreviations: *KS* ketosynthase, *AT* acyltransferase, *ACP* acyl-carrier protein, *KR* ketoreductase, *DH* dehydratase, *TE* termination. (Shen 2003)

doxorubicin biosynthesis in *Streptomyces peucetius* (Taguchi et al. 2000; Frankel and McCafferty 2004). The polypeptides involved in type II PKSs reactions typically consist of KS domain, chain length factor domain (CLF domain), ACP domain, KR domain, aromatase domain (ARO domain), and cyclase domain (CYC). The reaction mechanism of type II PKSs begins with the KS and CLF domain-containing heterodimers catalyzing the initiation and elongation of malonyl building blocks followed by ACP domain delivering the malonyl building blocks to KS-CLF domain. The collective action of these type II domains results in the production of poly-β-keto intermediates. The nascent polyketide chain thus produced by the action of KR, ARO, and CYC domains results in final polyketide production (Hertweck et al. 2007). Most of the known type II PKSs found in *Streptomyces* spand are extensively investigated; however, computational methods and genome mining have led to the identification of >150 putative type II PKS gene clusters in actinobacterial genomes (Ziemert et al. 2016).

Despite the detailed analysis of gene structure and reaction mechanism of type I and II PKSs, limited information is available on the three-dimensional organization of these biosynthetic complexes. To further manipulate the chemical and biological properties of delivered polyketides, further in-depth studies on 3D structures of PKS type I and II complexes are needed. Such efforts would facilitate precise understanding of substrate recognition, polyketide chain length, and cyclization specificity in these large multi-enzyme complexes (Shen 2003).

8.1.1 Type III PKSs

Type III PKSs have a simple architecture that is a homodimer of identical ketoacyl synthase (KS) monomeric domains (Fig. 8.1). The active site in each monomer catalyzes the entire reactions of priming, extension, and cyclization of acetate units into a polyketide chain. Primarily present in plants, type III PKSs are also found in bacteria and fungi as well. Type III PKS enzymes display multiple activities to catalyze varying numbers of iterative condensation reactions on a diverse set of much larger starter molecules. Polyketides produced by type III PKSs have applications in medicine, agriculture, and other fields, due to their notable structural diversity and biological activities (Yu et al. 2012). The first type III PKS system was discovered in plants in 1970 and chalcone synthase (CHS) is a typical and well-studied plant type III PKS involved in flavonoid biosynthesis. Flavonoids are major pigmentation compounds in plants. CHS is among the several type III PKSs that use 4-coumaroyl-CoA as the starter unit. CHS assisted by other synthetic enzymes such as chalcone isomerase and flavanone isomerase form a flavonoid scaffold to adapt reactions such as hydroxylation, O-methylation, glycosylation, acylation, prenylation, and conjugation resulting in the biosynthesis of more than 6000 naturally occurring flavonoids. In addition to pigmentation, flavonoids play a vital role in plant–microbe interaction, defense mechanism due to their antimicrobial property, UV protection, etc. (Austin and Noel 2003; Abe and Morita 2010).

A detailed review by Austin and Noel (2003) described the literature survey on chalcone synthase superfamily of type III PKSs from plants and bacteria. Other CHS-like type III PKSs reported in plants include benzalacetone synthase (BAS), octaketide synthase (OKS), styrylpyrone synthase (SPS), 4-coumaroyl triacetic acid synthase (CTAS), stilbene synthase (STS), acridone synthase, and valerophenone synthase (Yu et al. 2012). STS with a reaction mechanism similar to CHS is involved in the formation of resveratrol. CTAS synthesizes the same polyketide chain, but via a different reaction mechanism. OKS from *Aloe arborescens* has been shown to produce unnatural products such as C_{21}heptaketidechalcone, C_{19}hexaketidestilbene, C_{18}heptaketidephloroglucinol, and C_{16}hexaketide resorcinol. In the same organism, the enzyme pentaketide chromone synthase (PCS) produces an unnatural pentaketide 2,7-dihydroxy-5-methylchromone (Shi et al. 2009). In *Curcuma longa*, two novel type III PKSs, namely, diketide-CoA synthase (DCS) and curcumin synthase (CURS), have been reported and are involved in production of curcumin. In the same organism, two other CURSs, namely, CURS2 and CURS3, in addition to curcumin can synthesize demethoxycurcumin, bisdemethoxycurcumin, and demethoxycurcumin, respectively (Katsuyama et al. 2009). Type III PKS in *Oryza sativa* that under in vitro conditions synthesize curcuminoids are termed curcuminoid synthase (CUS). Resolved CUS structure has revealed the presence of the largest active site pocket among all reported plant type III PKSs (Katsuyama et al. 2007). A recent review by Lim et al. (2016) adds on the information about newly studied type III PKSs and possible synthetic potential in generating novel polyketides.

8.1.2 Domain Architecture in Type III PKSs

The main function of all available PKSs is to catalyze the incorporation of acetate units into a growing polyketide chain and thus possess a β-keto synthase (KS) domain and as their function is similar to fatty acid synthases(FASs), the same criteria as for

FASs is used to classify PKSs. Apart from KS domain, AT domain and ACP domain constitute the core domains in PKS enzymes. It is the arrangement and order of these domains that determine the diversity in polyketide production. Type I and II PKSs are believed to be evolved due to gene duplication or loss of function of FAS genes and exhibit structural similarity to FASs. In type I PKSs, multiple catalytic domains are organized as modules in a single protein, whereas in type II and type III PKSs, catalytic domains are observed in different protein subunits. In comparison to type I and II PKSs, type III PKSs perform multiple roles and are able to use a diverse set of starter molecules, but always maintain a simple homodimeric KS organization. The classical example of type III PKS, CHS contains a five-layer αβαβα core and each core domain comprises of three α-helices and five β-strands. Active site cavities are situated at the bottom of each αβαβα core and display a distinct bi-lobed architecture. In each monomer, a loop consisting of six residues separate the active sites between homodimers (Austin and Noel 2003). In few of the PKSs, similar to KAS III enzymes, a second series of loop forming an additional domain is observed, and the active site, in this case, is buried. Based on the substrate and synthesizing environment, dynamic mobility and conformational flexibility within the active site and tunnel determine the efficacy of PKS activity (Mizuuchi et al. 2008; Meslet-Cladière et al. 2013). A detailed review by Abe and Morita (2010) elaborated the structure and functional analyses of plant type III PKSs. Furthermore, crystallographic and site-directed mutagenesis studies enabled the structure-based and precursor-directed engineered biosynthesis of unnatural novel polyketides.

8.2 Bacterial Type III PKSs

Type III PKSs produce an array of molecules such as pyrones, chalcones, stilbenes, phloroglucinols, resorcinols, resorcinolic acids, acridones, quinones, etc. (Fig. 8.2). Traditionally assumed to be plant specific, presence of type III PKSs was described in bacteria and fungi in recent years. The first report on bacterial type III PKS emerged in 1999, and since then, several putative type III PKSs have been characterized. Bacterial type III PKSs fall into five groups primarily based on the structural aspects of the products they produce (Shimizu et al. 2017). In the actinomycetes *Streptomyces griseus*, a type III PKS, namely, RppA, was reported to be responsible for the production of 1,3,6,8-tetrahydroxynaphthalene (THN) that oxidizes spontaneously to flaviolin (Fig. 8.2) (Funa et al. 2001). Although RppA produced THN when malonyl-CoA was used as a substrate, with other C_4-C_8 aliphatic acyl-CoAs, synthesis of α-pyrones, phloroglucinols, 4-hydroxy-6-(2′,4′,6′-trioxotridecyl)-2-pyrone, and 3,6-dimethyl-4- hydroxy-2-pyrone was also produced. Chromosomal mutants for *RppA* were unable to produce THN, and structural analysis of RppA indicated the presence of conserved catalytic triad (consisting of three amino acid residues Cys138, His270, and Asn303). Besides *S. griseus*, enzymes similar to RppA have been reported in *Saccharopolysporaerythrea*, *Streptomyces coelicolor*, *Streptomyces toxytricini*, and *Sorangium cellulosum*. In *Pseudomonas fluorescens*, a type III PKS, PhlD, was shown to produce phloroglucinol product with three

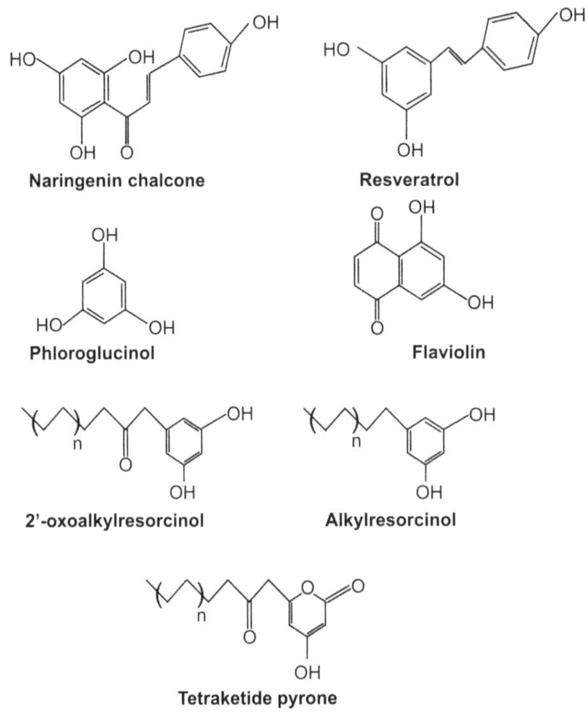

Fig. 8.2 Structural representation of aromatic type III polyketides produced by various type III PKSs

malonyl-CoAs as starter units. Interestingly this enzyme exhibited broad substrate specificity by accepting C_4–C_{12} aliphatic acyl-CoAs and phenylacetyl-CoA to produce triketide to heptaketideα-pyrones. In depth structural mutational analysis of PhlD revealed the presence of a buried tunnel extending out from the active site thereby assisting the binding of long-chain acyl-CoAs and possibly resulted in broad substrate specificity (Zha et al. 2006). Type III PKS DpgA from *Amycolatopsis orientalis* catalyzes the condensation reaction of four malonyl-CoAs to produce 3,5-dihydroxy phenylacetyl-CoA (DPA-CoA) and involved in the synthesis of 3,5-dihydroxyphenylglycine (Dpg), which is a vital component in vancomycin antibiotic family. For Dpg production in *A. orientalis*, four enzymes DpgA, DpgB, DpgC, and DpgD are required among which, DpgA is a type III PKS. All the three PKSs, RppA, PhlD, and DpgA, are capable of using malonyl-CoA as both the starter and extender units differing only in the number of individual units utilized and cyclization pattern (Wu et al. 2012). A novel type III polyketide synthase (RePKS) from *Rhizobium etli* produced a heptaketidepyrone using acetyl-CoA and six molecules of malonyl-CoA (Jeya et al. 2012a). Alkylresorcinol synthases, such as ArsB from *Azotobacter vinelandii*, *Streptomyces* resorcinol synthase A (SrsA) from *S. griseus*, and alkylpyrone synthases ArsB and ArsC from *A. vinelandii* represent the fourth group of type III microbial PKSs and are primarily involved in the biosynthesis of phenolic lipids representing alkyl resorcinols and alkylpyrones. In *A. vinelandii*, besides *arsB* and *arsC*, the *ars* operon harbors two other genes that encode two putative type I FASs, namely, ArsA and ArsD. FASs ArsA and ArsD catalyse the synthesis of long-chain fatty acids that are utilized by ArsB, which could ultimately yield alkylresorcinols (Funa et al. 2006). Similar to ArsB, SrsA from *S. griseus* yielded alkyl resorcinols by utilizing methyl malonyl-CoA and malonyl-CoA as starter molecules. The individual genes of the *srs* operon, namely, *srsA*, *srsB*, and *srsC*, when co-expressed yielded methylated products and various other alkylquinones (Funabashi et al. 2008). The final group member of the type III PKSs is represented by alkylpyrone synthases like ArsC and germicidin synthase (Gcs) from *A. vinelandii* and *S. coelicolor*, respectively. ArsC utilizes fatty acyl-CoAs as starter substrates to catalyze elongation with malonyl-CoA resulting in the yield of alkylpyrones. ArsB, a protein similar to ArsC due to its structural difference, is involved in the synthesis of alkylresorcinols. However, both the synthases eventually yield alkylpyrones. In *S. coelicolor*, germicidin biosynthesis is mediated by Gcs encoded by *sco*7221. Gcs catalyze elongation of β-ketoacyl-ACP with ethylmalonyl- or methylmalonyl-CoA followed by cyclization to yield triketide products (Chemler et al. 2012). In *Bacillus subtilis*, a type III PKS, BpsA, encoded by the *bpsA-bpsB* operon utilizes long-chain fatty acyl-coA thioesters as starter substrates and malonyl-CoA as an extender substrate to produce triketidepyrones, tetraketidepyrones, and alkylresorcinols (Nakano et al. 2009). In *Mycobacterium tuberculosis*, PKS18 is a new type III PKS that utilizes C_6–C_{20} aliphatic acyl-CoA esters as starter units to produce triketide or tetraketide by condensation reactions. Resolved crystal structure of PKS18 revealed the presence of a binding tunnel as similar to that of PhlD and known to accept broader substrates. Complete genome scanning of *S. coelicolor* has revealed the presence of several putative type III PKS genes; however, most of them are yet to be functionally characterized. PKS11 and SCO7671 from *S. coelicolor* functionally similar to PKS18 can utilize long-chain fatty acid CoA-esters either as such or even when bound to ACPs (Saxena et al. 2003). Alkylquinone biosynthesis in *Myxococcus xanthus* is confined to ftp gene cluster consisting of FtpA, FtpB, FtpC, FtpD, and FtpE. Herein, FtpA is a type III PKS that accepts acyl-ACP as starter unit to yield alkyl resorcylic acids, which subsequently by the action of FtpB and FtpE yield alkylquinones. Other bacterial type III PKSs with the capability to utilize other acyl-CoAs apart from malonyl-CoA as the extender units are Gcs (ethylmalonyl-CoA) and SrsA (methylmalonyl-CoA) (Hayashi et al. 2011). RpsA, a type III PKS from *Rhodospirillum centenum*, produced tetraketidepyranones from one C (10–14) fatty acyl-CoA unit, one malonyl-CoA unit, and two methylmalonyl-CoA units. The products were identified as 4-hydroxy-3-methyl-6-(1-methyl-2-oxoalkyl) pyran-2-one by NMR analysis. RpsA is the first bacterial type III PKS that preferred to incorporate two molecules of methylmalonyl-CoA as the extender substrate (Awakawa et al. 2013). A deep-sea-derived *Streptomyces somaliensis* SCSIO ZH66 type III PKS, vioA when expressed in heterologous marine strain, and *Streptomyces youssoufiensis* OUC681 produced methylated and nonmethylated violapyrones (Hou et al. 2018). RePKS from *Rhizobium etli* exhibited the highest catalytic efficiency toward malonyl-CoA (Jeya et al. 2012a).

8.3 Fungal Type III PKSs

Fungal type III PKSs have gained attention in recent years despite being well studied in plants and bacteria decades earlier. Fungi have contributed several natural products with biotechnological applications. Notable achievements in fungal type III characterization emerged following the genome sequencing of biotechnologically important fungi (Hashimoto et al. 2014). Search on genome sequences of *Aspergillus oryzae* and *Neurospora crassa* had revealed the presence of ORFs that could encode type III PKSs of which few have been functionally characterized already (Yu et al. 2010; Funa et al. 2007). *N. crassa* produces a functional type III PKS termed 2′-oxoalkylresorcylic acid synthase (ORAS). ORAS utilizes long-chain fatty acyl-CoAs and straight-chain saturated fatty acyl-CoAs ranging from C_4 to C_{20} as the starter units to synthesize triketide pyrones, tetraketide pyrones, tetraketide resorcylic acids, and pentaketide resorcylic acids. These resorcylic acids upon non-enzymatic decarboxylation yield resorcinols. In the industrially important fungi, *A. oryzae*, four CHS-like PKS genes (csyA, csyB, csyC, and csyD) are reported (Seshime et al. 2010a). Among these, CsyA specifically synthesizes fatty acyl-primed triketide and tetraketide pyrone compounds with hexanoyl-CoA and heptanoyl-SNAC as the starter units. Interestingly, expression of CsyA in *E. coli* has yielded polyketide products rendering it an explicit enzyme for the synthesis of pyrone compounds both in vivo and in vitro (Yu et al. 2010). CsyB from *A. oryzae M-2-3* accepts succinly-CoA and acetyl/malonyl-CoAs as starter units and produces csypyrone B1 (Seshime et al. 2010b). Type III PKS of *A. niger*, AnPKS, utilizes both short- and long-chain fatty acyl-CoA (C_2-C_{18}) and benzoyl-CoA as starter units to yield triketide and tetraketide pyrones(Li et al. 2011).

Botrytis cinerea type III PKS (BPKS) was also found to accept short-chain to long-chain acyl-CoAs (C_2–C_{18}) as starters with malonyl-CoA as the extender to generate tri- and tetraketide alkylpyrones (Jeya et al. 2012b). Moreover, BPKS is able to generate a hexaketide alkylresorcylic acid from C_{18} acyl-CoA. BPKS shares 65% identity with ORAS and similar to other type III PKSs that utilize long-chain acyl-CoAs as substrates, homology modeling studies show an acyl-binding tunnel in the active site. An-CsyA from *Aspergillus niger* NRRL328 synthesizes triketide pyrones, tetraketide pyrones, and pentaketide resorcinols from myriad acyl-CoAs as starter substrates and is a unique fungal type III PKS that can produce polyketides from malonyl-CoA as starter and extender substrates and produce tetraketide pyrones from short-chain fatty acyl-CoAs as starter substrates (Kirimura et al. 2016). Recently, two putative type III polyketide synthase genes (PKS) identified from *Sordariomycetes* fungi were characterized (Ramakrishnan et al. 2018). These two type III PKS genes from *Sordaria macrospora* (SmPKS) and *Chaetomium thermophilum* (CtPKS) shared 59.8% sequence identity. When expressed in *E. coli*, both the recombinant polyketide synthases could efficiently synthesize tri- and tetraketide pyrones, resorcinols, and resorcylic acids using various acyl-CoAs (C_4-C_{20}) as starter units.

8.4 Mining Type III PKSs

The field of genome mining for novel natural products has exploded during the past decade. Numerous literatures described the discovery of new natural products from microbes by genomics-guided approaches (Gomez-Escribano et al. 2016). Identification of newer type III PKSs could be performed either by searching DNA databases for genome sequences containing PKS genes or using catalytic domains in PKS proteins. Owing to the limitation in resolved structures of PKSs, protein-based homology search for newer PKSs is still understudied. Advancements in next-generation sequencing technology have enhanced the availability of newer whole genomes in public databases. Retrieval of PKS, putative PKS genes from DNA databases, holds much promise for the identification of new PKSs. Traditional pipeline to obtain new PKS genes is to first extract available PKS protein sequences from KEGG GENES or similar databases. Retrieved sequences should then be subjected to BLAST analysis with emphasis on elimination of PKS-like genes. Significant BLAST hits could then be studied for functional analysis to elucidate PKS activity. Conversely, bioinformatics tools for genome mining such as antiSMASH (Antibiotics and Secondary Metabolite Analysis Shell), NP.searcher, and NaPDos could be applied for direct PKS gene mining in bacterial genomes. antiSMASH follows BLAST-based comparison and as an added advantage takes into account of fragmented genomes and metagenomes and thus is regarded as a powerful tool for PKS identification. Dedicated database NORINE has a repository of PKS structures and can be utilized for protein-based queries. NRPS/PKS substrate predictor takes into account the AT domain in PKS, and thus available genomes could be scanned for AT domain initially to identify novel clusters which can further be refined for the search of other functional domains. PKSIII explorer could predict whether a query protein sequence is a type III PKS (Mallika et al. 2010). A novel type III PKS that synthesizes curcuminoids and polyphenols was identified by genome mining instead of classical genetic approach (Katsuyama et al. 2007). Mining of *S. coelicolor* genome revealed a cryptic type III PKS Gcs (SCO7221) shown to produce germicidins (Song et al. 2006). Muggia and Grube (2010) reported that BLAST search of 52 fungal genomes of *Pezizomycotina* revealed 37 putative type III CHS-like sequences in 31 species. Based on the alignment of known type III PKS genes, CODEHOP (COnsensus-DEgenerate Hybrid Oligonucleotide Primer) strategy type III PKS genes were predicted in fungal mycobionts (Muggia and Grube 2010). From 35 sequenced genomes, three putative type III PKSs from *Basidiomycetes* have been predicted by Lackner et al. (2012).

8.5 Structure-Based Engineering Approaches to Alter Substrate and Product Specificity

Crystal structures and site-directed mutagenesis studies have revealed amino acid residues that are critical for the structure and function of type III PKS enzymes. The crystal structures of representative type III PKSs are given in Fig. 8.3. The role of

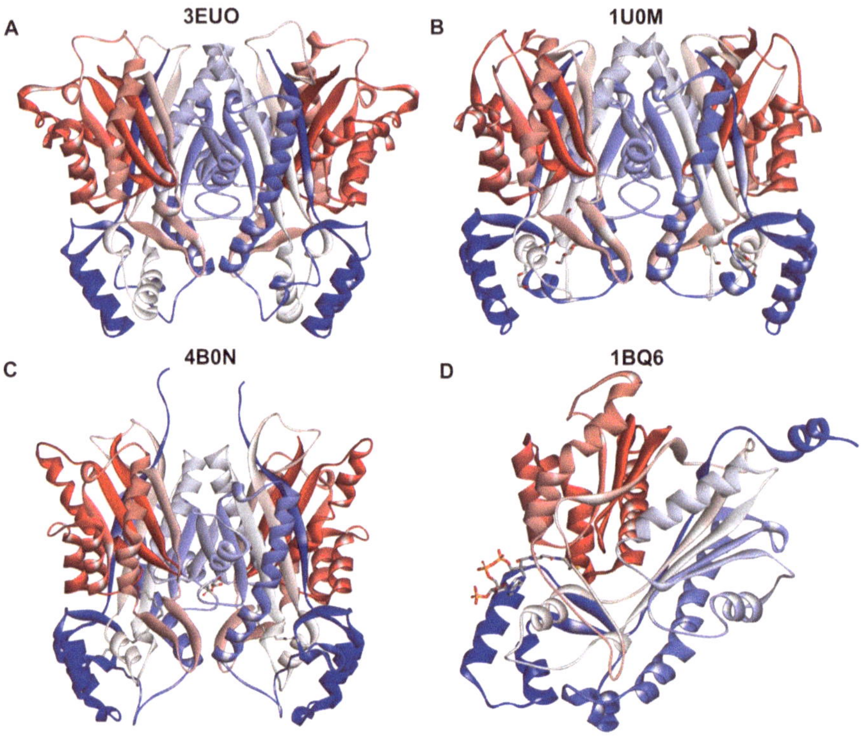

Fig. 8.3 A ribbon representation of the dimeric and monomeric crystal structures of various PKSs. (**a**) A dimeric structure of ORAS from *N. crassa* (3EUO) (Rubin-Pitel et al. 2008). (**b**) A dimeric structure of CHS from *Medicago sativa* (1UOM) (Austin et al. 2004). (**c**) A dimeric structure of PKS1 from *Ectocarpus siliculosus* (4B0N) (Meslet-Cladière et al. 2013). (**d**) A monomer structure of CHS2 from *Medicago sativa* (1BQ6) (Ferrer et al. 1999). Structures are colored with N-to-C terminal, and bound ligands are shown in stick model. Figures were generated using Discovery Studio Visualizer 4.0

the catalytic triad (Cys, His, and Asn) in the reaction mechanism of CHS has been well established based on the structural and functional studies of *Medicago sativa* CHS (Austin and Noel 2003). Catalytic cysteine (Cys164 numbered according to *M. sativa* CHS) acts as a nucleophile and attachment site for the starter-CoA. Previous studies on type III PKSs have shown that decreasing or increasing the number of CoA interactions may alter the kinetics of the association and dissociation of CoA thioesters and thus influence the fate of the type III PKS reaction intermediates (Austin and Noel 2003). The difference in product profile of type III PKSs arises from the variation in the starter substrate selection, the number of condensation steps, and in the mechanism of cyclization of reaction intermediates. The product profiles of type III PKSs from various sources are given in Table 8.1. Protein engineering studies provide insights on how the subtle variations in active site region

Table 8.1 Reported type III PKSs from various sources

PKSs	Product specificity	Source	Reference
Bacteria			
THNS	Tetrahydroxynaphthalene, pyrones, acylphloroglucinols	*Streptomyces griseus*	Funa et al. (2005)
PhlD	Phloroglucinols, pyrones	*Pseudomonas fluorescens* Pf-5	Achkar et al. (2005)
DpgA	Resorcinols	*Streptomyces orientalis*	Wu et al. (2012)
ArsB	Resorcinols	*Azotobacter vinelandii*	Miyanaga et al. (2008)
ArsC	Pyrones	*Azotobacter vinelandii*	Miyanaga et al. (2008)
MtbPKS18	Pyrones	*Mycobacterium tuberculosis*	Saxena et al. (2003)
Gcs	Germicidins, pyrones	*Streptomyces coelicolor*	Chemler et al. (2012)
Cpz6	Presulficidins	*Streptomyces* sp. MK730–62F2	Tang et al. (2013)
SCO7671	Pyrones	*Streptomyces coelicolor*	Thanapipatsiri et al. (2015)
DpyA	Dihydropyrones	*Streptomyces reveromyceticus*	Aizawa et al. (2014)
SrsA	Resorcylic acids, pyrones	*Streptomyces griseus*	Funabashi et al. (2008)
AgqA	Resorcinols	*Actinoplanes missouriensis*	Awakawa et al. (2011)
RpsA	Pyrones, resorcylic acids/resorcinols	*Rhodospirillum centenum*	Awakawa et al. (2013)
BpsA	Prones, resorcinols	*Bacillus subtilis*	Nakano et al. (2009)
Fungi			
ORAS	Resorcylic acids/resorcinols, pyrones	*Neurospora crassa*	Funa et al. (2007)
BPKS	Resorcylic acids/resorcinols, pyrones	*Botrytis cinerea*	Jeya et al. (2012a, b)
AnPKS	Resorcinols, pyrones, protocatechuic acid	*Aspergillus Niger*	Li et al. (2011)
CsyA	Dihydroxybenzoic acids, pyrones	*Aspergillus oryzae*	Seshime et al. (2010a)
CsyB	Pyrones, acetoacyl-CoA	*Aspergillus oryzae*	Seshime et al. (2010b)
SmPKS	Resorcylic acids/resorcinols, pyrones	*Sordaria macrospora*	Ramakrishnan et al. (2018)
CtPKS	Resorcylic acids/resorcinols, pyrones	*Chaetomium thermophilum*	Ramakrishnan et al. (2018)

(continued)

Table 8.1 (continued)

PKSs	Product specificity	Source	Reference
Other sources			
StlA	Resorcinols, pyrones	*Dictyostelium discoideum*	Ghosh et al. (2008)
StlB	Acylphloroglucinols	*Dictyostelium discoideum*	Austin et al. (2006)
EsiPKS1	(acyl)phloroglucinols, pyronestramenopeles	*Ectocarpus siliculosus*	Meslet-Cladière et al. (2013)
SbPKS	Pyrones	*Sargassum binderi*	Baharum et al. (2011)

could influence the product profiles of type III PKS enzymes. Type III PKSs exhibit remarkable substrate tolerance, accepting a series of nonphysiological substrate analogues to produce structurally unnatural novel polyketides (Abe 2008). Structure-based engineering of various plant type III PKSs has significantly expanded the catalytic potential of type III PKSs. Pentaketide chromone synthase (PCS) and octaketide synthase (OKS) from *Aloe arborescens* are novel plant-specific type III PKSs that produce 5,7-dihydroxy-2-methyl chromone and SEK4/SEK4b octaketides, respectively. Site-directed mutagenesis and crystallographic studies have shown that residue at 197 is responsible for chain length and product specificity. A triple mutant of PCS F80A/Y82A/M207G catalyzed the condensation of nine molecules of malonyl-CoA to produce a novel naphthopyrone. It is possible to control the starter molecule selection and the number of malonyl-CoA condensation to produce novel polyketides (Abe 2008). HsPKS1 from *Huperzia serrata* synthesize novel polyketide scaffolds (6.5.6-fused pyridoisoindole) based on precursor-directed and structure-based approaches. Further, S348G mutant altered the cyclization mechanism to produce dibenzoazepine by the condensation of 2-carbamoyl benzoyl-CoA with three malonyl-CoAs (Morita et al. 2011). Crystal structures of PKS18 and ORAS have been determined previously (Sankaranarayanan et al. 2004; Rubin-Pitel et al. 2008; Goyal et al. 2008). The composition of amino acid residues at the active site pocket and the substrate binding tunnel of representative type III PKSs, PKS18 (PDB: 1TED), from *M. tuberculosis*, CsyB (PDB: 3WXY) from *Aspergillus oryzae* (Mori et al. 2015), and PKS1 (PDB: 4B0N) from *Ectocarpus siliculosus* are shown in Fig. 8.4. Site-directed mutagenesis of *S. griseus*, RppA at Tyr position showed that Tyr is essential for selecting malonyl-CoA as starter substrate. Tyr224 and Ala305 were shown to be conserved in other RppAs from *S. antibioticus* and *S. lividans* (Funa et al. 2002). Structural study of THNS revealed a novel cavity that extends into an active site pocket, which substantiates the unusual catalytic activity of the enzyme (Austin et al. 2004). Homology modeling of PhlD from *Pseudomonas fluorescens* showed the presence of a buried tunnel extending out from the active site cavity to accommodate the binding of long-chain acyl-CoAs. Further, saturation mutagenesis altered the substrate specificity of PhlD (Zha et al. 2006). A key residue essential for the cyclization specificity of ArsB and ArsC from *A. vinelandii* was identified by site-directed mutagenesis study. ArsC mutant G284 W synthesized

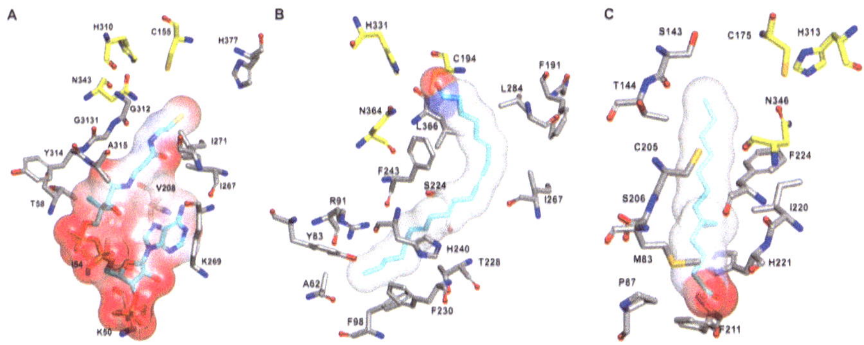

Fig. 8.4 Composition of amino acid residues at active site and substrate binding tunnel of various PKSs. (**a**) Active site of *Aspergillus oryzae* CsyB (3WXY) complexed with CoASH, (**b**) active site of brown algae PKS1 (4B0N) complexed with arachidonic acid (Meslet-Cladière et al. 2013), and (**c**) active site of *Mycobacterium* PKS (1TED) complexed with myristic acid (Sankaranarayanan et al. 2004). Active site and substrate binding tunnel residues are shown in stick model where active site residues are colored with yellow carbon, whereas tunnel residues are represented with element color. Bound ligands are shown in stick model and are shown in cyan color. Solvent solid surface around bound substrates is shown with a probe radius of 1.4 Å and colored with atom charge. Figures were generated using Discovery Studio Visualizer 4.0

alkyl resorcinol as a major product with alkyl pyrone as a minor product. Crystal structure analyses of G284 W revealed that the bulky side chain of tryptophan protrudes toward the space of active site cavity and causes a reduction in cavity volume leading to change in cyclization specificity from lactonization to aldol condensation (Satou et al. 2013). Mutagenic studies and crystal structure analyses showed insights into critical residues that are crucial for chain length specificity of PKS18 (Sankaranarayanan et al. 2004). A unique rearrangement of structural elements near the active site of the ORAS from *N. crassa* was shown by the crystal structure of wild-type and a mutant F252G. These changes reorient the critical residue essential for controlling cyclization and chain length. Similar to bacterial type III PKSs, ORAS possesses another active site thiol that may play a role in steering the polyketide reactivity (Rubin-Pitel et al. 2008). Further, residues that govern the cyclization specificity of PKSIIINc from *N. crassa* were determined by Parvez et al. (2018). The mutant C120S gained new cyclization specificity and produced acyl phloroglucinols in addition to wild-type products. Thus the catalytic potential and structurally simple nature of PKSs make them an excellent choice for further development of novel unnatural biocatalysts with novel catalytic functions.

8.6 Biotechnological Potential of Type III PKSs

Polyketides and polyketide derivatives play a novel role in the development of new therapeutics. Despite chemical synthesis offers flexibility for the synthesis of polyketide analogues, enzyme-based production offers a sustainable solution with better structural and chemical stability of the products. The pharmaceutical properties of turmeric are based on curcuminoids (curcumin, demethoxycurcumin, and bisdemethoxycurcumin) that are synthesized by type III PKS of *Curcuma longa* (Sharma et al. 2005). RppA of *S. griseus* is involved in the production of α-pyrones and phloroglucinols with anti-inflammatory and anti-HIV-1 properties. PhlD of *P. fluorescenes* synthesize phloroglucinol a vital precursor of antifungal drug 2,4-diacetyl phloroglucinol. DpgA of *Amycoltopsis mediterranei* yields 3,5-dihydroxyphenylacetic acid which is a precursor of (S)-3,5-dihydroxyphenylglycine involved in the biosynthesis of balhimycin. PKS11 and PKS18 of *M. tuberculosis* result in the production of tetraketide/triketide α-pyrones with clinical importance on drug development and unnatural polyketide synthesis. Gcs of *S. coelicolor* is involved in the production of germicidins, which are efficient inhibitors of spore germination. ArsB and ArsC of *A. vinelandii* catalyze the synthesis of alkylresorcinols with enzyme inhibitory and membrane insertion potential. RpsA of *R. centenum* yields alkyldimethyl pyrones and alkyldimethyl resorcylic acids capable of exhibiting diverse biological activities including anti-inflammatory and immune system modulatory effects.

8.7 Summary

Type III PKSs are the distinct group of PKSs compared to type I and II PKSs and have become the fascinating group of enzymes that possess a great potential to produce novel natural products. Recently, several new type III PKSs have been characterized from fungi. X-ray crystallographic studies and site-directed mutagenesis of these enzymes have revealed insights into the cyclization mechanism that altered the specificity of PKSs. Further analyses of these characterized/yet to be characterized type III PKSs would reveal more novel natural products.

References

Abe I (2008) Engineering of plant polyketide biosynthesis. Chem Pharm Bull 56(11):1505–1514. https://doi.org/10.5059/yukigoseikyokaishi.66.683

Abe I, Morita H (2010) Structure and function of the chalcone synthase superfamily of plant type III polyketide synthases. Nat Prod Rep 27(6):809–838. https://doi.org/10.1039/b909988n

Achkar J, Xian M, Zhao H, Frost JW (2005) Biosynthesis of phloroglucinol. J Am Chem Soc 127(15):5332–5333. https://doi.org/10.1021/ja042340g

Aizawa T, Kim SY, Takahashi S et al (2014) Alkyldihydropyrones, new polyketides synthesized by a type III polyketide synthase from *Streptomyces reveromyceticus*. J Antibiot (Tokyo) 67:819–823. https://doi.org/10.1038/ja.2014.80

Austin MB, Noel JP (2003) The chalcone synthase superfamily of type III polyketide synthases. Nat Prod Rep 20:79–110. https://doi.org/10.1039/b100917f

Austin MB, Izumikawa M, Bowman ME et al (2004) Crystal structure of a bacterial type III polyketide synthase and enzymatic control of reactive polyketide intermediates. J Biol Chem 279:45162–45174. https://doi.org/10.1074/jbc.M406567200

Austin MB, Saito T, Bowman ME et al (2006) Biosynthesis of *Dictyostelium discoideum* differentiation-inducing factor by a hybrid type I fatty acid-type III polyketide synthase. Nat Chem Biol 2(9):494–502. https://doi.org/10.1038/nchembio811

Awakawa T, Fujita N, Hayakawa M et al (2011) Characterization of the biosynthesis gene cluster for alkyl-O-dihydrogeranyl-methoxyhydroquinones in *Actinoplanes missouriensis*. Chembiochem 12(3):439–448. https://doi.org/10.1002/cbic.201000628

Awakawa T, Sugai Y, Otsutomo K et al (2013) 4-Hydroxy-3-methyl-6-(1-methyl-2-oxoalkyl) pyran-2-one synthesis by a type III polyketide synthase from *Rhodospirillumcentenum*. Chem Bio Chem 14(8):1006–1013. https://doi.org/10.1002/cbic.201300066

Baharum H, Morita H, Tomitsuka A et al (2011) Molecular cloning, modeling, and site-directed mutagenesis of type III polyketide synthase from *Sargassum binderi* (Phaeophyta). Mar Biotechnol 13:845–856. https://doi.org/10.1007/s10126-010-9344-5

Chemler JA, Buchholz TJ, Geders TW et al (2012) Biochemical and structural characterization of germicidin synthase: analysis of a type III polyketide synthase that employs acyl-ACP as a starter unit donor. J Am Chem Soc 134:7359–7366. https://doi.org/10.1021/ja2112228

Ferrer JL, Jez JM, Bowman ME et al (1999) Structure of chalcone synthase and the molecular basis of plant polyketide biosynthesis. Nat Struct Biol 6:775–784. https://doi.org/10.1038/11553

Frankel BA, McCafferty DG (2004) Profiling natural product biosynthesis. Chem Biol 11:290–291. https://doi.org/10.1016/j.chembiol.2004.03.007

Funa N, Funabashi M, Ohnishi Y, Horinouchi S (2005) Biosynthesis of hexahydroxyperylenequinone melanin via oxidative aryl coupling by cytochrome P-450 in *Streptomyces griseus*. J Bacteriol 187:8149–8155. https://doi.org/10.1128/JB.187.23.8149-8155.2005

Funa N, Ohnishi Y, Ebizuka Y et al (2001) Properties and substrate specificity of RppA , a chalcone synthase- related polyketide synthase in *Streptomyces griseus*. Biochemistry. https://doi.org/10.1074/jbc.M110357200

Funa N, Ohnishi Y, Ebizuka Y, Horinouchi S (2002) Alteration of reaction and substrate specificity of a bacterial type III polyketide synthase by site-directed mutagenesis. Biochem J 367:781–789. https://doi.org/10.1042/bj20020953

Funa N, Ozawa H, Hirata A et al (2006) Phenolic lipid synthesis by type III polyketide synthases is essential for cyst formation in *Azotobactervinelandii*. Proc Natl Acad Sci U S A 103:6356–6361. https://doi.org/10.1073/pnas.0511227103

Funa N, Awakawa T, Horinouchi S (2007) Pentaketideresorcylic acid synthesis by type III polyketide synthase from Neurosporacrassa. J Biol Chem 282:14476–14481. https://doi.org/10.1074/jbc.M701239200

Funabashi M, Funa N, Horinouchi S (2008) Phenolic lipids synthesized by type III polyketide synthase confer penicillin resistance on *Streptomyces griseus*. J Biol Chem 283(20):13983–13991. https://doi.org/10.1074/jbc.M710461200

Ghosh R, Chhabra A, Phatale PA et al (2008) Dissecting the functional role of polyketide synthases in *Dictyostelium discoideum*: biosynthesis of the differentiation regulating factor 4-methyl-5-pentylbenzene-1,3-diol. J Biol Chem 283(17):11348–11354. https://doi.org/10.1074/jbc.M709588200

Gomez-Escribano JP, Alt S, Bibb MJ (2016) Next generation sequencing of actinobacteria for the discovery of novel natural products. Mar Drugs 14:6–8. https://doi.org/10.3390/md14040078

Goyal A, Saxena P, Rahman A et al (2008) Structural insights into biosynthesis of resorcinolic lipids by a type III polyketide synthase in *Neurosporacrassa*. J Struct Biol 162(3):411–421. https://doi.org/10.1016/j.jsb.2008.02.009

Hashimoto M, Nonaka T, Fujii I (2014) Fungal type III polyketide synthases. Nat Prod Rep 31:1306–1317. https://doi.org/10.1039/c4np00096j

Hayashi T, Kitamura Y, Funa N et al (2011) Fatty acyl-AMP ligase involvement in the production of Alkylresorcylic acid by a *Myxococcusxanthus* type III polyketide synthase. Chembiochem 12:2166–2176. https://doi.org/10.1002/cbic.201100344

Hertweck C, Luzhetskyy A, Bechthold A (2007) Type II polyketide synthases: gaining a deeper insight into enzymatic teamwork. Nat Prod Rep 24:162–190. https://doi.org/10.1039/b507395m

Hou L, Huiming H, Li H, Wang S et al (2018) Overexpression of a type III PKS gene affording affording novel violapyrones with enhanced anti-influenza A virus activity. Microb Cell Factories 17:61. https://doi.org/10.1186/s12934-018-0908-9

Jeya M, Kim TS, Kumar Tiwari M et al (2012a) A type III polyketide synthase from *Rhizobium etli* condenses malonylCoAs to a heptaketidepyrone with unusually high catalytic efficiency. Mol Biosyst 8(12):3103–3106. https://doi.org/10.1039/c2mb25347j

Jeya M, Kim TS, Tiwari MK et al (2012b) The *Botrytis cinerea* type III polyketide synthase shows unprecedented high catalytic efficiency toward long chain acyl-CoAs. Mol Biosyst 8(11):2864–2867. https://doi.org/10.1039/c2mb25282a

Katsuyama Y, Matsuzawa M, Funa N et al (2007) In vitro synthesis of curcuminoids by type III polyketide synthase from *Oryza sativa*. J Biol Chem 282:37702–37709. https://doi.org/10.1074/jbc.M707569200

Katsuyama Y, Kita T, Funa N et al (2009) Curcuminoid biosynthesis by two type III polyketide synthases in the herb *Curcuma longa*. J Biol Chem 284(17):11160–11170. https://doi.org/10.1074/jbc.M900070200

Kirimura K, Watanabe S, Kobayashi K (2016) Heterologous gene expression and functional analysis of a type III polyketide synthase from *Aspergillusniger* NRRL 328. Biochem Biophys Res Commun 473(4):1106–1110. https://doi.org/10.1016/j.bbrc.2016.04.023

Lackner G, Misiek M, Braesel J, Hoffmeister D (2012) Genome mining reveals the evolutionary origin and biosynthetic potential of basidiomycetepolyketide synthases. Fungal Genet Biol. https://doi.org/10.1016/j.fgb.2012.09.009

Li J, Luo Y, Lee J-K, Zhao H (2011) Cloning and characterization of a type III polyketide synthase from *Aspergillusniger*. Bioorg Med Chem Lett 21(20):6085–6089. https://doi.org/10.1016/j.bmcl.2011.08.058

Lim YP, Go MK, Yew WS (2016) Exploiting the biosynthetic potential of type III polyketide synthases. Molecules 21:1–37. https://doi.org/10.3390/molecules21060806

Mallika V, Sivakumar KC, Jaichand S, Soniya EV (2010) Kernel based machine learning algorithm for the efficient prediction of type III polyketide synthase family of proteins. J Integr Bioinform 7(1):143. https://doi.org/10.2390/biecoll-jib-2010-143

Meslet-Cladière L, Delage L, Leroux CJ-J et al (2013) Structure/function analysis of a type III polyketide synthase in the brown alga *Ectocarpus siliculosus* reveals a biochemical pathway in phlorotannin monomer biosynthesis. Plant Cell 25:3089–3103. https://doi.org/10.1105/tpc.113.111336

Miyanaga A, Funa N, Awakawa T, Horinouchi S (2008) Direct transfer of starter substrates from type I fatty acid synthase to type III polyketide synthases in phenolic lipid synthesis. Proc Natl Acad Sci 105(3):871–876. https://doi.org/10.1073/pnas.0709819105

Mizuuchi Y, Shimokawa Y, Wanibuchi K et al (2008) Structure function analysis of novel type III polyketide synthases from *Arabidopsis thaliana*. Biol Pharm Bull 31:2205–2210. https://doi.org/10.1248/bpb.31.2205

Mori T, Yang D, Matsui T et al (2015) Structural basis for the formation of acylalkylpyrones from two beta-ketoacyl units by the fungal type III polyketide synthase CsyB. J Biol Chem 290:5214–5225. https://doi.org/10.1074/jbc.M114.626416

Morita H, Yamashita M, Shi S-P et al (2011) Synthesis of unnatural alkaloid scaffolds by exploiting plant polyketide synthase. Proc Natl Acad Sci 108(33):13504–13509. https://doi.org/10.1073/pnas.1107782108

Muggia L, Grube M (2010) Erratum type III polyketide synthases in lichen mycobionts. Fungal Biol 114(4):379–385. https://doi.org/10.1016/j.funbio.2010.03.001

Nakano C, Ozawa H, Akanuma G et al (2009) Biosynthesis of aliphatic polyketides by type III polyketide synthase and methyltransferase in *Bacillus subtilis*. J Bacteriol 191:4916–4923. https://doi.org/10.1128/JB.00407-09

Parvez A, Giri S, Bisht R, Saxena P (2018) New insights on cyclization specificity of fungal type III polyketide synthase, PKSIIINc in *Neurosporacrassa*. Indian J Microbiol 58:268–277. https://doi.org/10.1007/s12088-018-0738-9

Pfeifer A, Chaitan K (2001) Biosynthesis of polyketides in heterologous hosts. Microbiol Mol Biol Rev 65:106–118. https://doi.org/10.1128/MMBR.65.1.106-118.2001

Ramakrishnan D, Tiwari MK, Manoharan G et al (2018) Molecular characterization of two alkylresorcylic acid synthases from Sordariomycetes fungi. Enzyme Microb Technol 115:16–22. https://doi.org/10.1016/j.enzmictec.2018.04.006

Rimando AM, Baerson R (eds) (2007) Polyketides: biosynthesis, biological activity, and genetic engineering. American Chemical Society, Washington, DC

Rubin-Pitel SB, Zhang H, Vu T et al (2008) Distinct structural elements dictate the specificity of the type III pentaketide synthase from *Neurosporacrassa*. Chem Biol 15(10):1079–1090. https://doi.org/10.1016/j.chembiol.2008.08.011

Sankaranarayanan R, Saxena P, Marathe UB et al (2004) A novel tunnel in mycobacterial type III polyketide synthase reveals the structural basis for generating diverse metabolites. Nat Struct Mol Biol 11:894–900. https://doi.org/10.1038/nsmb809

Satou R, Miyanaga A, Ozawa H et al (2013) Structural basis for cyclization specificity of two *Azotobacter* type III polyketide synthases. J Biol Chem 288:34146–34157. https://doi.org/10.1074/jbc.M113.487272

Saxena P, Yadav G, Mohanty D et al (2003) A new family of type III polyketide synthases in *Mycobacterium tuberculosis*. J Biol Chem 278(45):44780–44790. https://doi.org/10.1074/jbc.M306714200

Seshime Y, Juvvadi PR, Kitamoto K et al (2010a) *Aspergillusoryzae* type III polyketide synthase CsyA is involved in the biosynthesis of 3,5-dihydroxybenzoic acid. Bioorg Med Chem Lett 20(16):4785–4788. https://doi.org/10.1016/j.bmcl.2010.06.119

Seshime Y, Juvvadi PR, Kitamoto K et al (2010b) Identification of csypyrone B1 as the novel product of *Aspergillusoryzae* type III polyketide synthase CsyB. Bioorg Med Chem 18(12):4542–4546. https://doi.org/10.1016/j.bmc.2010.04.058

Sharma RA, Gescher AJ, Steward WP (2005) Curcumin: the story so far. Eur J Cancer 41(13):1955–1968

Shen B (2003) Polyketide biosynthesis beyond the type I, II and III polyketide synthase paradigms. Curr Opin Chem Biol 7:285–295. https://doi.org/10.1016/S1367-5931(03)00020-6

Shi SP, Wanibuchi K, Morita H et al (2009) Enzymatic formation of unnatural novel chalcone, stilbene, and benzophenone scaffolds by plant type III polyketide synthase. Org Lett 11:551–554. https://doi.org/10.1021/ol802606w

Shimizu Y, Ogata H, Goto S (2017) Type III polyketide synthases: functional classification and phylogenomics. Chembiochem 18:1048–1049. https://doi.org/10.1002/cbic.201700230

Song L, Barona-Gomez F, Corre C et al (2006) Type III polyketide synthase β-ketoacyl-ACP starter unit and ethylmalonyl-CoA extender unit selectivity discovered by *Streptomyces coelicolor* genome mining. J Am Chem Soc 128(46):14754–14755. https://doi.org/10.1021/ja065247w

Taguchi T, Itou K, Ebizuka Y et al (2000) Chemical characterisation of disruptants of the *Streptomyces coelicolor*A3(2) actVI genes involved in actinorhodin biosynthesis. J Antibiot (Tokyo) 53:144–152. https://doi.org/10.7164/antibiotics.53.144

Tang X, Eitel K, Kaysser L et al (2013) A two-step sulfation in antibiotic biosynthesis requires a type III polyketide synthase. Nat Chem Biol 9(10):610–615. https://doi.org/10.1038/nchembio.1310

Thanapipatsiri A, Claesen J, Gomez-Escribano J-P et al (2015) A *Streptomyces coelicolor* host for the heterologous expression of type III polyketide synthase genes. Microb Cell Factories 14:145. https://doi.org/10.1186/s12934-015-0335-0

Wu HC, Li YS, Liu YC et al (2012) Chain elongation and cyclization in type III PKS DpgA. Chem Bio Chem 3(6):862–871. https://doi.org/10.1002/cbic.201200051

Yu D, Zeng J, Chen D, Zhan J (2010) Characterization and reconstitution of a new fungal type III polyketide synthase from Aspergillusoryzae. Enzyme Microb Technol 46:575–580. https://doi.org/10.1016/j.enzmictec.2010.02.011

Yu D, Xu F, Zeng J, Zhan J (2012) Type III polyketide synthases in natural product biosynthesis. IUBMB Life 64:285–295. https://doi.org/10.1002/iub.1005

Zha W, Rubin-Pitel SB, Zhao H (2006) Characterization of the substrate specificity of PhlD, a type III polyketide synthase from *Pseudomonas fluorescens*. J Biol Chem 281(42):32036–32047. https://doi.org/10.1074/jbc.M606500200
Ziemert N, Alanjary M, Weber T (2016) The evolution of genome mining in microbes – a review. Nat Prod Rep 33:988–1005. https://doi.org/10.1039/C6NP00025H

Production of a Variety of Industrially Significant Products by Biological Sources Through Fermentation

9

Manoj Kumar Enamala, Divya Sruthi Pasumarthy,
Pavan Kumar Gandrapu, Murthy Chavali,
Harika Mudumbai, and Chandrasekhar Kuppam

Abstract

Fermentation is an attractive field in the area of bioprocess engineering where it represents a promising and environmentally friendly option to find an alternative for the well-established chemical process which is being used currently. With the various fermentation techniques available, various microorganisms are utilized to an extent in producing a wide range of products which range from the production of biofuels from various sources of algae, bacteria, yeasts, etc. Apart from the various renewable sources, microorganisms are even capable to synthesize a wide range of fermented food products from various biological sources. In this chapter, we have discussed various advantages of fermented products as well as explained the various renewable energy sources which are the most needed for the survival of mankind.

M. K. Enamala
Bioserve Biotechnologies (India) Private Limited, Hyderabad, Telangana, India

D. S. Pasumarthy
ITC Spices (ITC Agri Business Division), Guntur, Andhra Pradesh, India

P. K. Gandrapu
CP Aqua Culture Indian Pvt Ltd, Tangutur Mandal, Prakasam District, Andhra Pradesh, India

M. Chavali
MCETRC, Tenali, Guntur 522201, Andhra Pradesh, India

Shree Velagapudi Ramakrishna Memorial College (SVRMC; Autonomous), Guntur, Andhra Pradesh, India

H. Mudumbai
Montessori Mahila Kalasala, Vijayawada, Andhra Pradesh, India

C. Kuppam (✉)
Green Processing, Bioremediation and Alternative Energies (GPBAE) Research Group, Faculty of Environment and Labour Safety, Ton Duc Thang University, Ho Chi Minh City, Vietnam

P. K. Arora (ed.), *Microbial Technology for the Welfare of Society*, Microorganisms for Sustainability 17,
https://doi.org/10.1007/978-981-13-8844-6_9

Keywords
Syngas · Electrofermentation · Bacteria · Algae · Biohydrogen

9.1 Introduction

Fermentation is an anaerobic pathway which is used by many microorganisms for breaking down of glucose into its respective components (Reddy et al. 2011a; Mohan et al. 2013; Chandrasekhar et al. 2015b; Abbasiliasi et al. 2017). Fermentation can also be defined as a process of growing a culture in a nutrient medium and converting that into desired feed or product of our choice. Fermentation can also be defined as a large-scale growth of microorganisms under aerobic or anaerobic conditions in a large container stated as fermenter or bioreactor carried out under optimized growth parameters like temperature, pH, and optimized nutrient medium (Stanbury et al. 1995; Kumar et al. 2017). The term fermentation is a derivative from the Latin term "fervere" meaning to boil. In a biochemical sense, a process by which microbes produce energy from organic compounds is referred to as fermentation. The fermentation process includes a wide range of aspects like the conversion of carbohydrates into alcohols or acids, conversion of sugar to alcohol using yeast, used for preservation of foods like yogurt, kimchi, sour foods, etc. (Tamang et al. 2015; Singh et al. 2017). The major advantages in the fermented foods are that it improves the shelf life of the food, reduces toxicity, and imparts desirable flavor to the food. In the fermentation processes, the contribution of a variety of enzymes and organic catalysts is generated by a variety of microorganisms and thus produces chemical transformations (Fig. 9.1). In recent years, ultrasound technology has been employed for monitoring fermentation process; these parameters depend upon the various intrinsic as well as extrinsic factors. In today's industry generally, 20 or 40 kHz is sufficient for the microorganisms to develop acoustic cavitation as a standard technique for the microorganisms to release the

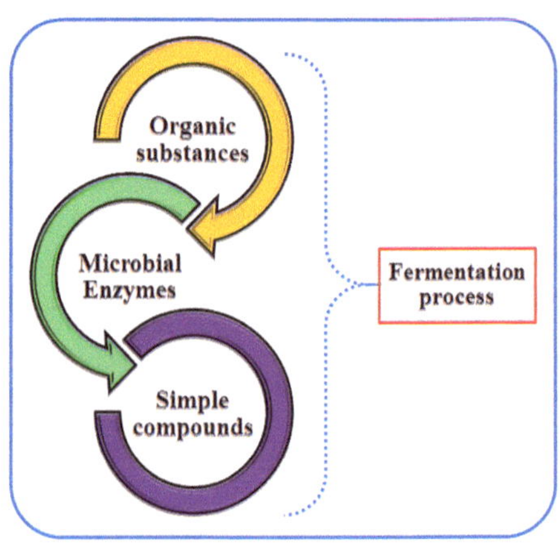

Fig. 9.1 Schematic representation of fermentation process

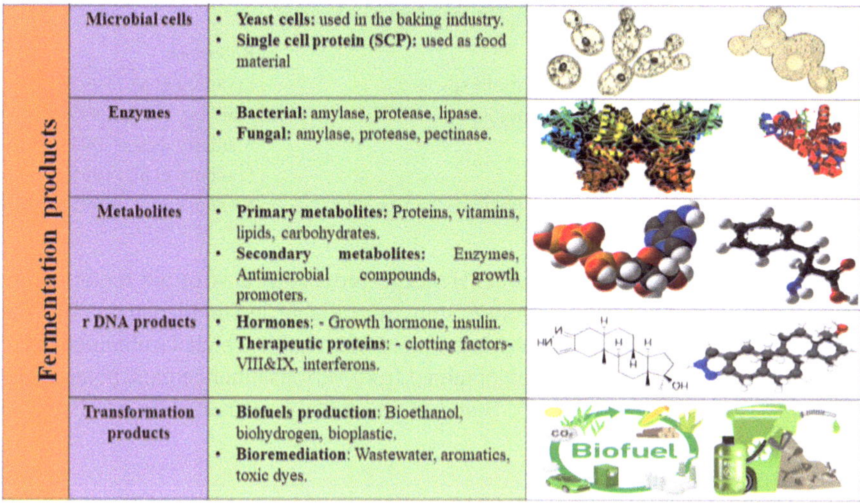

Fig. 9.2 Various commercially fermentation products utilized by the numerous industries

products from their cell walls (Ojha et al. 2017). Various commercial fermentation products utilized by the numerous industries are represented in Fig. 9.2.

Bioenergy is derived from various sources like bacterial, fungal, yeasts, plant sources, etc. Among them algae have gained a prominent interest in the production of biofuels. This green fuel is the world's cleanest form of fuel which can emit low pollutants and cause less damage to the environment (Enamala et al. 2018).

9.2 Design and Selection of Fermentation Media

9.2.1 Carbon Source

Carbon (C) is considered to be a rich energy source for various microbes which plays a vital role in the generation of primary and secondary metabolites (Mohan and Chandrasekhar 2011a; Reddy et al. 2011b; Venkata Mohan et al. 2011; Kiran Kumar et al. 2012). The rate at which the C source is utilized frequently influences the formation of intermediate metabolites as well as the biomass formation. The nature of C source also affects product formation in terms of both quality and quantity. Since the usage of methanol is not possible as a C source because of its toxicity to the cells and low flash points, therefore the cost as well as dynamics of the C source needs to be measured (Singh et al. 2017). The product of the fermentation mainly depends upon the C source and also determines the choice of C source. The main fermentation product will determine the choice of C source (Gürtekin 2014). In many of the fermentation process of single cell proteins as well as ethanol production, the vending cost of the product will be determined mainly by the

Table 9.1 List of various carbohydrate nutrients and its sources

Nutrient	Source	Reference
Sucrose	Sugarcane, sugar beet molasses	Stanbury et al. (1995)
Glucose	Corn sugar, starch, cellulose	Stanbury et al. (1995)
Lactose	Milk whey	Stanbury et al. (1995)
Fats	Vegetable oil	Stanbury et al. (1995)
Starch	Maize grains, cereals, potatoes, and cassava	Stanbury et al. (1995)

availability and cost of the C source. Hence to reduce the cost of the fermentation process one should test upon various C sources available and finalize upon one. The most common sources of carbohydrates used are starch which is obtained from maize grains; carbohydrate is even obtained from various other cereals, potato, cassava, etc. (Stanbury et al. 1995). Some of the widely used sources are listed in Table 9.1.

9.2.2 Nitrogen Source

The selection of nitrogen (N) source also plays an important role in the formation of intermediate metabolites as well as product. The microorganisms can consume both organic and inorganic sources of N. Use of certain amino acids increases the productivity and also the synthesis of secondary metabolites can be inhibited by unsuitable amino acids. Sometimes these N molecules have also shown some inhibitory effects on the growth of the microorganisms (Singh et al. 2017).

The choice of fermentation media is an important factor in getting a high yield of the product. The quality and quantity is an important factor as it also provides nutrients as well as energy for the microorganisms. The major component present in the fermentation medium should compose of C and N which are the sole components in providing energy to the microorganisms (Pandit et al. 2018a, b). The poor choice of the nutrient medium will obviously lead to poor yield of the product. In today's market there are several kinds of media available which include brain heart infusion, de Man, Rogosa, and Sharpe (MRS) media, sodium lactate (NaLa), and trypticase soy broth yeast extract (TSBYE). The use of these kinds of media becomes uneconomical because of their high price, specific preparation steps, and long incubation time. Other N sources like peptone, yeast extract, and beef extract are expensive, and at the final stage, the proteins could effect upon downstream processing techniques for extraction and purification.

9.2.2.1 Examples of Nitrogen Sources Used

The various N sources used in the fermentation media are categorized based upon the availability of N in two different forms known as an inorganic and organic form of N. The inorganic N includes ammonia gas, ammonium salts, and nitrates. The functions of this inorganic Nis to control pH whereas the organic N is used in the form of amino acid, protein, urea, or as a yeast extract in complex media. In comparison with inorganic N sources, organic N sources are most commonly used because they are enriched with amino acids which are certainly essential for the growth of microorganisms. Organic N sources promote faster cell growth than inorganic N sources. The examples for various other complex nitrogenous compounds include corn steep liquor, soya meal, peanut meal, and cotton seeds. Nitrate is converted to an ammonium ion with an enzyme known as nitrate reductase and is repressed in the presence of ammonia (Stanbury et al. 1995).

9.2.3 Role of Fermentation Media

The fermentation media contain high amounts of nutrients and broth for microorganisms and even desired optimum conditions which enhance the bacterial growth and product yield. During this process, the fermentative microorganisms will be benefitted from the metabolic process and result in excessive consumption of nutrients and lead to partial degradation of fermentation media. The amount of substrate present in the medium should not exceed the inhibitory levels since it will inhibit the production of vital products like proteins, enzymes, alcohols, etc. By the addition of an excess amount of sugar or salt to the media will inhibit the metabolism and results in the death of microorganisms. The excess amount of substrate will increase the osmotic pressure and influence enzymes present in the cells. Consequently, excreting this excess substrate in the form of reserve food material will be harmless to the cell. An ideal fermentation media should possess the following qualities (Stanbury et al. 1995):

- It should produce a maximum biomass/product.
- The media should create fewer problems when sterilized.
- The fermentation media should not react with our desired product.
- The materials for fermentation should be readily available all year round.
- It should possess the maximum rate of product formation.

9.2.4 Role of Carbon and Nitrogen Sources

In the fermentation process, both the C and N sources are equally important in the metabolism of an organism. When C and N sources exist together and the concentration of N is higher, then the microorganisms change their source to C. Glucose is considered as major C source for the majority of microorganisms, which uses this C source at a rapid pace for its growth and energy conversion because of its rapid uptake by the cells of microorganisms. The C source is readily utilized by the microorganisms at a higher N concentration and then the concentration of C becomes limited in the following stages which suggest that the yield of biomass would decrease with increase in N concentration. The N source is required in all processes including the biological growth, synthesis of proteins and nucleic acid production (Abbasiliasi et al. 2017).

9.2.5 Role of Metal Sources

Phosphorus (P) is the compound whose synthesis mainly occurs inside the DNA, RNA, and ATP in the form of sugar phosphates. This P is digested in its organic form where the phosphate ion is esterified. When the cell becomes nonfunctional, then the unreacted P is liberated through hydrolysis. In the media, various metals like magnesium (Mg), P, potassium (K), sulfur (S), chlorine (Cl), and calcium (Ca) are necessary ingredients because they require certain concentrations in the medium. Other essential metals like cobalt (Co), copper (Cu), iron (Fe), manganese (Mn), molybdenum (Mo), and zinc (Zn) are regularly present as an impurity in other main ingredients. The minor compounds need to be added purposefully (Stanbury et al. 1995).

9.3 Monitoring Fermenter and Contamination During the Fermentation Process

9.3.1 Fermentation Monitoring During Fermentation

The primary objective of industrial fermentation is to provide and establish a viable method by increasing product yields and reduce the costs of operating. The monitoring of the fermentation system has drastically changed with changes in terms of various parameters like bioprocess instrumentation, bioprocess control, and modeling. The control of these fermenters is categorized into biological and engineering point. In the biological perspective, the study is based upon the kinds of substrates which are synthesized by a variety of organisms both aerobic and anaerobic. In the engineering point of view, it depends upon the various reactor types, and operation is based upon various modes like batch fermentation, fed-batch fermentation, and continuous fermentation. When the different kinds of substrates are used, there should be continuous monitoring of the fermenter to avoid contamination (Montague et al. 1989). Contamination is defined as the migration of an unknown species of microorganism along with known microorganism and resulting in the growth of unknown species thereby affecting the growth of the known microorganisms and also affecting the quality of the product (Yang et al. 2015). Microbial fermentations are used in many of the various types of biological products like antibiotics, amino acids, pharmaceuticals, enzymes, etc. Microbial contaminations are one of the most important problems faced by the fermentation industry. To control these contaminations, effective and continuous monitoring is needed. The contaminants must be monitored by the biochemical tests and various online methods available to monitor the process. The tests done by biochemical analysis are time-consuming, and hence the monitoring of the contamination should be done during the fermentation processes itself. During this process contamination by external sources like bacterial and human errors causes the product to deteriorate, and loss of expenses is incurred for the production company. However, with contamination, all the efforts and many precautions taken to protect the fermenter will go to waste (Elmroth et al. 1990). Consequently, to avoid these problems, fermentation tanks should be monitored when fermenter is in process.

9.3.2 Online Monitoring of Fermentation

There are several monitoring instruments to monitor fermenters which are available nowadays. Today's electronic instrumentation provides a calculated, complicated, and consistent measurement of signals (electrical signals) at a very low price. In the industrial fermentations, measurements will be influenced considerably by the ease of use, robustness, and reliability. The most important sensors for control of the fermentation process are the least consistent sensors which are used in the chemical control process. The best online growth-related measurements from many available sources are based upon the carbon monoxide (CO) and oxygen (O_2) production which are measured continuously with the infrared, paramagnetic, and zirconia analyzers which lead to discrete analyses. Mass spectroscopy is becoming one of the more cost-effective which is used for the generation of data on fast control purposes. The other online measuring devices that include HPLC, GC, and GC-MS are being used in measuring the analytical measurements. These techniques give quick results if the readings are taken regularly. But if this is put into practice the samples are collected occasionally which makes the fermentation parameters to be adjusted manually. The measurements provided by sensors are not much linear and simple to the fermentation process variable of interest. The study of the measurement and the state variable typically specifies that the measured variable is an intricate function of several factors. Various new techniques for determination of biomass like infrared optical fiber light scattering detection, online fluorescence probes for NAD(P)H, may all parade good associations under appropriate calibration circumstances but are intricate functions of both bio- and physicochemical effects (Montague et al. 1989).

9.4 Fermentation Process in Microalgae and Its Evolution

Many foods which we eat today are based upon the fermentation process carried out by the microorganisms whereas we find very less fermented food items from algae. The fermented food products from these photosynthetic species need to be developed. The major polysaccharides present in the algae like fucoidan, galactan, cellulose, and hemicellulose are unsuitable substrates for fermentation process due to the fact that this algae fermentation is still in the development stage. If these polysaccharides are saccharified with the cellular enzymes then it could open up the idea of the possibility for obtaining algal fermented foods and algal fermentation can be more significant.

9.5 Production of Bioethanol Through Algae Fermentation

Microalgae and cyanobacteria are rich sources of carbohydrates. The various carbohydrates present in them are starch, glycogen, and cellulose. In the process of bioethanol production, starch and cellulose are considered as an imperative feedstock. For the production of bioethanol, certain yeasts (*Saccharomyces*) as well as bacteria (genus *Zymomonas*) are used in the mixed consortium. Cyanophycean starch is produced as an energy storage compound by cyanobacteria from glycogen which is a polymer of glucose. Mild operating conditions are needed for the fermentation process. The production of bioethanol from microalgae can be processed in two different procedures (de Farias Silva and Bertucco 2016):

- Bioethanol by dark fermentation.
- Bioethanol production by photofermentation.

9.5.1 Bioethanol by Dark Fermentation

Alteration of organic substrates into hydrogen (H_2) is referred to as dark fermentation process (Mohan and Chandrasekhar 2011b; Chandrasekhar et al. 2015a). Some of the microalgae species are *C. reinhardtii*, *Chlamydomonas moewusii*, *C. vulgaris*, *Oscillatoria limnetica*, *Oscillatoria limosa*, *Gleocapsa alpícola*, *Cyanothece* sp., *Chlorococcum littorale*, *and Spirulina* sp. *Synechococcus* sp. is skilled of expelling bioethanol by the intracellular process through the cell walls in the absence of light. The microalgae are able to synthesize ethanol forcefully through the process of the metabolic fermentative process when stored during the dark phase and are favored by carbohydrate accumulation in the cells through photosynthesis. Alternate routes for the production of bioethanol should be rediscovered as this process has some disadvantages.

9.5.2 Bioethanol Production by Photofermentation

Photofermentative process is a usual mechanism of altering sunlight (solar energy) into fermentation products through the metabolic pathways. Many cyanobacterial processes are genetically modified according to the conditions and then tested as a fermentative organism. Among these cyanobacterial strains *Synechocystis* sp. PCC 6803, *Synechococcus elongatus* sp. PCC 7992, *Synechococcus* sp. PCC 7002, and *Anabaena* sp. PCC 7120 have gained high reputation in synthesizing and genetically modified and have been well characterized. Calvin cycle helps in fixation of C which forms phosphoglycerate which is then converted into pyruvate by the two key enzymes pyruvate decarboxylase and alcohol dehydrogenase. Therefore, the photofermentation process has key steps of photosynthesis and fermentation; each of these steps has key factors which control the effectiveness of this process.

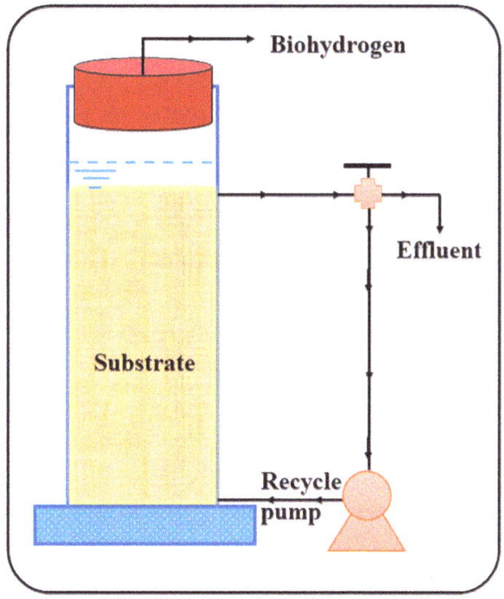

Fig. 9.3 Schematic illustration of hydrogen-producing bioreactor

9.5.3 Production of Hydrogen Through Fermentation

In today's generation, fossil fuels are depleting, and hence a search for an alternative source of energy and fuel system is underway. In this regard algae, the photosynthetic organisms, are gaining a great demand in the years to come which makes them a potential source for the generation of biofuels (Enamala et al. 2018). Production of H_2 through dark fermentation has gained prominence during the recent years. The H_2gas is one of the cleanest gases and is used as an alternative fuel to drive vehicles. The H_2 can be used in conventional gasoline engines with modifications to generate energy via combustion. The combustion of H_2 does not produce any harmful effects as this gas when burnt does not emit greenhouse gases like carbon dioxide (CO_2), Methane (CH_4), etc. Hence for production of this cleanest gas, many of the researchers have used the process known as dark fermentation where H_2 is produced in the absence of light Fig. 9.3. Moreover, the substrates used in this process are also renewable substrates like stillage, sludge, pomace, stalks, bagasse, etc. Moreover keeping in mind about the environmental concern organic waste materials are best sources for the choice of substrates when compared to the costlier substrates like sugar, starch, and other carbohydrate substances used (Wong et al. 2014).

9.5.4 Significant Advantages of Hydrogen as an Energy Source (Jianxiong and Version 2008)

- The H_2 does not cause environmental pollution as it releases water vapor (H_2O) which is a non-greenhouse gas.
- The H_2 can be easily stored as a metal hydride.

- The burning of H_2 in automobiles is 50% more efficient than gasoline.
- The H_2 batteries are considered to be the future power supply for automobiles.
- The H_2 energy is nontoxic.
- Practically H_2 is considered as a clean energy source.

9.6 Microalgae Hydrogen Production Through Fermentation

There are many ways where the biohydrogen (Bio-H_2) production can be carried out, and they are discussed below.

9.6.1 Direct Photolysis

In this process, there is a direct dissociation of H_2O into H_2 and O_2 in the presence of light. Microalgae use this light as a source for carrying out photosynthesis. The direct photolysis has been reported only in microalgae, and the process takes place by channeling of electron resultant from splitting of light-driven H_2O from the photosystem II directly to produce hydrogenase. Herein course, solar energy is directly transformed into H_2 through photosynthetic reactions. Algae split H_2O molecules into H_2 and O_2 through the process of photosynthesis (Oey et al. 2016).

9.6.1.1 Limitations of Hydrogen Production

- Efficiency for the conversion of solar energy by the photosynthetic organisms.
- Separation of H_2 and O_2 from the oxidation of H_2O.
- Cost of bioreactor construction and design.

9.6.2 Indirect Biophotolysis

This process involves the division of H_2 and O_2 into separate molecules which are mediated by CO_2 fixation. Cyanobacteria have exclusive uniqueness in performing this reaction since they possess key enzymes known as hydrogenase and nitrogenase which carry various metabolic functions in order to produce H_2. The mutant strains of *A. variabilis* have attained a production rate up to the order of 0.355 mmol H_2 per liter (Gürtekin 2014).

9.6.2.1 Advantages

- It can produce higher H_2 yields based upon the separation of H_2 and O_2 molecules (Sharma and Arya 2017).

9.6.2.2 Disadvantages

- The continuous light source is needed if the production is on a large scale.
- Adenosine triphosphate requirement should be sufficient for the requirement of nitrogenase (Sharma and Arya 2017).

9.6.3 Factors Influencing the Biohydrogen Production

Microalgae and macroalgae possess a very good amount of various sources like N, P, and microelements which are necessary for the growth of H_2-producing bacteria (HPB). The algae biomass with high concentrations of C is considered as the best sources in the dark fermentation process as the HPB can hydrolyze sugar faster than proteins. Fermentation of H_2 in the dark is a process which necessitates little energy and several types of organic wastes. The various common pathways involved in the production of H_2 include the acetate pathway and butyrate pathways (Sivagurunathan et al. 2017). In general, 1 mole of glucose gives 4 moles of H_2 through these pathways. The main source for the production of Bio-H_2 depends upon either pure or mixed culture HPB. When the sugar substrate is fermented with pure cultures, the maximum yield about 3.5–3.8 mol H_2 per mole glucose equal can be attained. In order to get optimized results, one needs to optimize various parameters like temperature and pH, and genetic engineering techniques like manipulation of genes can be applied for an increase of H_2 yield. This genetic manipulation should associate with various kinds of substrates being used for instance if the same strain has to be used against different substrates then the manipulation should be done accordingly. In the production of Bio-H_2, mixed cultures of bacteria are easier to use since they do not require sterile conditions and they are simpler to operate. The *Clostridia* and *Bacillus* are the most used bacterial cultures for degrading a wide range of organic and complex substrates. Methanogens are considered as the H_2-consuming species which can be eradicated using various heat treatment methods, pH treatments, etc. Apart from the inoculum sources selected, the H_2 yield does not require to entirely depend upon these mixed consortiums of bacteria. The operational parameters like temperature, pH, and organic loading rate effect significantly on the performance and efficiency of HPB (Sambusiti et al. 2015) (Table 9.2).

9.7 Production of a Variety of Industrially Significant Products by Biological Sources Through Fermentation

The fermentation of foods is the most common way of preserving consumable foods whereby improving the nutritional value of these foods. Fermented foods are those which are subjected to the action of microorganisms or enzymes into necessary changes in foods. In the chemical point of view, fermentation is catalyzed with enzymes in living cells. The most famous examples of fermentation are the use of lactic acid bacteria for the production of cheese and other dairy and yeasts for alcoholic fermentation. Currently, WHO has approved for introducing live microorganisms into the foods in sufficient amounts which benefits the health of the living beings. These probiotics are used as dietary supplements. The commercially available probiotic cultures belong to the genera *Bifidobacterium* and *Lactobacillus*. Member of this bacterium tends to grow well in the milk (Ouwehand and Röytiö 2014). These microorganisms are even capable of fermenting soybeans and its various products. Some of the certain bacterial species like *Bacillus subtilis*, lactic

Table 9.2 Biohydrogen potentials of various algae

Species name	Pretreatment method	Biohydrogen production potential (mL H_2 g^{-1})	References
Laminaria japonica	Thermal	58.6–107.5[a]	Sambusiti et al. (2015)
Laminaria japonica	Thermo-acidic (HCl)	9.5–163.1[a]	Sambusiti et al. (2015)
Chlorella vulgaris	Ultrasonic	31.9–37.9	Sambusiti et al. (2015)
Arthrospira platensis	Ultrasonication	55.9[a]	Sambusiti et al. (2015)
Chlamydomonas reinhardtii	Sonication	60	Sambusiti et al. (2015)
C. pyrenoidosa	Microwave heating (centrifuged and spry dried)	12.6[a]	Sambusiti et al. (2015)
Chlorella vulgaris	Acidic HCl (wet algae)	13.6–36.5	Sambusiti et al. (2015)
Chlorella vulgaris	Acidic HCl ultrasonic (wet algae)	24.2–41.6	Sambusiti et al. (2015)
Scenedesmus obliquus	Thermal (autoclave)	90.3[a]	Sambusiti et al. (2015)
Scenedesmus obliquus (wet)	Thermal (autoclave)	45.1	Sambusiti et al. (2015)
C. pyrenoidosa	Steam heating	13.1[a]	Sambusiti et al. (2015)
Lipid-extracted Scenedesmus	Thermo-alkaline	33.5–40.8[a]	Sambusiti et al. (2015)
Chlamydomonas reinhardtii	Thermal (autoclave)þ acid	1160	Sambusiti et al. (2015)
C. pyrenoidosa	Steam heating with diluted acid	97.2[a]	Sambusiti et al. (2015)
Chlamydomonas reinhardtii	Methanol	980	Sambusiti et al. (2015)
Arthrospira platensis	Ultrasonication and enzymatic	82.4[a]	Sambusiti et al. (2015)
Chlamydomonas reinhardtii	Enzymatic	1100	Sambusiti et al. (2015)
Arthrospira platensis	Bead milling	38.5[a]	Sambusiti et al. (2015)

[a]Dry alga

acid bacteria, and the fungal species like *Aspergillus* sp. and *Rhizopus* sp. are capable and are considered to be the key players in fermenting the soybean products. During the fermentation of soybeans, peptides as well as certain biopeptides are released from soybeans which results in the hydrolysis of soya bean proteins. During the fermentation process, the microbial cultures release bioactive peptides by the hydrolysis of soybean proteins, and they exhibit various properties like anti-hypertensive, antimicrobial, antioxidant, anti-diabetic, and anticancer activities (Sanjukta and Rai 2016).

9.7.1 Advancement of Industrially Important Microbial Strains by Genome Shuffling

The fast-rising population, quick depletion of fossil fuels, and ecological contamination have initiated many researchers to put in efforts in increasing various biotechnological products within the limited facility. Microorganisms are used to convert the substrates into products through several biochemical pathways or through the enzymatic pathways. Obtaining products with high yield and with natural microorganisms decreases the content of the final product thereby allowing researchers to come up with a new idea known as genome shuffling.

Genome shuffling is defined as the recombination of desired phenotypes in the parent strains by the combinatorial or recursive combination. This is an attractive and auspicious technology for quick development of the whole genome of the bacterial strains. This technology has achieved in expressing a single strain into several recombinants by bringing about the recombination in the entire genome. The major advantage of this technique is that in this only the desired characteristics of the industrial strains can be improved instead of manipulating the entire genome of an organism. The major difference which lies between the classical recombination and the currently involved techniques is that sexual and whole genome is being synthesized or evolved which does not happen in the classical shuffling system. The various technical aspects involved in this process include:

- Preparation of parental library
- Protoplast fusion
- Selection of the desired phenotype

The genome shuffling technique can be applied in various applications like utilization for better substrate yield, better product yield, and increasing in the tolerance of an organism toward particular component (temperature, pH, etc.) (Magocha et al. 2018).

Fermented foods have various functional properties in imparting health benefits in the presence of microorganisms. The functional properties include antimicrobial, antioxidant, peptide production etc. have listed out various microbes which are involved in imparting health benefits. Despite the advantages gained from fermented foods, they also contain hazardous effects on the health of human beings. The reason for this is the presence of biogenic amines in fermented foods. These biogenic amines have typical characteristics in their structure which makes them disturb the health of a living being. These biogenic amines are the nitrogenous compounds which are mainly formed by decarboxylation of amino acids, or these can be also formed by amination and transamination of aldehydes and ketones. The *Enterobacteriaceae* and *enterococci* are the main producers of biogenic amines in

Table 9.3 A few various fermented foods which are available in India

Name of the food	Name of the microorganism	Name of the substrate used	Reference
Naan	*Yeasts and lactic acid bacteria*	Wheat	Tamang et al. (2015)
Paneer	*Lactic acid bacteria*	Milk	Tamang et al. (2015)
Kinema	*Bacillus subtilis, Enterococcus, yeasts*	Soybeans	Tamang et al. (2015)
Mishti dahi	*Lactic acid bacteria*	Milk	Tamang et al. (2015)
Ghee	*Lactococcus lactis* subsp. *cremoris*	Cow milk	Tamang et al. (2015)
Lassi	*Lactobacillus acidophilus, Streptococcus thermophilus*	Cow milk	Tamang et al. (2015)
Idli	*Lb. delbrueckii, Lb. fermenti, Lb. coryniformis, Ped*	Black gram dal	Tamang et al. (2015)
Jilebi	*Sacch. bayanus, Lb. fermentum, Lb. buchneri, Lact. lactis*	Wheat flour	Tamang et al. (2015)
Dosa	*Leuc. mesenteroides, Lb. delbrueckii, Lb. fermenti, Ent. faecalis*	Black gram dal	Tamang et al. (2015)

foods. The remedy to remove such biogenic amines is to introduce lactic acid bacteria starter cultures and ferment them with cabbage. The lactic acid bacteria include *L. casei* subsp. *casei, L. plantarum, and L. curvatus.* Some of the various Indian traditional fermented foods are listed in Table 9.3. Tamang and his team (Tamang et al. 2015) have described detailed fermented foods available worldwide. In this paper, a few of them have been listed.

9.8 Integrating Approaches for the Production of Various Renewable Energy Sources

The Bio-H_2 system was integrated with fermentation and bioelectrolysis systems. The in situ extraction methods and novel tube-shaped microbial electrolysis cells (MEC) improved the overall Bio-H_2 production efficiency (Kadier et al. 2015 2016a, b, 2017, 2018a, b; Sivagurunathan et al. 2017). The by-products obtained like hydrogen, CO_2, and volatile fatty acids (VFAs) were utilized to enhance the bioelectrochemical process. The production of Bio-H_2 from wastewater has various benefits in treating waste treatment as well as producing biofuel (Chandrasekhar et al. 2017; Deval et al. 2017; Kakarla et al. 2017; Patel et al. 2017; Sivagurunathan et al. 2018; Venkata Mohan et al. 2019;). Lin and his team (2018) have suggested to use the palm connected biomass and some lignocellulosic waste are great sources for the production of biofuels. Also, these wastes should be verified in experimental scale plants before taking them to the industrial application (Lin et al. 2018).

9.8.1 Production of Syngas

Syngas comprises three different gases mainly H_2, CO, and CO_2 which can be produced from the gasification of biomass. This syngas can also be used as a byproduct in the other industries like petroleum refining, steam milling another process which involves combustion and partial oxidation. The chemical composition of syngas usually contains 30–60% CO, 25–30% H_2, 5–15% CO_2, and 0–5% CH_4. It also comprises a smaller quantity of other gases. Syngas fermentation with the combination of acetate and a mixed consortium of bacteria is also a promising field in the field of biotechnology for the production of renewable sources of energy. Liu and his team (2018) have conducted an experiment, and they have found that *Clostridium* and *Acetobacterium* genera are found well in the generation of this syngas and moreover these are enriched in initial pH of 9 which shows that CO and H_2 are converted into acetate. They could also show that pH 9 and temperature of 20 °C were suitable for the VFA (acetate) production (Liu et al. 2018). In industrial application, this syngas is produced in the fixed bed and fluidized bed gasifiers, and the gas effluent is always cleaned up before it is fed to the reactor. The future research on syngas should focus on the following terms in order to have better availability of this product for the coming generations:

- Development of novel models which enhances gas to liquid mass transfer.
- Operational parameters need to be optimized.
- Different kinds of bioreactors need to be studied in producing various different bioproducts (Asimakopoulos et al. 2018) (Table 9.4).

Table 9.4 Some of the various products synthesized available for the production of gases through fermentation

Microorganism	Substrate used	Products synthesized	Reference
Acetobacterium woodii	$CO_2 + H_2$	Acetate	Dürre and Eikmanns (2015)
Sporomusa ovata	CO_2 + electricity	Acetate, 2-oxobutyrate	Dürre and Eikmanns (2015)
Ideonella sp. *O-1*	$CO_2 + H_2 + O_2$	Poly-D-3-hydroxybutyric acid	Dürre and Eikmanns (2015)
Ralstonia eutropha	$CO_2 + H_2 + O_2$	Pentadecane/heptadecene, botryococcene	Dürre and Eikmanns (2015)
Synechocystis sp. *PCC6803*	CO_2 + light + O_2	Isobutanol, 3-methyl-1-butanol, ethanol, lactate	Dürre and Eikmanns (2015)
Synechococcus elongates	CO_2 + light + O_2	2,3-Butanediol, isobutyraldehyde	Dürre and Eikmanns (2015)
Methylotrophus capsulatus	$CH_4 + O_2$	Methanol	Dürre and Eikmanns (2015)

9.8.2 Production of Current Through Electrofermentation of Yeast and Algae

A microbial cell is defined as a device which converts the chemical energy of organic matter into the flow of electrons and thus generating current at the cathode end. Usually anaerobic bacteria are used at the anode for the oxidation of organic matter, and at the anode, the generation of electrons takes place which results in the generation of current (Chandrasekhar and Venkata Mohan 2014a, b; Chandrasekhar and Ahn 2017; Chandrasekhar 2019; Chandrasekhar et al. 2018; Pandit et al. 2019). The yeast extract added at the end acts as a mediator for electron transfer for the algae present at the cathode (Gouveia et al. 2014; Saratale et al. 2017). The Nafion and montmorillonite are the best sources of electrolytes which are used in the MFC (Ejaz et al. 2017).

9.8.3 Production of Oils Through the Fermentation Process of Yeast and Algae

Nowadays, the struggle for petroleum products is getting a huge demand because fossil fuels are getting depleted and then there will be not any further production of fuels if these are exhausted. Some trials are also being done to utilize the plant sources in the production of these oils to be used in vehicles. If done so then there would be a huge scarcity for the entire world if all the plant sources are used up in the generation of fuels. Hence then the human mankind will starve to death. Hence scientists are currently carrying out various experiments to produce fuels through biological sources. Among them, algae have gained a prominent interest in the generation of biofuel. These microalgae species can accumulate more amount of lipids; almost 80% of their dry weight (Enamala et al. 2018). Apart from algae certain oleaginous organisms like bacterial sources; yeast sources are also gaining prominence in the field of production of single cell oils.

Heterotrophic oleaginous microorganisms convert the C source into storage lipids which later can be converted into oils and which can be used for the transportation industry. The major advantage of this process is that these oleaginous microorganisms can be grown in any season and it does not depend upon the seasonal variation.

Cryptococcus curvatus yeast is one of the most prominent yeast species which is considered as the most prominent producer of oils in its category. This species is easy to cultivate in bioreactors with high growth rates and even high lipid yield is noticed for a cultivation period of 6 days. This species is able to convert various C sources into fatty acids. The *Phaeodactylum tricornutum* is considered as the most prominent species in the category of microalgae. Its major advantage is that it can grow with high growth rates and can accumulate lipids to high specific contents and can produce polyunsaturated fatty acid like eicosapentaenoic acid. Studying the process of production of oils by integrating yeast and algae has shown that when both of them are combined and fermented together for the production of oils, one would get all the potential benefits from this process, and moreover this fermentation process is considered to be an environmental free process (Dillschneider et al. 2014).

9.8.4 Electrofermentation

The fermentation process has played an important role in productions of various food metabolites, transformation, and conservation throughout the history of mankind. Fermentation technology at industrial levels has various constraints like:

- Fermentations need to rely on a particular substrate.
- Culture media are highly specific and are optimized for various microbial strains.
- pH should be controlled with the buffer solutions or by adding acid/base depending upon the nature of the solution.

Electrofermentation is a newly developed technique which combines fermentation and bioelectrochemical systems (Schievano et al. 2016; Kumar et al. 2018; Moscoviz et al. 2018). It is defined as the transformation of organic molecules by microbes which are improved and are mediated by electron transfer. Electrodes are used as substitutes for chemical donors or acceptors to stabilize the fermentation environment. This particular process is utilized toward waste biomass utilization in the generation of various biorefineries (Kumar et al. 2018). The various functions of electrofermentation include:

- Optimize fermentation metabolisms by controlling imbalance which happens due to substrate purity, pH conditions, by-product accumulation, etc.
- Increase ATP synthesis and increase the yield of microbial biomass.
- Target proteins are extracted by selected membranes.

The major advantage of this process includes the production of biochemicals with improved selectivity, increases C efficiency, and enhances microbial growth (Schievano et al. 2016). On an overall basis with the help of electrofermentation production of electricity from waste, sources seem to be an attractive option (Chandrasekhar and Venkata Mohan 2012; Kadier et al. 2015, 2016b, 2018a, b; Pandit et al. 2017). In this process, a survey of various materials and microbes is required to study on an extensive basis. Integrating this process with various other methods like biodegradation leads to a successful utilization of waste biorefineries for a suitable and sustainable economy.

9.9 Conclusion

The fermented foods have gained global attention from a long time ago, and the tradition is still followed. These fermented foods have possessed various benefits over normal foods. Fermented foods have proven to cure many diseases like chronic diseases, neurological diseases, anti-inflammatory diseases, etc. Moreover, the recent development in the genome shuffling has taken these fermented foods to a new height where one can modify the microorganisms to their choice of interest and inculcate them in the starter culture of the food of their choice. Globally the food eating style has also changed where traditional food style is being changed into a modern commercial style where we can find most of the junk foods which impart a great loss to the health of human lifestyle. The consumption of junk food is not limited to a particular country; it is applicable globally. The benefits of the fermented foods must be taught to all the common people who are unaware of the benefits.

The renewable energy sources are also being developed at a faster pace which provides clean and green fuels for the society. These fermented renewable energy sources are being researched with various microorganisms like bacteria, algae, fungi, yeasts, etc., which are making them as potential sources in the field of renewable energy as well as fermented foods. Hence in the years to come, fermentation will gain much more prominence, and many more fermented food products and renewable sources will be emerging as the global leaders which will help in improvising the lives of human beings.

References

Abbasiliasi S et al (2017) Fermentation factors influencing the production of bacteriocins by lactic acid bacteria: a review. RSC Adv 7(47):29395–29420. https://doi.org/10.1039/C6RA24579J

Asimakopoulos K, Gavala HN, Skiadas IV (2018) Reactor systems for syngas fermentation processes: a review. Chem Eng J 348(May):732–744. https://doi.org/10.1016/j.cej.2018.05.003

Chandrasekhar K (2019) Chapter 3.5 - effective and nonprecious cathode catalysts for oxygen reduction reaction in microbial fuel cells. In: Mohan SV, Varjani S, Pandey A (eds) Microbial electrochemical technology. Elsevier, pp 485–501

Chandrasekhar K, Ahn Y (2017) Effectiveness of piggery waste treatment using microbial fuel cells coupled with elutriated-phased acid fermentation. Bioresour Technol. https://doi.org/10.1016/j.biortech.2017.08.021

Chandrasekhar K, Venkata Mohan S (2012) Bio-electrochemical remediation of real field petroleum sludge as an electron donor with simultaneous power generation facilitates biotransformation of PAH: effect of substrate concentration. Bioresour Technol 110:517–525. https://doi.org/10.1016/j.biortech.2012.01.128

Chandrasekhar K, Venkata Mohan S (2014a) Bio-electrohydrolysis as a pretreatment strategy to catabolize complex food waste in closed circuitry: function of electron flux to enhance acidogenic biohydrogen production. Int J Hydrog Energy 39(22):11411–11422. https://doi.org/10.1016/j.ijhydene.2014.05.035

Chandrasekhar K, Venkata Mohan S (2014b) Induced catabolic bio-electrohydrolysis of complex food waste by regulating external resistance for enhancing acidogenic biohydrogen production. Bioresour Technol 165(C):372–382. https://doi.org/10.1016/j.biortech.2014.02.073

Chandrasekhar K, Amulya K, Mohan SV (2015a) Solid phase bio-electrofermentation of food waste to harvest value-added products associated with waste remediation. Waste Manag (New York, NY) 45:57–65. https://doi.org/10.1016/j.wasman.2015.06.001

Chandrasekhar K, Lee Y-J, Lee D-W (2015b) Biohydrogen production: strategies to improve process efficiency through microbial routes. Int J Mol Sci 16:8266–8293. https://doi.org/10.3390/ijms16048266

Chandrasekhar K, Pandit S, Kadier A, Dasagrandhi C, Velpuri J (2017) Biohydrogen production: integrated approaches to improve the process efficiency. In: Kalia VC, Kumar P (eds) Microbial applications Vol.1: bioremediation and bioenergy. Springer International Publishing, Cham, pp 189–210

Chandrasekhar K, Kadier A, Kumar G, Nastro RA, Jeevitha V (2018) Challenges in microbial fuel cell and future scope. In: Das D (ed) Microbial fuel cell: a bioelectrochemical system that converts waste to Watts. Springer International Publishing, Cham, pp 483–499

de Farias Silva CE, Bertucco A (2016) Bioethanol from microalgae and cyanobacteria: a review and technological outlook. Process Biochem. Elsevier Ltd 51(11):1833–1842. https://doi.org/10.1016/j.procbio.2016.02.016

Deval AS, Parikh HA, Kadier A, Chandrasekhar K, Bhagwat AM, Dikshit AK (2017) Sequential microbial activities mediated bioelectricity production from distillery wastewater using bio-electrochemical system with simultaneous waste remediation. Int J Hydrog Energy 42(2):1130–1141

Dillschneider R et al (2014) Combination of algae and yeast fermentation for an integrated process to produce single cell oils. Appl Microbiol Biotechnol 98(18):7793–7802. https://doi.org/10.1007/s00253-014-5867-4

Dürre P, Eikmanns BJ (2015) C1-carbon sources for chemical and fuel production by microbial gas fermentation. Curr Opin Biotechnol 35:63–72. https://doi.org/10.1016/j.copbio.2015.03.008

Ejaz H et al (2017) Electrochemical integration of algal biodiesel production and yeast based fermentation for power production. J Biotechnol Res 8(January):1944–328548

Elmroth I et al (1990) Detection of microbial contamination in fermentation processes: mass spectrometric determination of gram-negative bacteria in Leuconostoc mesenteroides cultures. Biotechnol Bioeng 35(8):787–792. https://doi.org/10.1002/bit.260350806

Enamala MK et al (2018) Production of biofuels from microalgae – a review on cultivation, harvesting, lipid extraction, and numerous applications of microalgae. Renew Sustain Energy Rev 94(May):49–68. https://doi.org/10.1016/j.rser.2018.05.012

Gouveia L et al (2014) Effect of light on the production of bioelectricity and added-value microalgae biomass in a photosynthetic alga microbial fuel cell. Bioresour Technol 154:171–177. https://doi.org/10.1016/j.biortech.2013.12.049

Gürtekin E (2014) Biological hydrogen production methods. AKADEMIK Platform (c):463–471

Jianxiong R, Version D (2008) Bio-hydrogen production by dark fermentation from organic wastes and residues. Dawei Liu

Kadier A, Simayi Y, Chandrasekhar K, Ismail M, Kalil MS (2015) Hydrogen gas production with an electroformed Ni mesh cathode catalysts in a single-chamber microbial electrolysis cell (MEC). Int J Hydrog Energy 40(41):14095–14103

Kadier A, Sahaid M et al (2016a) Recent advances and emerging challenges in microbial electrolysis cells (MECs) for microbial production of hydrogen and value-added chemicals. Renew Sustain Energy Rev 61:501–525. https://doi.org/10.1016/j.rser.2016.04.017

Kadier A, Simayi Y et al (2016b) A comprehensive review of microbial electrolysis cells (MEC) reactor designs and configurations for sustainable hydrogen gas production. Alex Eng J 55:427–443. https://doi.org/10.1016/j.aej.2015.10.008

Kadier A, Chandrasekhar K, Kalil MS (2017) Selection of the best barrier solutions for liquid displacement gas collecting metre to prevent gas solubility in microbial electrolysis cells. Int J Renewable Energy Technology 8(2):93–103

Kadier A, Jiang Y, Lai B, Rai PK, Chandrasekhar K, Mohamed A, Kalil MS (2018a) Biohydrogen production in microbial electrolysis cells from renewable resources. In: Bioenergy and biofuels. CRC Press, pp 331–356

Kadier A et al (2018b) Bioelectrochemistry Surpassing the current limitations of high purity H 2 production in microbial electrolysis cell (MECs): strategies for inhibiting growth of methanogens. Bioelectrochemistry 119:211–219. https://doi.org/10.1016/j.bioelechem.2017.09.014

Kakarla R, Kuppam C, Pandit S, Kadier A, Velpuri J (2017) Algae—the potential future fuel: challenges and prospects. In: Kalia VC, Kumar P (eds) Microbial applications Vol. 1: bioremediation and bioenergy. Springer International Publishing, Cham, pp 239–251

Kiran Kumar A et al (2012) Endocrine disruptive estrogens role in electron transfer: bio-electrochemical remediation with microbial mediated electrogenesis. Bioresour Technol 104:547–556. https://doi.org/10.1016/j.biortech.2011.10.037

Kumar G et al (2017) A comprehensive overview on light independent fermentative hydrogen production from wastewater feedstock and possible integrative options. Energy Convers Manag 141:390–402. https://doi.org/10.1016/j.enconman.2016.09.087

Kumar P et al (2018) Electro-fermentation in aid of bioenergy and biopolymers. Energies 11:343. https://doi.org/10.3390/en11020343

Lin CY et al (2018) Fermentative biohydrogen production and its byproducts: a mini review of current technology developments. Renew Sustain Energy Rev 82(November 2017):4215–4220. https://doi.org/10.1016/j.rser.2017.11.001

Liu C et al (2018) The effects of pH and temperature on the acetate production and microbial community compositions by syngas fermentation. Fuel 224(March):537–544. https://doi.org/10.1016/j.fuel.2018.03.125

Magocha TA et al (2018) Improvement of industrially important microbial strains by genome shuffling: current status and future prospects. Bioresour Technol 257(February):281–289. https://doi.org/10.1016/j.biortech.2018.02.118

Mohan SV, Chandrasekhar K (2011a) Self-induced bio-potential and graphite electron accepting conditions enhances petroleum sludge degradation in bio-electrochemical system with simultaneous power generation. Bioresour Technol 102(20):9532–9541. https://doi.org/10.1016/j.biortech.2011.07.038

Mohan SV, Chandrasekhar K (2011b) Solid phase microbial fuel cell (SMFC) for harnessing bioelectricity from composite food waste fermentation: influence of electrode assembly and buffering capacity. Bioresour Technol 102(14):7077–7085. https://doi.org/10.1016/j.biortech.2011.04.039

Mohan SV et al (2013) Biohydrogen production from wastewater. In: Biohydrogen, pp 223–257. https://doi.org/10.1016/B978-0-444-59555-3.00010-6

Montague GA, Morris AJ, Ward AC (1989) Fermentation monitoring and control: a perspective. Biotechnol Genet Eng Rev 7(1):147–188. https://doi.org/10.1080/02648725.1989.10647858

Moscoviz R, Trably E, Bernet N (2018) Electro-fermentation triggering population selection in mixed-culture glycerol fermentation. Microb Biotechnol 11(1):74–83. https://doi.org/10.1111/1751-7915.12747

Oey M et al (2016) Challenges and opportunities for hydrogen production from microalgae. Plant Biotechnol J 14(7):1487–1499. https://doi.org/10.1111/pbi.12516

Ojha KS et al (2017) Ultrasound technology for food fermentation applications. Ultrason Sonochem 34:410–417. https://doi.org/10.1016/j.ultsonch.2016.06.001

Ouwehand AC, Röytiö H (2014) Probiotic fermented foods and health promotion. In: Advances in fermented foods and beverages: improving quality, technologies and health benefits. Elsevier Ltd. https://doi.org/10.1016/B978-1-78242-015-6.00001-3

Pandit S et al (2017) Basic principles of microbial fuel cell: technical challenges and economic feasibility. In: Microbial applications, vol 1. Bioremediation and Bioenergy, pp 165–188. https://doi.org/10.1007/978-3-319-52666-9_8

Pandit S, Sarode S, Chandrasekhar K (2018a) Fundamentals of bacterial biofilm: present state of art. In: Kalia VC (ed) Quorum sensing and its biotechnological applications. Springer Singapore, Singapore, pp 43–60

Pandit S, Sarode S, Sargunaraj F, Chandrasekhar K (2018b) Bacterial-mediated biofouling: fundamentals and control techniques. In: Kalia VC (ed) Biotechnological applications of quorum sensing inhibitors. Springer Singapore, Singapore, pp 263–284

Pandit S, Chandrasekhar K, Jadha DA, Ghangrekar MM, Das D (2019) Contaminant removal and energy recovery in microbial fuel cells. in: Microbial biodegradation of xenobiotic compounds, (Ed.) Y.-C. Chang, Vol. 1, CRC Press. Published January 30, 2019 pp. 76

Patel V, Pandit S, Chandrasekhar K (2017) Basics of Methanogenesis in anaerobic digester. In: Kalia VC (ed) Microbial applications Vol. 2: biomedicine, agriculture and industry. Springer International Publishing, Cham, pp 291–314

Reddy MV, Chandrasekhar K, Mohan SV (2011a) Influence of carbohydrates and proteins concentration on fermentative hydrogen production using canteen based waste under acidophilic microenvironment. J Biotechnol 155(4):387–395. https://doi.org/10.1016/j.jbiotec.2011.07.030

Reddy MV et al (2011b) Aerobic remediation of petroleum sludge through soil supplementation: microbial community analysis. J Hazard Mater 197:80–87. https://doi.org/10.1016/j.jhazmat.2011.09.061

Sambusiti C et al (2015) Algae as promising feedstocks for fermentative biohydrogen production according to a biorefinery approach: a comprehensive review. Renew Sust Energ Rev 44:20–36. https://doi.org/10.1016/j.rser.2014.12.013

Sanjukta S, Rai AK (2016) Production of bioactive peptides during soybean fermentation and their potential health benefits. In: Trends in food science and technology, vol 50. Elsevier Ltd, pp 1–10. https://doi.org/10.1016/j.tifs.2016.01.010

Saratale RG et al (2017) Bioelectrochemical systems using microalgae – a concise research update. Chemosphere 177:35–43. https://doi.org/10.1016/j.chemosphere.2017.02.132

Schievano A et al (2016) Electro-fermentation – merging electrochemistry with fermentation in industrial applications. Trends Biotechnol 34(11):866–878. https://doi.org/10.1016/j.tibtech.2016.04.007

Sharma A, Arya SK (2017) Hydrogen from algal biomass: a review of production process. Biotechnol Rep 15(February):63–69. https://doi.org/10.1016/j.btre.2017.06.001

Singh V et al (2017) Strategies for fermentation medium optimization: an in-depth review. Front Microbiol 7(Jan). https://doi.org/10.3389/fmicb.2016.02087

Sivagurunathan P et al (2017) A comprehensive review on two-stage integrative schemes for the valorization of dark fermentative effluents. Crit Rev Biotechnol 0(0):1–15. https://doi.org/10.1080/07388551.2017.1416578

Sivagurunathan P, Kadier A, Mudhoo A, Kumar G, Chandrasekhar K, Kobayashi T, Xu K (2018) Nanomaterials for biohydrogen production. Nanomaterials: biomedical, environmental, and engineering applications. 2018, p. 217. https://doi.org/10.1002/9781119370383.ch9.

Stanbury et al (1995) Media for industrial fermentation. Principles Ferment Technol:93–122. https://doi.org/10.1016/B978-0-08-099953-1/00004-1

Tamang J et al (2015) Microorganisms in fermented foods and beverages. In: Health benefits of fermented foods and beverages. CRC Press, Boca Raton, pp 1–110. https://doi.org/10.1201/b18279-2

Venkata Mohan SR et al (2011) Bioremediation of petroleum sludge under anaerobic microenvironment: influence of biostimulation and bioaugmentation. Environ Eng Manag J 10(11):1609–1616

Venkata Mohan S, Chiranjeevi P, Chandrasekhar K, Babu PS, Sarkar O (2019) Chapter 11 - Acidogenic biohydrogen production from wastewater. In: Pandey A, Mohan SV, Chang J-S, Hallenbeck PC, Larroche C (eds) Biohydrogen, Second edn. Elsevier, pp 279–320

Wong YM, Wu TY, Juan JC (2014) A review of sustainable hydrogen production using seed sludge via dark fermentation. Renew Sustain Energy Rev 34:471–482. https://doi.org/10.1016/j.rser.2014.03.008

Yang J, Chen X, Jin H (2015) Online prediction for contamination of chlortetracycline fermentation based on Dezert-Smarandache theory. Chin J Chem Eng 23(6):1009–1016. https://doi.org/10.1016/j.cjche.2014.06.043

Molecular Diagnosis of Acute and Chronic Brucellosis in Humans

10

Maryam Dadar, Youcef Shahali, and Gamal Wareth

Abstract

Human brucellosis is a zoonosis distributed in many countries around the world. It represents an important public health problem in areas where *Brucella* infection is primarily enzootic in cattle, sheep, goat, and swine populations. More than 500,000 new human cases are reported annually worldwide, whereas the number of undetected cases is believed to be considerably higher. Both microbiological and serological methods are commonly used for diagnosis and characterization of *Brucella* infections. However, DNA-based assays such as conventional PCR, real-time PCR, MLST, as well as MLVA have provided better sensitivity when compared to bacteriological tests and much more specificity than the traditional serological methods. In addition, PCR assays allow the fast detection of *Brucella* infections and detection of relapse and facilitate the post-treatment follow-up of the patients. The improvement and optimization of molecular techniques along with their accessibility to most clinical laboratories allow the rapid diagnosis of the disease and better control of occupational risks for laboratory personnel while handling live *Brucella* spp. Therefore, PCR-based methods represent applicable and very efficient methods for the diagnosis of acute brucellosis at the earlier time of disease,

The original version of this chapter was revised: The chapter was originally published with an additional affiliation for the authors "Maryam Dadar and Youcef Shahali" which has been removed now. The correction to this chapter is available at https://doi.org/10.1007/978-981-13-8844-6_16

M. Dadar (✉) · Y. Shahali
Razi Vaccine and Serum Research Institute (RVSRI), Agricultural Research, Education and Extension Organization (AREEO), Karaj, Iran

G. Wareth
Friedrich-Loeffler-Institut, Institute of Bacterial Infections and Zoonoses, Jena, Germany

Faculty of Veterinary Medicine, Benha University, Toukh, Egypt

P. K. Arora (ed.), *Microbial Technology for the Welfare of Society*, Microorganisms for Sustainability 17,
https://doi.org/10.1007/978-981-13-8844-6_10

exploring predictive biomarkers for the post-treatment control as well as for monitoring the course of the disease evolution for the early recognition of relapses.

Keywords
Human brucellosis · Acute and chronic infection · PCR · Real-time PCR

10.1 Introduction

Brucellosis is one of the most common zoonoses worldwide. The disease widely affects domestic and wild animals causing tremendous economic losses among livestock; it provokes abortion, stillbirth, metritis, mastitis, and placental retention in females and orchitis and arthritis in males. Infertility may be seen in both sexes. Additionally, prolonged antibiotic therapy is required in humans (Ducrotoy et al. 2014). According to the World Health Organization (WHO), brucellosis is classified among the top seven world neglected zoonotic diseases and has social, economic, and public health impact on tropical and subtropical regions of the world. The disease in human is primarily an occupational disease that affects individuals who have close contact with the infected animals such as veterinarians, abattoir workers, farmers, and laboratory workers. Approximately more than 500,000 new cases occur every year (Pappas et al. 2006). In human, *Brucella* can cause systemic infections with an acute, subacute, or chronic course (Franco et al. 2007). The disease is associated with abortion, orchitis, acute renal failure, endocarditis, splenic abscess, spondylitis, arthritis, and encephalitis (Dagli et al. 2011; Zhong et al. 2013). *Brucella* has always been in the focus of military decision-makers because it can easily be transmitted via aerosols. The fear still exists to use *Brucella* as possible biological (B-) agent in biological crimes, biological warfare, and biological terrorism (Neubauer 2010). Up to date, genus *Brucella* included 12 accepted species; each species was named based on primary host species and antigenic characteristics. The 'classical' six species are *B. melitensis*, *B. abortus*, *B. suis*, *B. canis*, *B. ovis*, and *B. neotomae* which are primarily isolated from small ruminants, bovines, pigs, dogs, sheep, and desert wood rats, respectively (Corbel and Brinley-Morgan 1984). Another two species of marine origin were described, *B. ceti* which have been isolated from dolphins and whales and *B. pinnipedialis* isolated from seals. *B. microti* was isolated from the common vole *Microtus arvalis* (Foster et al. 2007; Scholz et al. 2008), and *B. inopinata* was isolated from a breast implant wound of a North American female patient (Scholz et al. 2010). Recently, *Brucella papionis* was described from an isolate in baboons (*Papio* spp.) (Whatmore et al. 2014), and *B. vulpis* was recovered from red foxes (Whatmore et al. 2016). Among 12 known brucellae, only *B. melitensis*, *B. abortus*, *B. suis* (except bv 2), and in rare cases *B. canis* are human pathogen, while the zoonotic potential of the rest are not proved yet. The genus *Brucella* is transmitted either by direct contact with infected animal's excreta or aborted materials or through the consumption of potentially contaminated unpasteurized dairy products (Wareth et al. 2014). *Brucella* could escape from the host immune system, infect any organs in the body, and survive in the intracellular organisms, thereby inducing prolonged relapses, morbidity, and chronic condition (Shehabi et al. 1990; Young 1995). Due to the wide range of clinical symptoms, brucellosis may mimic other infectious and noninfectious conditions and, therefore, impeding the accurate and rapid diagnosis of the disease. Nowadays, diagnosis of brucellosis is commonly based on serological and microbiological laboratory tests, although bacterial isolation is restricted by the lack of a suitable commercial blood culture system as well as slow growing characteristics of this organism (Yagupsky 1999). In the early stages of the disease, the diagnostic value of serological test is limited because of serological cross-reactions, low sensitivity, and the inability to distinguish between inactive and active infection due to antibody persistence after therapy. Furthermore, conventional serological tests and blood cultures showed critical restrictions for the detection of *Brucella* strains during relapsing, chronic, and focal complications of the disease (Mitka et al. 2007). This fact is of particular importance as the earlier detection of this microorganism could lead to a more effective treatment to prevent complications and relapses. For this purpose, the technical development of more sensitive and rapid detection methods is required to improve the diagnostic of brucellosis. The present chapter aimed to highlight recent technical advances and the molecular approaches for accurate and fast diagnosis of chronic and acute human brucellosis to improve treatment efficiencies.

10.2 Acute and Chronic Brucellosis in Humans

Human brucellosis is known in many developing countries such as the Mediterranean region, Central Asia, Near East countries, northern and eastern Africa, India, Central and South America, and Mexico. It is also referred to as Mediterranean fever, undulant fever, and Malta fever that infects humans by the consumption of contaminated milk, dairy products, and meat from domestic livestock (cattle, camels, goats, sheep, pigs, and water buffalo), as well as close contact with their carcasses and secretions. The symptoms of brucellosis are nonspecific such as high fever, arthralgia of the large joints, and myalgia (Doganay and Aygen 2003). Recently, the clinical feature of brucellosis has changed due to consumption of antibiotic at the suggestion of the pharmacist or the self-medication of patients with fever. This led to difficulties in the diagnosis of brucellosis and has decreased the rate of positive blood culture (Araj 1999). Four different forms of brucellosis, namely, acute, subacute, chronic, and relapse, could be observed (Young 2006). The symptoms of patients with acute brucellosis usually include weight loss, chills, sweating, fever, headache, back pain, arthralgia, anorexia, and fatigue. Subacute brucellosis cases typically present less severe symptoms compared to the acute form. Chronic infections also cause emotional lability, malaise, depression, nervousness, as well as generalized musculoskeletal pain (Galinska and Zagórski 2013; Young 2006). The relapse of brucellosis is very commonly reported because of partial, incomplete, or inappropriate antibiotic treatment as well as deficient immunologic response that is sometimes difficult to diagnose (Ariza et al. 1995). The disease often becomes chronic, and clinical manifestation could persist for more than16 months, and the majority of patients experience relapse, even after appropriate treatments (Buzgan et al. 2010).

10.3 Evaluation of Different PCR Methods for Early Detection of Acute and Chronic Brucellosis

The nonspecific and heterogeneous symptoms of brucellosis always highlight the necessity of laboratory confirmation either by isolation of pathogen or by demonstration of specific antibody and/or detection of DNA (Buzgan et al. 2010). The most specific method for isolation and identification of *Brucella* spp. is blood culture during feverish condition. However, this approach presents several important limitations, including low sensitivity (Ariza et al. 1995; Young 1991; Colmenero et al. 1996), particularly in cases with focal complications in patients with long-term symptoms and previously treated (Sanjuan-Jimenez et al. 2017; Buzgan et al. 2010); time-consuming even with the use of modern semiautomated approaches – this method needs several days or sometimes 1–2 weeks for blood culture results; as well as hazardous (Al Dahouk and Nöckler 2011). On the other hand, serological methods present also several limitations. The high titres of specific antibodies are required for the indirect identification of *Brucella* infections; this approach could not be effective for the early diagnosis of brucellosis as well as for treatment follow-up (Al Dahouk and Nöckler 2011) because the antibody takes 2–3 weeks to appear in the blood. Moreover, the serological tests are not able to clearly differentiate individuals with a history of brucellosis from those either presenting an active infection without clinical manifestation or with immune memory due to numerous exposures to the etiologic agent. Thus, the sole recognition of anti-*Brucella* antibodies does not support the evidence for the active pathogen presence (Al Dahouk and Nöckler 2011; Young 1991), and serologic test results are unable to differentiate acute and chronic patients.

In this regard, PCR-based methods appeared to be specific and sensitive diagnostic tools for the surveillance and detection of acute and chronic brucellosis which both constitute important prerequisites for effective control and prevention programmes (Gemechu et al. 2011; Boeri et al. 2018). The sensitivity and accuracy of the PCR tests are 100% in patients with acute infection presenting a positive blood culture as well as for chronic patients with no bacteraemia in two or more peripheral blood cultures (Sanjuan-Jimenez et al. 2017; Mitka et al. 2007). This aspect is critically important because 22% of patients show clinical forms which persist more than 1 month and 25.5% present focal forms with a lower number of circulating microorganisms (Sanjuan-Jimenez et al. 2017). The high sensitivity of molecular approaches is likely associated with its capability to detect 10 fg of bacterial DNA (approximately two bacteria) in 1 ml of peripheral blood sample from patients with clinical brucellosis (Queipo-Ortuño et al. 1997; Sanjuan-Jimenez et al. 2017). Moreover, different matrices such as whole blood, serum, urine, synovial or pleural fluid, cerebrospinal fluid, different tissues, and pus are applicable for the *Brucella* detection, although the preferred clinical samples are sera and whole blood samples (Zerva et al. 2001; Queipo-Ortuño et al. 1997, 2005). It has been reported that buffy coat is the optimal clinical specimen for detection of acute brucellosis via PCR methods due to haem compounds, whereas the use of anticoagulants may have inhibitory effect on PCR (Mitka et al. 2007).

In addition, the use of PCR methods is of critical importance in patients with negative serological results despite clinical signs allowing the rapid confirmation of potential *Brucella* infection (Asaad and Alqahtani 2012). Thus, PCR-based tests represent suitable alternative methods for the diagnosis of acute and chronic brucellosis. Asaad and Alqahtani (2012) reported that the PCR method correctly diagnosed acute brucellosis patients with 100% specificity and 70.4% sensitivity (Asaad and Alqahtani 2012). In accordance, another molecular study performed on 200 acute brucellosis patients demonstrated 99% of PCR-positive outcomes among infected individuals (Mitka et al. 2007), whereas Surucuoglu et al. (2009) reported 91.2% of positive cases among 36 patients (Surucuoglu et al. 2009). The variability in the detection percentage of PCR-based assays is partly associated with the lack of standardization and uniformity in the PCR protocols among studies, including the sample volume, optimal clinical specimen, target and primer sequences, extraction method, storage conditions of the samples, or experimental design (Mantur and Mangalgi 2004; Mitka et al. 2007). In addition, Asaad and Alqahtani (2012) showed positive rate of 11.1% for PCR tests performed according to a 31 kDa membrane protein (BCSP31) in the patients with chronic disease. They speculated that declined load of bacteria in the blood of chronic brucellosis patients might lead to lower rate of PCR results.

10.4 Molecular Identification of *Brucella* spp.

The first studies on *Brucella* spp. at molecular level had been conducted in early 1968 using DNase agar and a filter method (Hoyer and McCullough 1968a, b). These molecular studies revealed that *B. abortus*, *B. melitensis*, *B. suis*, and *B. canis* are all members of a single genus. Few years later, the close genetic relation between different species was confirmed by using the modern DNA–DNA hybridization method (Verger et al. 1985). A genetic similitude of more than 80% was identified in all studied species and biovars (De Ley et al. 1987). Since then, the near genetic relation of different *Brucella* species was further confirmed by polymerase chain reaction (PCR) technology and DNA sequencing method (Gopaul et al. 2008). It is now well established that *B. abortus*, *B. melitensis*, *B. suis*, and *B. canis* spp. have the same *recA*- and 16S ribosomal (r)RNA gene sequences and show similarity in the majority of housekeeping genes (only a few nucleotide substitutions) (Scholz and Vergnaud 2013; Scholz et al. 2008). The whole genome sequencing of *Brucella* spp. was published for the first time for *B. melitensis* 16 M in 2002 (DelVecchio et al. 2002), followed by *B. suis* strain 1330 (Osterman and Moriyon 2006). The results of comparative genome analyses demonstrated that these genomes present above 99% of typical nucleotide identity.

These findings were of overwhelming importance as some *Brucella* spp. such as *B. inopinata* and *B. microti* are phenotypically distant from other members of this genus because of faster growth and rapid metabolic activity (Scholz et al. 2008, 2010). Therefore, these species were initially misrecognized by phenotypic approaches such as the VITEK (both manufactured by BioMérieux) and API® NE (non-enteric) as members of the genus *Ochrobactrum*, the closest phylogenetic neighbour to *Brucella* (Scholz et al. 2010). For this reason, comprehensive molecular analyses are recommended to confirm species with unconventional phenotypes. Using molecular approaches, an atypical isolate of *Brucella* strain (BO2) was identified from a lung biopsy specimen collected from a 52-year-old Australian patient with chronic pneumonia. Molecular and phenotypic analyses placed strain BO2 as a novel lineage of *B. inopinata* (Tiller et al. 2010).

The possibility of whole genome sequence analysis have smoothed the way for the use of promising molecular methods such as multilocus sequence typing (MLST), multilocus sequence analysis (MLSA), single-nucleotide polymorphism (SNP) analysis, and multiple-locus variable number tandem repeats analysis (MLVA) (Scholz and Vergnaud 2013), thereby enabling the differentiation of *Brucella* spp. at species, at biovar, and even at individual strain levels.

10.5 Commonly Used PCR Methods for *Brucella* Identification

Molecular analysis of *Brucella* organisms and likely new *Brucella* species will support a better understanding of the host specificity, pathogenicity, and evolution of this genus (Scholz and Vergnaud 2013). PCR-based methods could also be used to confirm the results obtained from other conventional detection approaches. Numerous studies at the early 1990s were carried out in order to detect and discriminate different *Brucella* spp. using PCR methods (Fekete et al. 1990, 1992; Baily et al. 1992; Herman and De Ridder 1992). Fekete and his colleagues amplified a fragment of 635-bp belonging to a gene coding an outer membrane protein from *B. abortus* strain 19 (Fekete et al. 1990). Several studies have reported PCR as an effective method for the early detection of relapses and for the post-treatment follow-up as well as a rapid and accurate diagnosis tool of acute brucellosis (Wang et al. 2014).

The sequencing of different proteins provides the accessibility for various primers to amplify specific sequences of the genus *Brucella*, although no agreement was reached regarding the most suitable and applicable methods (Romero et al. 1995; Romero and Lopez-Goñi 1999). Nowadays, PCR-based methods are commonly applied in the diagnosis of human and animal brucellosis (Yu and Nielsen 2010), and more than 200 reports of laboratory diagnosis of human brucellosis using PCR-based assays have been published including nonspecific high-resolution typing assays, specific high-resolution assays, HOOF-Prints and subsequent selections of loci for multiple-locus variable number tandem repeats analysis, multilocus sequence typing of single-nucleotide polymorphisms, real-time polymerase chain reaction assays of single-nucleotide polymorphisms, and whole genome sequencing and global analysis of single-nucleotide polymorphisms (Scholz and Vergnaud 2013; Scholz et al. 2008).

10.5.1 Conventional PCR

Several PCR tests are increasingly used for the direct identification of human brucellosis because of their high specificity and sensitivity when compared to serological and blood culture methods, respectively (Araj 2010; Scholz and Vergnaud 2013; Rodríguez-Lázaro et al. 2017). One of the most important characteristics of the PCR methods that highlight its value is the ability to confirm the diagnosis of acute brucellosis (Asaad and Alqahtani 2012). Numerous target genes have been selected for identification of the genus *Brucella*, including the 16S rRNA (Herman and De Ridder 1992; Romero et al. 1995; Nimri 2003), insertion element IS711 (Ouahrani et al. 1993; Osterman and Moriyon 2006; Cloeckaert et al. 2000), the 16S-23S internal transcribed spacer region (ITS) (Rijpens et al. 1996), the region of coding sequence for recA protein (Scholz et al. 2008), outer membrane protein coding for omp2 (omp2a, omp2b) gene (Leal-Klevezas et al. 1995; Bardenstein et al. 2002; Barua et al. 2016), and the coding sequence for genus-specific PCR assays that target the bcsp31 gene associated with a 31 kDa immunogenic outer membrane protein conserved among all *Brucella* spp. (Sanjuan-Jimenez et al. 2017; Bounaadja et al. 2009; Ohtsuki et al. 2008).

In 1988, the gene encoding for the membrane protein of 31 kDa that was specific for *B. abortus* has been cloned by Mayfield et al. (Mayfield et al. 1988). Few years later, Baily et al. designed a 223-bp fragment of this gene which was conserved in all species and biovars of the genus *Brucella* in order to evaluate *Brucella* cultures obtained from clinical samples via PCR methods (Baily et al. 1992). Another study showed that the amplification of the design primers, according to species-specific polymorphism in the rRNA 16S-23S spacer region, consistently produced five bands characteristic of the *Brucella* spp. These bands were common to all 25 examined strains (Fox et al. 1998). Since then, this method has been applied for the identification of *Brucella* strains at the genus level, while carbohydrate profiling was capable to differentiate *B. canis* from the other *Brucella* spp. (Fox et al. 1998). Mitka et al. have performed four different PCR assays for the diagnosis of acute and relapsing forms of brucellosis with (I) bcsp31 (223-bp-sequence of the conserved gene bcsp31 encoding the immunogenic outer membrane protein of 31 kDa); (II) a 193-bp sequence of the gene omp2 encoding a 26 kDa outer membrane protein of *B. abortus* in all *Brucella* species; (III) a 282-bp sequence of different region of the gene omp2; and (IV) a 1029-bp sequence of bp26 gene of *B. melitensis* 16 M (also named Omp28), which is identified as an immunodominant antigen. All four tests showed excellent specificity (100%) and high diagnostic sensitivity ranging from 95.5% to 100% in acute infections, depending on the type of specimen and the PCR assay as the I and III PCRs appeared to be slightly more sensitive than II and IV PCRs (Mitka et al. 2007) (Table 10.1). In addition, the diagnostic sensitivity for the target genes of I and II PCRs were evaluated as 10–100 fg and 25–250 fg, respectively. PCRs III and IV have been particularly suggested for the diagnosis of acute brucellosis because of their high detection limits and sensitivity (Mitka et al. 2007).

Table 10.1 Different primers used for *Brucella* spp. detection and differentiation

Brucella spp.	Target	Primer	PCR-based method/ product	Sample	Stage of brucellosis	References
B. abortus *B. melitensis* *B. suis* *B. canis*	16S–23S rRNA Ribosomal DNA interspace region	KF5: 5′-GAAGTCGTAACAAGG-3′ KF6: 5′-CAAGCATCCATCGT-3′	PCR /1530, 916, 701, 399, and 316-bp	Bacteria	Acute	Fox et al. (1998)
Brucella spp.	BCSP31[a]	B4: 5′-TGGCTCGGTTGCCAATATCAA-3′ B5: 5′-CGCGCTTGCCTTTCAGGTCTG-3′	PCR (223-bp)	Blood	Acute, chronic	Baily et al. (1992); Queipo-Ortuño et al. (1997), Mitka et al. (2007), Asaad and Alqahtani (2012), and Maas et al. (2007)
Brucella spp.	BCSP31[a]	B4: 5′-TGGCTCGGTTGCCAATATCAA-3′ B5: 5′-CGCGCTTGCCTTTCAGGTCTG--3′	Real-time PCR (223-bp)	Blood	Acute, chronic	Queipo-Ortuño et al. (2005)
B. melitensis	IS711	EFQ1: 5′-TGTTTCGGCTCAGAATAATCC A-3′, ERQ2: 5′-GCATGCGCTATGATCTGGTTA C-3′ Specific probe sequence (STqME) 5′-6-carboxy-fluorescein (FAM)-AGCTTACCCGCCAATCTTCGCCCXT-3′	Real-time PCR (251-bp)	Blood	Acute, chronic	Navarro et al. (2006)
Brucella spp.	16S-23S ITS	Primers	Real-time PCR (147-bp)	Blood	Acute, chronic	Kattar et al. (2007)
		Bru ITS-S: 5′- TGCCTGTTCTGTATGAAATCGT-3′				Kattar et al. (2007)
		Bru ITS-A: 5′- GCAGAAAGACCAGCTTCTCGA-3′				Kattar et al. (2007)
		Probes				Kattar et al. (2007)

	OMP25	ITS_FL: CTTGCTCAAGCCTTGCATAATGATTGA-F	(141-bp)			Kattar et al. (2007)
		ITS_LC: LC Red640-TGTTTAACCGCCATCACCGATTGTA-p				Kattar et al. (2007)
		Bru 25-S: 5′- GGTTATTCCTGGGCCAAGAA –3′				Kattar et al. (2007)
		Bru 25-A: 5′- AGCCGTGAGGTACGGCATA-3′				Kattar et al. (2007)
		Probes				
	OMP31	Melit FL+: AGGGCTTTGAAGGCTCGCTGCGT-F	(281-bp)			Kattar et al. (2007)
		Melit-LC+: LC Red640-CCCGCGTTGGCTACGACCTG-p				Kattar et al. (2007)
		Bru-31-F: 5′- TGGTAAGGTCAAGTCTGCGTT-3′				Kattar et al. (2007)
		Bru-31-R: 5′- CTTCTTCATTCCGTGTTCGTG-3′				Kattar et al. (2007)
		Probes				
		Bru31-FL: TGAGAGCAAGGTCAATTTCCACACTG-F				Kattar et al. (2007)
		Bru31-LC: LC Red640-CGCGTCGGTCTGAACTACAAGTTC-p				Kattar et al. (2007)
Brucella spp.	BCSP31[a]	B4: 5′- TGGCTCGGTTGCCAATATCAA-3′ B5: 5′- CGCGCTTGCCTTTCAGGTCTG--3′ Specific probe (5’6FAM-ACTCCAGAGCGCCCGACTTGATCG-DB 3′)	Real-time PCR (223-bp)	Blood	Acute, chronic	Gwida et al. (2011) and Probert et al. (2004)
Brucella spp.	IS711	Not determined	Real-time PCR (207-bp)	Blood	Acute, chronic	Alsayed and Monem (2012)
Brucella spp.	omp2[b]	JPF: 5′- GCGCTCAGGCTGCCGACGCAA-3′ JPR: 5′- ACCAGCCATTGCGGTCGGTA--3′	PCR (193-bp)	Blood	Acute, chronic	Mitka et al. (2007)

(continued)

Table 10.1 (continued)

Brucella spp.	Target	Primer	PCR-based method/ product	Sample	Stage of brucellosis	References
Brucella spp.	omp2[b]	P1: 5′- TGGAGGTCAGAAATGAAC-3′ P2: 5′- GAGTGCGAAACGAGCGC-3′	PCR 282-bp	Blood	Acute, chronic	Mitka et al. (2007)
Brucella spp.	bp26[c]	26A: 5′- GCCCCTGACATAACCCGCTT –3′ 26B: 5′- GAGCGTGACATTTGCCGATA-3′	PCR (1029-bp)	Blood	Acute, chronic	Mitka et al. (2007)
Brucella spp.	Vdcc	5′- GTGGCGATCTTGTCCG-3′ 5′- ACGGCGATGGATTTCCG-3′	Real-time PCR (67-bp)	Blood	Acute	Piranfar et al. (2015) and Winchell et al. (2010)
B. melitensis	Int-hyp	5′- GTGGCGATCTTGTCCG-3′ 5′- ACGGCGATGGATTTCCG--3′	(125-bp)			Piranfar et al. (2015) and Winchell et al. (2010)
B. abortus	Glk	5′- GACCTCTTCGCCACCTATCTGG-3′ 5′- CCTTGTGCGGGGCCTTGTCCT-3′	(164-bp)			Piranfar et al. (2015) and Winchell et al. (2010)
B. melitensis	Acetyl-CoA acetyltransferase gene	5′-GCTCGACACAAAGGGCCA-3′ 5′-CAAGCGTGGTCTGGCGA-3′ Probe -CCGAGATACAAA-MGB	Real-time PCR (223-bp)	Blood	Acute, chronic	Kaden et al. (2017)

[a]BCSP31, encoding an immunogenic 31 kDa outer membrane protein of *B. abortus*
[b]omp2 encoding a 26 kDa outer membrane protein of *B. abortus*
[c]bp26 encoding the BP26 protein of *B. melitensis*, also named Omp28

10.5.2 Real-Time PCR

Real-time PCR (RT-PCR) assay is a very useful method for the evaluation of *Brucella* nucleic acids in individual blood samples. Currently, this approach becomes more accessible for microbiological laboratories because of decreasing prices of the thermocyclers machines and the required reagents (Ginzinger 2002). RT-PCR has been successfully adapted for the rapid and accurate identification of *Brucella* species in clinical samples, targeting the genes coding bcsp31 (Colmenero et al. 2005; Queipo-Ortuño et al. 2008, 2005), omp25, omp31 (Kattar et al. 2007), and IS711 (Cerekci et al. 2011; Zhang et al. 2013) and 16S-23S internal transcribed spacer region (ITS) (Newby et al. 2003).

Queipo-Ortuño and colleagues used bcsp31 biomarker for a fast SYBR Green I LightCycler-based real-time PCR assay (LC-PCR). The authors reported the technical simplicity of this method as well as its ability to confirm *Brucella* infections in doubtful or negative serological test results or in serum samples from patients with suspected brucellosis symptoms as an efficient alternative to blood cultures. In this study, the sensitivity of LC-PCR for *Brucella* spp. detection in sera of suspected brucellosis patients reached 93.3%. The positive likelihood ratios and specificity were 94.6% (Queipo-Ortuño et al. 2005). When applied to a total of 62 serum samples from active brucellosis patients and 65 negative control samples, this approach showed 91.9% sensitivity and 95.4% specificity (Queipo-Ortuño et al. 2005). Another study reported that real-time PCR, targeting the genus-specific biomarker bcsp31, was positive for 146 confirmed brucellosis patients and represented a reliable clinical predictor for brucellosis (Njeru et al. 2016). Moreover, brucellosis suspected patients with serum antibody titres of more than 1/80 have been analysed with multiplex real-time PCR and high-resolution melt analysis curve from human blood and showed a more rapid and convenient closed-tube method with specificity of 56% for *B. melitensis* and 44.4% for *B. abortus* (Piranfar et al. 2015; Winchell et al. 2010). Recently, a single real-time PCR has been implemented for the specific identification of all biovars of *B. melitensis* with high sensitivity (1.25 GE/μl) and high specificity (100%), thereby recommending its use as a reliable and rapid detection method in all clinical laboratories across Sweden (Kaden et al. 2017).

Surucuoglu et al. developed a TaqMan real-time PCR assay to improve the diagnosis of brucellosis from patients' sera. They have reported specificity, sensitivity, and positive and negative predictive values of this method as 100%, 88%, 100%, and 83%, respectively (Surucuoglu et al. 2009). The use of RT-PCR in combination with ELISA and/or blood culture has also been recommended if the agglutination test is negative, improving by this way the diagnosis of brucellosis (Alsayed and Monem 2012).

In several other studies, RT-PCR has been reported as a rapid, sensitive, and specific method for the diagnosis of human brucellosis (Winchell et al. 2010; Navarro-Martínez et al. 2008; Colmenero et al. 2011). The specificity of the primers as well as the amount of interfering immunoglobulin G extracted with the template DNA from serum samples also could determine the efficiency of RT-PCR (Wang et al. 2014). Three real-time PCR tests for diagnosis of human brucellosis were evaluated at genus level with hybridization primers and probes from omp25, omp31, and 16S-23S ITS and revealed that real-time PCR with 16S-23S ITS primers and its probes were the most sensitive. The clinical sensitivity, specificity, and positive and negative predictive values of this approach reached 66.7%, 99.7%, 94.1%, and 97.6% (Kattar et al. 2007).

10.5.3 Multiplex PCR

Multiplex PCR tests have been used in numerous studies for the diagnosis of human brucellosis (Wang et al. 2014). Lübeck et al. (2004) developed a multiplex PCR test for *Brucella* diagnostics based on the perosamine synthetase gene (Lübeck et al. 2004). In another study, El Kholy and his coworkers also reported a multiplex PCR test via two types of primers (B4/B5 and JPF/JPR) for the diagnosis of active human brucellosis in Egypt (El Kholy et al. 2009). The authors demonstrated the high accuracy, sensitivity, and specificity of this method as a promising alternative to culture tests for the diagnosis of human brucellosis. In addition, a multiplex PCR test can be applied to simultaneously identify different *Brucella* spp. present in human samples. For example, a multiplex PCR assay was developed to detect four important species of the genus *Brucella* in a single test (Imaoka et al. 2007). For this purpose, four pairs of primers targeting bcsp31, omp2b, omp2a, and omp31 genes specific to each *Brucella* spp. have been amplified allowing the accurate detection and differentiation of *Brucella* species. Multiplex PCR appeared to be an accurate and rapid method to trace sources of infection and could be of great value for the early detection of human brucellosis. Concerning *Brucella spp.*, AMOS-PCR (*B. abortus*, *B. melitensis*, *B. ovis*, and *B. suis*-PCR) was the first species-specific multiplex PCR which is based on the polymorphism arising from species-specific localization of the insertion sequence IS711 in the *Brucella* chromosome and used to identify and differentiate *B. abortus* bv 1, 2, and 4, *B. melitensis*, *B. ovis*, and *B. suis* bv 1 (Bricker and Halling 1994). Another multiplex PCR assay called Bruce-ladder multiplex PCR assay has been reported for diagnosis of *Brucella* even at the biovar level through the combination of various primers. For instance, a 19-primer multiplex PCR was able to simultaneously detect *B. microti*, *B. ceti*, *B. pinnipedialis*, and *B. neotomae* as well as *B. abortus* biovars 1, 2, and 4 from biovars 3, 5, 6, 9 (Huber et al. 2009). Two pairs of primers derived from IS711 have also been used in nested PCR and led to higher sensitivity and specificity in the detection of human brucellosis (Al-Nakkas et al. 2005).

10.6 Interpretation of Molecular Data in Acute and Chronic *Brucella* Infections

For rapid and reliable diagnosis of acute and chronic brucellosis, the simple detection of *Brucella* at genus level by PCR remains adequate to start the conventional antibiotic therapy for human against the majority of *Brucella* species. After completing the treatment, if the result of PCR-based analysis remains positive, the treatment should be continued with a more effective therapeutic regimen in order to prevent relapse (Mitka et al. 2007). To differentiate patients re-exposed to *Brucella* spp. from those subjected to a relapse of the initial infection, the subtyping of bacteria at strain level can be useful (Al Dahouk et al. 2005). For this purpose, multiple-locus variable number tandem repeats analysis based on 16 markers (MLVA-16) was performed as adapted by Al Dahouk et al. (2007a) and Le Fleche et al. (2006) and has demonstrated its high diagnostic value (Al Dahouk et al. 2007b; Jiang et al. 2011; Kadanali et al. 2009). The assay comprises two panels; panel 1 is used to trace back the geographic origin of strains, while panel 2 is useful for outbreak investigation. It has been shown that the diagnosis of a primary or relapse treatment failure can be performed through the evaluation of the degree of identity of MLVA-16 *Brucella* marker for strains isolated from the brucellosis patients before and after the first-line therapy as a reliable indication for the prolongation of antibiotic therapy. The identification of divergent genotypes in the *Brucella* isolates may be indicative of re-infection and conventional antibiotics could be administered as an efficient therapeutic approach (Al Dahouk et al. 2013). It is however important to note that the detection of *Brucella* DNA in the course of the disease has poor clinical significance considering the fact that *Brucella* DNA-negative patients can also relapse (Navarro et al. 2006). The molecular identification of the *Brucella* DNA can confirm the acute or chronic form of brucellosis but could also occur in asymptomatic subjects with a history of brucellosis (Castaño and Solera 2009). In the majority of brucellosis patients completing successful clinical cure, *Brucella* DNA persists in sera for weeks to years, even in the absence of any chronic or relapse symptoms (Maas et al. 2007; Navarro et al. 2006; Vrioni et al. 2008). This phenomenon might be explained by the survival and persistence of *Brucella* in human macrophages and the higher sensitivity of modern real-time PCR assays for the detection of non-viable or phagocytosed microorganisms. Quantitative real-time PCR could thus represent the tool of choice to differentiate active and past *Brucella* infections, allowing the interpretation of different levels of DNA persistence in brucellosis patients (Queipo-Ortuño et al. 2008). Therefore, considering the significant sensitivity of PCR-based assays, there is always a possibility to amplify nonviable *Brucella* or phagocytosed bacteria, and this probability should be taken into account in the interpretation of any positive PCR outcomes.

10.7 Advantages and Restrictions of PCR-Based Approaches

The sensitivity of each PCR assays is comparable to traditional blood cultures and serological assays in acute brucellosis that showed a combined sensitivity reaching 96% (Mitka et al. 2007). However, non-PCR traditional methods require optimal sample storage/handling conditions and are time-consuming. Moreover, these methods should often be carried out in combination to obtain reliable results. Some laboratories use nonstandardized and non-uniform serological tests that commonly lead to false-positive and false-negative results (Zerva et al. 2001), contributing to the lack of specificity and sensitivity reported for some serological methods (Al et al. 2003). In addition, the sample volume applied to PCR assays is much smaller than those required for bacteriological tests, representing an important practical advantage in clinical laboratory practice. Thus, PCR-based tests might be applied as sensitive and rapid diagnostic tools for identification of acute brucellosis (Mitka et al. 2007). One of the most important characteristics of the PCR-based tests is its ability to detect *Brucella* spp. at the earlier stages of the disease. This represents a crucial point enabling earlier antibiotic treatment and thereby decreasing the rates of focal disease as well as the risk of relapse (Ariza et al. 1995). PCR assays are also very applicable, specific, sensitive, and relatively inexpensive tools, with high positive and negative predictive potency in order to improve the efficacy of treatments (Mitka et al. 2007). The diagnosis of relapses is very difficult and inefficient by conventional tests, and PCR represents a reliable alternative for predicting chronic and relapse course of the disease (Queipo-Ortuño et al. 1997). DNA preparation by the boiling method for the diagnosis of human brucellosis is a rapid, simple, cost-effective, and reproducible technique that can be achieved on other types of clinical samples, such as cerebrospinal fluid and urine. However, the DNA extraction by boiling lysis did not prevent the occurrence of some inhibitors in tested sera, including immunoglobulin G (IgG) (Queipo-Ortuño et al. 2008). Higher dilutions of different samples could decrease the presence of inhibitory factors, although this may lead to a decreased concentration of the target DNA in clinical samples, sometimes below the detection limits. Approximately five bacteria per reaction can be identified through well-conducted *Brucella* real-time PCR methods (Al Dahouk et al. 2013; Scholz et al. 2007). Real-time PCR tests are easily standardized and induced minimum risk of infection in laboratory workers. This method also allows the differentiation among seropositive, active, and inactive states of the *Brucella* infections (Wang et al. 2014). Previously, AMOS PCR targeting the IS711 DNA represented the most common PCR test to differentiate *B. abortus*, *B. melitensis*, *B. ovis*, and *B. suis* species (Bricker and Halling 1994). However, this assay was unable to detect *B. canis* and *B. neotomae* as well as *Brucella* species isolated from marine mammals and some biovars of *B. abortus* and *B. suis*. Therefore, a new IS711 chromosomal specific location was designed for *Brucella* isolates from a marine mammal in order to identify and classify these species according to the same principle of AMOS PCR (Maquart et al. 2008; Zygmunt et al. 2010). In addition, Bruce-ladder PCR is known as a multiplex PCR test that could successfully differentiate *Brucella* isolates from marine mammals and all six classical species (Lopez-Goñi et al. 2008; Mayer-Scholl et al. 2010). These molecular tests now allow the accurate identification of *Brucella* isolates at species level (except some strains belonging to *B. suis* and *B. canis* species) without expensive laboratory equipment (Koylass et al. 2010). The use of MLSA represents a promising way to characterize the global population picture of *Brucella* and appeared to be highly suitable for accurate epidemiological analysis (Whatmore et al. 2007). Recently, VNTR has been presented as a highly effective method to detect and confirm the relapse of brucellosis infection (Al Dahouk et al. 2005; Kattar et al. 2008) and allow the detection of various genotypes with different pathogenic profiles (Nöckler et al. 2009) in human brucellosis outbreaks (Lucero et al. 2008, 2010; Valdezate et al. 2007).

We cannot ignore that conventional PCR requires the cultivation of clinical samples and use of colonies in test, but RT-PCR solved the problem because it is able to detect *Brucella* DNA extracted directly from clinical samples. False-negative PCR results could be obtained after treatment because of the small number of circulating bacteria and the absence of the target DNA in serum (Morata et al. 2001). Besides, high genomic similarity between different *Brucella* species makes the design of species-specific real-time PCR methods difficult (Kaden et al. 2017). For instance, it has been reported that the primer pairs have important effect on the efficiency of multiplex PCR, and the presence of more than one primer pair increases risks of primer-dimers and nonspecific products (Brownie et al. 1997). Therefore, the multiplex PCR need laborious optimization.

It worth to mention that the protein-based assays such as matrix-assisted laser desorption ionization time of flight mass spectrometry (MALDI-TOF MS) is another promising method for the detection of *Brucella* isolates in clinical laboratories. This method appeared to be cost-effective, precise, and rapid when compared to conventional phenotypic methods or molecular test (Carbonnelle et al. 2011; Ferreira et al. 2010; Singhal et al. 2015) but requires adequate laboratory equipment. MALDI-TOF MS has been recently applied in the brucellosis diagnosis and is capable to detect *Brucella* at genus level, directly from blood culture bottles and culture plates (Colmenero et al. 2011; Singhal et al. 2015). This promising technique is not yet commonly applied as few laboratories are equipped with this device and the accessibility to comprehensive protein profile databases is still limited (Al Dahouk et al. 2013). Other important drawbacks of this method are the lack of standardized protocols as well as the impossibility for detection of new isolates given the restricted database for peptide mass fingerprints for different strains of *Brucella* spp. (Singhal et al. 2015). The appropriate use of these molecular tests should finally lead to rapid and accurate identification of internationally or nationally dispersed clusters, identification of emerging and novel strains, identification of transmission chains, and recognition of outbreaks sources.

10.8 Summary

Among *Brucella* spp., *B. melitensis*, *B. abortus*, *B. suis*, and, to a lesser extent, *B. canis* represent important pathogenic species for humans (Araj 2010). In addition, *B. ceti* or *B. pinnipedialis* have also been responsible for rare cases of human brucellosis (Brew et al. 1999; Hoover-Miller et al. 2017). Additionally, a novel *B. inopinata* strain (BO1) has been detected from a woman subjected to a prosthetic breast implant infection (De et al. 2008), and only little is known about the pathogenicity of *B. microti*. The source of *Brucella* infections highly depends on the geographical region and related endemic species (Pappas et al. 2006). The alarming situation of human brucellosis incidence could be correlated with the nonspecific clinical manifestation of brucellosis, shortcomings in laboratory diagnosis, and low awareness of the disease in non-endemic countries. The number of infected animals within defined regions is directly attributed to the number of human cases. For example, the reported human incidence of the brucellosis in the United States is low, but it is estimated that only 4–10% of patients are in fact identified (Colmenero et al. 1996). The diagnosis of brucellosis needs to be supported by fast and safe laboratory tests such as PCR-based assay due to nonspecific symptoms (Mitka et al. 2007).

Moreover, the number of bacteria in clinical samples might widely vary depending on antibiotic pre-treatment, acute and chronic stages of the disease, and storage/handling conditions of biological specimens (Al et al. 2003). In addition, the time of detection is contrarily related with the concentration of viable bacteria in the blood specimen (Yagupsky 1999). The PCR-based methods according to the 16S rRNA amplify a DNA fragment common to all *Brucella* spp., although some cross-reaction with members of the closely related genus *Ochrobactrum* may occur (Scholz et al. 2006, 2008). The IS711 target became the chosen region for general identification of *Brucella* spp. because of the presence of multiple copies and its exclusive occurrence in *Brucella* spp., thereby revealing high sensitivity and allowing direct testing on clinical samples (Bounaadja et al. 2009; Halling et al. 1993). Other molecular biomarkers such as omp2 (Leal-Klevezas et al. 1995), recA (Scholz et al. 2008), or the 16S-23S intergenic transcribed sequence (Rijpens et al. 1996) were also able to effectively detect *Brucella* spp. (Godfroid et al. 2011).

A large number of patients suffer from brucellosis worldwide because their infection remains misdiagnosed or not appropriately treated (McDermott and Arimi 2002). Reliable and rapid diagnosis is critical for effective treatment as well as to prevent complications and relapses. PCR alone or in combination with the serology and bacteriology is the fast and effective method allowing the accurate diagnosis of acute brucellosis as well as the early detection of relapses and post-treatment follow-up using reliable biomarkers. This method is more sensitive than the microbial test alone and allows earlier detection of the microorganism. The first step in PCR designing is the appropriate choice of the DNA target to be amplified in the *Brucella* spp. Therefore, to avoid false-positive results and cross-reaction with the genomes of human or other bacteria, the primers should be designed according to a well-conserved specific target of the *Brucella* genome. In optimal circumstances, the high sensitivity of the molecular tests in pure cultures and clinical samples is able to detect even low amounts of DNA related to *Brucella* spp. through clearly visible amplified products in agarose gel. Despite considerable advances in molecular diagnosis of brucellosis, there is still an urge to design PCR-based techniques with high specificity to better characterize and differentiate DNA from the six classic species and six new classified categories of the *Brucella* genus as well as against DNA from a high range of facultative intracellular bacteria and different microorganisms phylogenetically related to *Brucella* that produce similar bacteraemia in humans. DNA amplification from a peripheral whole blood specimens could be used for diagnosis of human brucellosis through PCR assay, although DNA amplification could be done from other different biological specimens such as joint fluid, bone marrow, urine, cerebrospinal fluid, abscesses, and tissues (Morata et al. 1998). The DNA sample for PCR can be extracted and purified from the bacteria or added directly to the reaction mixture. Some substances are able to inhibit the PCR action including host DNA, immunoglobulins, EDTA, phenol, proteinase K, porphyrins, and heparin (Sanjuan-Jimenez et al. 2017). The detection limit of molecular tests remains a critical issue given that the number of bacteria in clinical samples of focal forms of brucellosis is usually low. Furthermore, the direct extraction of *Brucella* DNA in patients with brucellosis is challenging due to inhibitory effect of matrix components and the low number of bacteria in clinical samples. Therefore, the basic methods for sample preparation need to concentrate the bacterial DNA template and remove inhibitory effects of interfering substances. Regarding molecular PCR tests using the IS711 insertion element of *Brucella* as a DNA target, the method sensitivity has been improved given that IS711 is present in *Brucella* chromosomes with multiple copies. In recent years, the well-conserved coding sequences for bcsp31 protein (Bounaadja et al. 2009), a 31 kDa immunogenic outer membrane protein common to all *Brucella* spp., became the most common target for molecular-based tests in clinical microbiological laboratories. This genus-specific PCR test is sufficient for the molecular detection of acute and chronic human brucellosis (Al Dahouk and Nöckler 2011). Moreover, this approach allowing the reliable detection of *Brucella* at genus level can help to prevent the false-negative outcomes in brucellosis patients affected by uncommon species and biovars. For the subsequent diagnosis confirmation and the distinction between closely related species, 16S rRNA gene sequencing can be used. Differential PCR-based tests such as species-specific real-time PCR assays or the conventional Bruce-ladder PCR also could be applied for confirmatory recognition and differentiation of *Brucella* species.

References

Al Dahouk S, Nöckler K (2011) Implications of laboratory diagnosis on brucellosis therapy. Expert Rev Anti-Infect Ther 9(7):833–845

Al Dahouk S, Tomaso H, Nöckler K, Neubauer H, Frangoulidis D (2003) Laboratory-based diagnosis of brucellosis – a review of the literature. Part I: techniques for direct detection and identification of Brucella spp. Clin Lab 49(9–10):487–505

Al Dahouk S, Hagen RM, Nöckler K, Tomaso H, Wittig M, Scholz HC, Vergnaud G, Neubauer H (2005) Failure of a short-term antibiotic therapy for human brucellosis using ciprofloxacin. Chemotherapy 51(6):352–356

Al Dahouk S, Fleche PL, Nockler K, Jacques I, Grayon M, Scholz HC, Tomaso H, Vergnaud G, Neubauer H (2007a) Evaluation of Brucella MLVA typing for human brucellosis. J Microbiol Methods 69(1):137–145. https://doi.org/10.1016/j.mimet.2006.12.015

Al Dahouk S, Le Flèche P, Nöckler K, Jacques I, Grayon M, Scholz HC, Tomaso H, Vergnaud G, Neubauer H (2007b) Evaluation of Brucella MLVA typing for human brucellosis. J Microbiol Methods 69(1):137–145

Al Dahouk S, Sprague L, Neubauer H (2013) New developments in the diagnostic procedures for zoonotic brucellosis in humans. Rev Sci Tech 32(1):177–188

Al-Nakkas A, Mustafa AS, Wright SG (2005) Large-scale evaluation of a single-tube nested PCR for the laboratory diagnosis of human brucellosis in Kuwait. J Med Microbiol 54(8):727–730

Alsayed Y, Monem F (2012) Brucellosis laboratory tests in Syria: what are their diagnostic efficacies in different clinical manifestations? J Infect Dev Countries 6(06):495–500

Araj GF (1999) Human brucellosis: a classical infectious disease with persistent diagnostic challenges. Clin Lab Sci 12(4):207

Araj GF (2010) Update on laboratory diagnosis of human brucellosis. Int J Antimicrob Agents 36:S12–S17

Ariza J, Corredoira J, Pallares R, Viladrich PF, Rufi G, Pujol M, Gudiol F (1995) Characteristics of and risk factors for relapse of brucellosis in humans. Clin Infect Dis 20(5):1241–1249

Asaad AM, Alqahtani JM (2012) Serological and molecular diagnosis of human brucellosis in Najran, southwestern Saudi Arabia. J Infect Public Health 5(2):189–194

Baily G, Krahn J, Drasar B, Stoker N (1992) Detection of Brucella melitensis and Brucella abortus by DNA amplification. Tropical Med Int Health 95(4):271–275

Bardenstein S, Mandelboim M, Ficht TA, Baum M, Banai M (2002) Identification of the Brucella melitensis vaccine strain rev. 1 in animals and humans in Israel by PCR analysis of the PstI site polymorphism of its omp2 gene. J Clin Microbiol 40(4):1475–1480

Barua A, Kumar A, Thavaselvam D, Mangalgi S, Prakash A, Tiwari S, Arora S, Sathyaseelan K (2016) Isolation & characterization of Brucella melitensis isolated from patients suspected for human brucellosis in India. Indian J Med Res 143(5):652

Boeri E, Wanke M, Madariaga M, Teijeiro M, Elena S, Trangoni M (2018) Comparison of four polymerase chain reaction assays for the detection of Brucella spp. in clinical samples from dogs. Vet World 11(2):201–208. Abstract

Bounaadja L, Albert D, Chénais B, Hénault S, Zygmunt MS, Poliak S, Garin-Bastuji B (2009) Real-time PCR for identification of Brucella spp.: a comparative study of IS711, bcsp31 and per target genes. Vet Microbiol 137(1–2):156–164

Brew S, Perrett L, Stack J, MacMillan A, Staunton N (1999) Human exposure to Brucella recovered from a sea mammal. Vet Rec 144(17):483

Bricker B, Halling S (1994) Differentiation of Brucella abortus bv. 1, 2, and 4, Brucella melitensis, Brucella ovis, and Brucella suis bv. 1 by PCR. J Clin Microbiol 32(11):2660–2666

Brownie J, Shawcross S, Theaker J, Whitcombe D, Ferrie R, Newton C, Little S (1997) The elimination of primer-dimer accumulation in PCR. Nucleic Acids Res 25(16):3235–3241

Buzgan T, Karahocagil MK, Irmak H, Baran AI, Karsen H, Evirgen O, Akdeniz H (2010) Clinical manifestations and complications in 1028 cases of brucellosis: a retrospective evaluation and review of the literature. Int J Infect Dis 14(6):e469–e478

Carbonnelle E, Mesquita C, Bille E, Day N, Dauphin B, Beretti J-L, Ferroni A, Gutmann L, Nassif X (2011) MALDI-TOF mass spectrometry tools for bacterial identification in clinical microbiology laboratory. Clin Biochem 44(1):104–109

Castaño MJ, Solera J (2009) Chronic brucellosis and persistence of Brucella melitensis DNA. J Clin Microbiol 47(7):2084–2089

Cerekci A, Kılıç S, Bayraktar M, Uyanık M, Yaşar E, Esen B (2011) Comparison of conventional methods and real-time multiplex polymerase chain reaction for identification and typing of Brucella isolates of human origin. Mikrobiyoloji bulteni 45(3):392–400

Cloeckaert A, Grayon M, Grepinet O (2000) An IS711 element downstream of thebp26 gene is a specific marker of Brucella spp. isolated from marine mammals. Clin Diagn Lab Immunol 7(5):835–839

Colmenero J, Reguera J, Martos F, Sanchez-De-Mora D, Delgado M, Causse M, Martín-Farfán A, Juarez C (1996) Complications associated with Brucella melitensis infection: a study of 530 cases. Medicine 75(4):195–211

Colmenero J, Queipo-Ortuno M, Reguera J, Baeza G, Salazar J, Morata P (2005) Real time polymerase chain reaction: a new powerful tool for the diagnosis of neurobrucellosis. J Neurol Neurosurg Psychiatry 76(7):1025–1027

Colmenero JD, Clavijo E, Morata P, Bravo MJ, Queipo-Ortuño MI (2011) Quantitative real-time polymerase chain reaction improves conventional microbiological diagnosis in an outbreak of brucellosis due to ingestion of unpasteurized goat cheese. Diagn Microbiol Infect Dis 71(3):294–296

Corbel M, Brinley-Morgan W (1984) Genus Brucella: En: section 4. Gram negative aerobic rods and coccid Bergey's manual, John Wiley & Sons, Inc., in association with Bergey's Manual Trust, pp 377–387. https://doi.org/10.1002/9781118960608.gbm00807

Dagli O, Dokur M, Guzeldag G, Ozmen Y (2011) Acute renal failure due to Brucella melitensis. J Infect Dev Ctries 5(12):893–895

De Ley J, Mannheim W, Segers P, Lievens A, Denijn M, Vanhoucke M, Gillis M (1987) Ribosomal ribonucleic acid cistron similarities and taxonomic neighborhood of Brucella and CDC group Vd. Int J Syst Evol Microbiol 37(1):35–42

De BK, Stauffer L, Koylass MS, Sharp SE, Gee JE, Helsel LO, Steigerwalt AG, Vega R, Clark TA, Daneshvar MI (2008) Novel Brucella strain (BO1) associated with a prosthetic breast implant infection. J Clin Microbiol 46(1):43–49

DelVecchio VG, Kapatral V, Redkar RJ, Patra G, Mujer C, Los T, Ivanova N, Anderson I, Bhattacharyya A, Lykidis A (2002) The genome sequence of the facultative intracellular pathogen Brucella melitensis. Proc Natl Acad Sci 99(1):443–448

Doganay M, Aygen B (2003) Human brucellosis: an overview. Int J Infect Dis 7(3):173–182

Ducrotoy M, Bertu W, Ocholi R, Gusi A, Bryssinckx W, Welburn S, Moriyón I (2014) Brucellosis as an emerging threat in developing economies: lessons from Nigeria. PLoS Negl Trop Dis 8(7):e3008. https://doi.org/10.1371/journal.pntd.0003008

El Kholy AA, Gomaa HE, El Anany MG, Abdel Rasheed E (2009) Diagnosis of human brucellosis in Egypt by polymerase chain reaction. EMHJ – East Mediterr Health J 15(5):1068–1074. http://www.who.int/iris/handle/10665/117734

Fekete A, Bantle J, Halling S, Sanborn MR (1990) Rapid, sensitive detection of Brucella abortus by polymerase chain reaction without extraction of DNA. Biotechnol Tech 4(1):31–34

Fekete A, Bantle JA, Halling SM (1992) Detection of Brucella by polymerase chain reaction in bovine fetal and maternal tissues. J Vet Diagn Investig 4(1):79–83

Ferreira L, Castaño SV, Sánchez-Juanes F, González-Cabrero S, Menegotto F, Orduña-Domingo A, González-Buitrago JM, Muñoz-Bellido JL (2010) Identification of Brucella by MALDI-TOF mass spectrometry. Fast and reliable identification from agar plates and blood cultures. PLoS One 5(12):e14235

Foster G, Osterman BS, Godfroid J, Jacques I, Cloeckaert A (2007) Brucella ceti sp. nov. and Brucella pinnipedialis sp. nov. for Brucella strains with cetaceans and seals as their preferred hosts. Int J Syst Evol Microbiol 57(11):2688–2693

Fox KF, Fox A, Nagpal M, Steinberg P, Heroux K (1998) Identification of Brucella by ribosomal-spacer-region PCR and differentiation of Brucella canis from other Brucella spp. pathogenic for humans by carbohydrate profiles. J Clin Microbiol 36(11):3217–3222

Franco M, Mulder M, Gilman R, Smits H (2007) Human brucellosis. Lancet Infect Dis 7:775–786

Galinska EM, Zagórski J (2013) Brucellosis in humans-etiology, diagnostics, clinical forms. Ann Agric Environ Med 20(2):233–238

Gemechu MY, Gill JPS, Arora AK, Ghatak S, Singh DK (2011) Polymerase chain reaction (PCR) assay for rapid diagnosis and its role in prevention of human brucellosis in Punjab, India. Int J Prevent Med 2(3):170

Ginzinger DG (2002) Gene quantification using real-time quantitative PCR: an emerging technology hits the mainstream. Exp Hematol 30(6):503–512

Godfroid J, Scholz H, Barbier T, Nicolas C, Wattiau P, Fretin D, Whatmore A, Cloeckaert A, Blasco J, Moriyon I (2011) Brucellosis at the animal/ecosystem/human interface at the beginning of the 21st century. Prev Vet Med 102(2):118–131

Gopaul KK, Koylass MS, Smith CJ, Whatmore AM (2008) Rapid identification of Brucella isolates to the species level by real time PCR based single nucleotide polymorphism (SNP) analysis. BMC Microbiol 8(1):86

Gwida MM, El-Gohary AH, Melzer F, Tomaso H, Rösler U, Wernery U, Wernery R, Elschner MC, Khan I, Eickhoff M (2011) Comparison of diagnostic tests for the detection of Brucella spp. in camel sera. BMC Res Notes 4(1):525

Halling SM, Tatum FM, Bricker BJ (1993) Sequence and characterization of an insertion sequence, IS711, from Brucella ovis. Gene 133(1):123–127

Herman L, De Ridder H (1992) Identification of Brucella spp. by using the polymerase chain reaction. Appl Environ Microbiol 58(6):2099–2101

Hoover-Miller A, Dunn J, Field C, Blundell G, Atkinson S (2017) Seroprevalence of Brucella antibodies in harbor seals in Alaska, USA, with age, regional, and reproductive comparisons. Dis Aquat Org 126(1):1–12

Hoyer BH, McCullough NB (1968a) Homologies of deoxyribonucleic acids from Brucella ovis, canine abortion organisms, and other Brucella species. J Bacteriol 96(5):1783–1790

Hoyer BH, McCullough NB (1968b) Polynucleotide homologies of Brucella deoxyribonucleic acids. J Bacteriol 95(2):444–448
Huber B, Scholz HC, Lucero N, Busse H-J (2009) Development of a PCR assay for typing and subtyping of Brucella species. Int J Med Microbiol 299(8):563–573
Imaoka K, Kimura M, Suzuki M, Kamiyama T, Yamada A (2007) Simultaneous detection of the genus Brucella by combinatorial PCR. Jpn J Infect Dis 60(2/3):137
Jiang H, Fan M, Chen J, Mi J, Yu R, Zhao H, Piao D, Ke C, Deng X, Tian G (2011) MLVA genotyping of Chinese human Brucella melitensis biovar 1, 2 and 3 isolates. BMC Microbiol 11(1):256
Kadanali A, Ozden K, Altoparlak U, Erturk A, Parlak M (2009) Bacteremic and nonbacteremic brucellosis: clinical and laboratory observations. Infection 37(1):67–69
Kaden R, Ferrari S, Alm E, Wahab T (2017) A novel real-time PCR assay for specific detection of Brucella melitensis. BMC Infect Dis 17(1):230
Kattar MM, Zalloua PA, Araj GF, Samaha-Kfoury J, Shbaklo H, Kanj SS, Khalife S, Deeb M (2007) Development and evaluation of real-time polymerase chain reaction assays on whole blood and paraffin-embedded tissues for rapid diagnosis of human brucellosis. Diagn Microbiol Infect Dis 59(1):23–32
Kattar MM, Jaafar RF, Araj GF, Le Flèche P, Matar GM, Rached RA, Khalife S, Vergnaud G (2008) Evaluation of a multilocus variable-number tandem-repeat analysis scheme for typing human Brucella isolates in a region of brucellosis endemicity. J Clin Microbiol 46(12):3935–3940
Koylass MS, King AC, Edwards-Smallbone J, Gopaul KK, Perrett LL, Whatmore AM (2010) Comparative performance of SNP typing and 'Bruce-ladder' in the discrimination of Brucella suis and Brucella canis. Vet Microbiol 142(3–4):450–454
Le Fleche P, Jacques I, Grayon M, Al Dahouk S, Bouchon P, Denoeud F, Nockler K, Neubauer H, Guilloteau LA, Vergnaud G (2006) Evaluation and selection of tandem repeat loci for a Brucella MLVA typing assay. BMC Microbiol 6:9. https://doi.org/10.1186/1471-2180-6-9
Leal-Klevezas DS, Martínez-Vázquez IO, Lopez-Merino A, Martínez-Soriano JP (1995) Single-step PCR for detection of Brucella spp. from blood and milk of infected animals. J Clin Microbiol 33(12):3087–3090
Lopez-Goñi I, Garcia-Yoldi D, Marin C, De Miguel M, Munoz P, Blasco J, Jacques I, Grayon M, Cloeckaert A, Ferreira A (2008) Evaluation of a multiplex PCR assay (Bruce-ladder) for molecular typing of all Brucella species, including the vaccine strains. J Clin Microbiol 46(10):3484–3487
Lübeck PS, Skurnik M, Ahrens P, Hoorfar J (2004) A multiplex PCR-detection assay for Yersinia enterocolitica serotype O: 9 and Brucella spp. based on the perosamine synthetase gene. In: The genus Yersinia. Springer, New York, pp 451–454
Lucero N, Ayala S, Escobar G, Jacob N (2008) Brucella isolated in humans and animals in Latin America from 1968 to 2006. Epidemiol Infect 136(4):496–503
Lucero N, Corazza R, Almuzara M, Reynes E, Escobar G, Boeri E, Ayala S (2010) Human Brucella canis outbreak linked to infection in dogs. Epidemiol Infect 138(2):280–285
Maas KS, Mendez M, Zavaleta M, Manrique J, PÍA M, Mulder M, Bonifacio N, Castaneda ML, Chacaltana J, Yagui E (2007) Evaluation of brucellosis by PCR and persistence after treatment in patients returning to the hospital for follow-up. Am J Trop Med Hyg 76(4):698–702
Mantur BG, Mangalgi SS (2004) Evaluation of conventional castaneda and lysis centrifugation blood culture techniques for diagnosis of human brucellosis. J Clin Microbiol 42(9):4327–4328
Maquart M, Fardini Y, Zygmunt MS, Cloeckaert A (2008) Identification of novel DNA fragments and partial sequence of a genomic island specific of Brucella pinnipedialis. Vet Microbiol 132(1–2):181–189
Mayer-Scholl A, Draeger A, Göllner C, Scholz HC, Nöckler K (2010) Advancement of a multiplex PCR for the differentiation of all currently described Brucella species. J Microbiol Methods 80(1):112–114
Mayfield JE, Bricker BJ, Godfrey H, Crosby RM, Knight DJ, Hailing SM, Balinsky D, Tabatabai LB (1988) The cloning, expression, and nucleotide sequence of a gene coding for an immunogenic Brucella abortus protein. Gene 63(1):1–9

McDermott JJ, Arimi S (2002) Brucellosis in sub-Saharan Africa: epidemiology, control and impact. Vet Microbiol 90(1–4):111–134

Mitka S, Anetakis C, Souliou E, Diza E, Kansouzidou A (2007) Evaluation of different PCR assays for early detection of acute and relapsing brucellosis in humans in comparison with conventional methods. J Clin Microbiol 45(4):1211–1218

Morata P, Queipo-Ortuño MI, de Dios Colmenero J (1998) Strategy for optimizing DNA amplification in a peripheral blood PCR assay used for diagnosis of human brucellosis. J Clin Microbiol 36(9):2443–2446

Morata P, Queipo-Ortuno M, Reguera J, Miralles F, Lopez-Gonzalez J, Colmenero J (2001) Diagnostic yield of a PCR assay in focal complications of brucellosis. J Clin Microbiol 39(10):3743–3746

Navarro E, Segura JC, Jesús Castaño M, Solera J (2006) Use of real-time quantitative polymerase chain reaction to monitor the evolution of Brucella melitensis DNA load during therapy and post-therapy follow-up in patients with brucellosis. Clin Infect Dis 42(9):1266–1273

Navarro-Martínez A, Navarro E, Castaño MJ, Solera J (2008) Rapid diagnosis of human brucellosis by quantitative real-time PCR: a case report of brucellar spondylitis. J Clin Microbiol 46(1):385–387

Neubauer H (2010) Brucellosis: new demands in a changing world. Prilozi 31(1):209–217

Newby DT, Hadfield T, Roberto FF (2003) Real-time PCR detection of Brucella abortus: a comparative study of SYBR green I, 5′-exonuclease, and hybridization probe assays. Appl Environ Microbiol 69(8):4753–4759

Nimri LF (2003) Diagnosis of recent and relapsed cases of human brucellosis by PCR assay. BMC Infect Dis 3(1):5

Njeru J, Melzer F, Wareth G, El-Adawy H, Henning K, Pletz MW, Heller R, Kariuki S, Fèvre E, Neubauer H (2016) Human brucellosis in febrile patients seeking treatment at remote hospitals, northeastern Kenya, 2014–2015. Emerg Infect Dis 22(12):2160

Nöckler K, Maves R, Cepeda D, Draeger A, Mayer-Scholl A, Chacaltana J, Castañeda M, Espinosa B, Castillo R, Hall E (2009) Molecular epidemiology of Brucella genotypes in patients at a major hospital in Central Peru. J Clin Microbiol 47(10):3147–3155

Ohtsuki R, Kawamoto K, Kato Y, Shah M, Ezaki T, Makino SI (2008) Rapid detection of Brucella spp. by the loop-mediated isothermal amplification method. J Appl Microbiol 104(6):1815–1823

Osterman B, Moriyon I (2006) International committee on systematics of prokaryotes; subcommittee on the taxonomy of Brucella: minutes of the meeting, 17 September 2003, Pamplona, Spain. Int J Syst Evol Microbiol 56(5):1173–1176

Ouahrani S, Michaux S, Widada JS, Bourg G, Tournebize R, Ramuz M, Liautard J-P (1993) Identification and sequence analysis of IS6501, an insertion sequence in Brucella spp.: relationship between genomic structure and the number of IS6501 copies. Microbiology 139(12):3265–3273

Pappas G, Papadimitriou P, Akritidis N, Christou L, Tsianos EV (2006) The new global map of human brucellosis. Lancet Infect Dis 6(2):91–99

Piranfar V, Sharif M, Hashemi M, Vahdati AR, Mirnejad R (2015) Detection and discrimination of two Brucella species by multiplex real-time PCR and high-resolution melt analysis curve from human blood and comparison of results using RFLP. Iran J Basic Med Sci 18(9):909

Probert WS, Schrader KN, Khuong NY, Bystrom SL, Graves MH (2004) Real-time multiplex PCR assay for detection of Brucella spp., B. abortus, and B. melitensis. J Clin Microbiol 42(3):1290–1293

Queipo-Ortuño MI, Morata P, Ocón P, Manchado P, Colmenero JD (1997) Rapid diagnosis of human brucellosis by peripheral-blood PCR assay. J Clin Microbiol 35(11):2927–2930

Queipo-Ortuño M, Colmenero J, Reguera J, Garcia-Ordonez M, Pachón M, Gonzalez M, Morata P (2005) Rapid diagnosis of human brucellosis by SYBR green I-based real-time PCR assay and melting curve analysis in serum samples. Clin Microbiol Infect 11(9):713–718

Queipo-Ortuño M, Colmenero J, Bravo M, Garcia-Ordonez M, Morata P (2008) Usefulness of a quantitative real-time PCR assay using serum samples to discriminate between inactive, serologically positive and active human brucellosis. Clin Microbiol Infect 14(12):1128–1134

Rijpens NP, Jannes G, Van Asbroeck M, Rossau R, Herman L (1996) Direct detection of Brucella spp. in raw milk by PCR and reverse hybridization with 16S-23S rRNA spacer probes. Appl Environ Microbiol 62(5):1683–1688

Rodríguez-Lázaro D, López-Enríquez L, Ocampo-Sosa AA, Muñoz P, Blasco JM, Marín C, Hernández M (2017) Evaluation of eryC as a molecular marker for the quantitative detection of Brucella Spp. by real-time PCR in food samples. Food Anal Methods 10(5):1148–1155

Romero C, Lopez-Goñi I (1999) Improved method for purification of bacterial DNA from bovine milk for detection of Brucella spp. by PCR. Appl Environ Microbiol 65(8):3735–3737

Romero C, Gamazo C, Pardo M, Lopez-Goñi I (1995) Specific detection of Brucella DNA by PCR. J Clin Microbiol 33(3):615–617

Sanjuan-Jimenez R, Colmenero JD, Morata P (2017) Lessons learned with molecular methods targeting the BCSP-31 membrane protein for diagnosis of human brucellosis. Clin Chim Acta 469:1–9

Scholz H, Vergnaud G (2013) Molecular characterisation of Brucella species. Rev Sci Tech 32(1):149–162

Scholz HC, Tomaso H, Dahouk SA, Witte A, Schloter M, Kämpfer P, Falsen E, Neubauer H (2006) Genotyping of Ochrobactrum anthropi by recA-based comparative sequence, PCR-RFLP, and 16S rRNA gene analysis. FEMS Microbiol Lett 257(1):7–16

Scholz HC, Pfeffer M, Neubauer H, Tomaso H (2007) Evaluation of genus-specific and species-specific real-time PCR assays for the identification of Brucella spp. Clin Chem Lab Med 45(11):1464–1470

Scholz HC, Al Dahouk S, Tomaso H, Neubauer H, Witte A, Schloter M, Kämpfer P, Falsen E, Pfeffer M, Engel M (2008) Genetic diversity and phylogenetic relationships of bacteria belonging to the Ochrobactrum–Brucella group by recA and 16S rRNA gene-based comparative sequence analysis. Syst Appl Microbiol 31(1):1–16

Scholz HC, Nöckler K, Göllner C, Bahn P, Vergnaud G, Tomaso H, Al Dahouk S, Kämpfer P, Cloeckaert A, Maquart M (2010) Brucella inopinata sp. nov., isolated from a breast implant infection. Int J Syst Evol Microbiol 60(4):801–808

Shehabi A, Shakir K, El-Khateeb M, Qubain H, Fararjeh N, Shamat ARA (1990) Diagnosis and treatment of 106 cases of human brucellosis. J Infect 20(1):5–10

Singhal N, Kumar M, Kanaujia PK, Virdi JS (2015) MALDI-TOF mass spectrometry: an emerging technology for microbial identification and diagnosis. Front Microbiol 6:791

Surucuoglu S, El S, Ural S, Gazi H, Kurutepe S, Taskiran P, Yurtsever SG (2009) Evaluation of real-time PCR method for rapid diagnosis of brucellosis with different clinical manifestations. Pol J Microbiol 58(1):15–19

Tiller RV, Gee JE, Frace MA, Taylor TK, Setubal JC, Hoffmaster AR, De BK (2010) Characterization of novel Brucella strains originating from wild native rodent species in North Queensland, Australia. Appl Environ Microbiol 76(17):5837–5845

Valdezate S, Cervera I, Hernandez P, Navarro A, Saez Nieto J (2007) Characterisation of human outbreaks of brucellosis and sporadic cases by the use of hyper-variable octameric oligonucleotide fingerprint (HOOF) variable number tandem repeats. Clin Microbiol Infect 13(9):887–892

Verger J-M, Grimont F, Grimont PA, Grayon M (1985) Brucella, a monospecific genus as shown by deoxyribonucleic acid hybridization. Int J Syst Evol Microbiol 35(3):292–295

Vrioni G, Pappas G, Priavali E, Gartzonika C, Levidiotou S (2008) An eternal microbe: Brucella DNA load persists for years after clinical cure. Clin Infect Dis 46(12):e131–e136

Wang Y, Wang Z, Zhang Y, Bai L, Zhao Y, Liu C, Ma A, Yu H (2014) Polymerase chain reaction–based assays for the diagnosis of human brucellosis. Ann Clin Microbiol Antimicrob 13(1):31

Wareth G, Melzer F, Elschner MC, Neubauer H, Roesler U (2014) Detection of Brucella melitensis in bovine milk and milk products from apparently healthy animals in Egypt by real-time PCR. J Infect Dev Ctries 8(10):1339–1343. https://doi.org/10.3855/jidc.4847

Whatmore AM, Perrett LL, MacMillan AP (2007) Characterisation of the genetic diversity of Brucella by multilocus sequencing. BMC Microbiol 7(1):34

Whatmore AM, Davison N, Cloeckaert A, Al Dahouk S, Zygmunt MS, Brew SD, Perrett LL, Koylass MS, Vergnaud G, Quance C (2014) Brucella papionis sp. nov., isolated from baboons (Papio spp.). Int J Syst Evol Microbiol 64(12):4120–4128

Whatmore AM, Koylass MS, Muchowski J, Edwards-Smallbone J, Gopaul KK, Perrett LL (2016) Extended multilocus sequence analysis to describe the global population structure of the genus Brucella: phylogeography and relationship to biovars. Front Microbiol 7:2049

Winchell JM, Wolff BJ, Tiller R, Bowen MD, Hoffmaster AR (2010) Rapid identification and discrimination of Brucella isolates by use of real-time PCR and high-resolution melt analysis. J Clin Microbiol 48(3):697–702

Yagupsky P (1999) Detection of Brucellae in blood cultures. J Clin Microbiol 37(11):3437–3442

Young EJ (1991) Serologic diagnosis of human brucellosis: analysis of 214 cases by agglutination tests and review of the literature. Rev Infect Dis 13(3):359–372

Young EJ (1995) An overview of human brucellosis. Clin Infect Dis 21(2):283–289

Young EJ (2006) Brucella spp. In: Gillepsie SH, Hawkey PM (eds) Principles and practice of clinical bacteriology, 2nd edn. Wiley, Chichester, pp 265–271

Yu WL, Nielsen K (2010) Review of detection of Brucella spp. by polymerase chain reaction. Croat Med J 51(4):306–313

Zerva L, Bourantas K, Mitka S, Kansouzidou A, Legakis N (2001) Serum is the preferred clinical specimen for diagnosis of human brucellosis by PCR. J Clin Microbiol 39(4):1661–1664

Zhang B, Wear DJ, Stojadinovic A, Izadjoo M (2013) Sequential real-time PCR assays applied to identification of genomic signatures in formalin-fixed paraffin-embedded tissues: a case report about Brucella-induced osteomyelitis. Mil Med 178(1):88–94

Zhong Z, Yu S, Wang X, Dong S, Xu J, Wang Y, Chen Z, Ren Z, Peng G (2013) Human brucellosis in the People's Republic of China during 2005–2010. Int J Infect Dis 17(5):e289–e292. https://doi.org/10.1016/j.ijid.2012.1012.1030. Epub 2013 Feb 1023

Zygmunt MS, Maquart M, Bernardet N, Doublet B, Cloeckaert A (2010) Novel IS711-specific chromosomal locations useful for identification and classification of marine mammal Brucella strains. J Clin Microbiol 48(10):3765–3769

Arbuscular Mycorrhizal Fungi Remediation Potential of Organic and Inorganic Compounds

11

Thinhinane Fecih and Hafida Baoune

Abstract

Industrialization has led to the serious issue of soil contamination caused by both organic and inorganic pollutants which are persistent and could generate dangerous human diseases. In the last few years, several researches about the dangerous effects of these compounds on biodiversity and human health were carried out to find and to develop new techniques that reduce or eliminate pollution without causing negative effects on the environment. Mycorrhizoremediation is one of the biological solutions that can solve the problem of pollution; using mycorrhizal fungi in bioremediation can not only remove pollutants but also enhance host plant development and even help synergically the growth of non-mycorrhizal plants. This chapter will highlight the potential of arbuscular mycorrhizal fungi (AMF) inocula on bioremediation and methods of inoculum production and successful application and commercialization.

Keywords

Bioremediation · Mycorrhizoremediation · Arbuscular mycorrhizal fungi (AMF) · Organic and inorganic pollutants

T. Fecih
Laboratoire de Bio-Resources Sahariennes FNSV, Université Kasdi Merbah Ouragla, Ouargla, Algeria

H. Baoune (✉)
Laboratoire de protection des écosystème en zones arides et semi-arides, FNSV, Université Kasdi Merbah Ouragla, Ouargla, Algeria

P. K. Arora (ed.), *Microbial Technology for the Welfare of Society*, Microorganisms for Sustainability 17,
https://doi.org/10.1007/978-981-13-8844-6_11

Abbreviations

ACC	1-Aminocyclopropane Carboxylate Deaminase
AMF	Arbuscular Mycorrhizal Fungi
EcM	Ectomycorrhiza
HM	Heavy Metals
PAHs	Poly Aromatic Hydrocarbons
PGPR	Plant Growth-Promoting Rhizobacteria
PIPs	Persistent Inorganic Pollutants
POPs	Persistent Organic Pollutants
UNEP	United Nations Environmental Pollution

11.1 Introduction

Chemicals and inorganic compounds manufacturies are unavoidably a pollution source of our aquatic and terrestrial environments; the persistence of these pollutants in the nature constitutes a serious threat to human health and environment. Physicochemical and biological strategies have been widely used to remediate these sites (Diane 2016). Symbiotic interactions particularly mycorrhizal associations appear to be consequential for exploitation in bioremediation in polluted ecosystems (Giasson et al. 2008; Singh and Ward 2010). The use of mycorrhizae in bioremediation is termed mycorrhizoremediation (Whitacre 2015); their capacity to remediate polluted areas has been widely explored (Bertold 2012). Mycorrhizae increase plant development and enhance their resistance to different environmental stresses (Souza 2015), such as soilborne pathogen attack, aggressive agricultural weeds, drought and salinity stress, as well as the presence of toxic pollutants and desertification (Bertold 2012). Furthermore, mycorrhizae have been well known as a key indicator for soil contamination and omnipresent and furnish a direct link between root hair and the substrate. They also ensure the viability and the growth of certain plants and transfer to them elements and contaminants from soil (Bhalerao 2013).

Many reasons were attributed to the extensive production of arbuscular mycorrhizal fungi (AMF) inocula in the last few years: their favorable effect on the health and growth of plants, restoration of ecosystems, disease management, and phytoremediation (Siddiqui et al. 2008). In addition, the awareness of the society about the necessity to replace agrochemicals which constitute a severe threat for the biodiversity (especially soil microorganisms) by other biological solutions is of paramount importance for a sustainable agriculture and forestry (Ahmad et al. 2011).

In this chapter, the role of arbuscular mycorrhizal fungi (AMF) in the bioremediation of contaminated sites, the effect of their inoculation on host plants and the ecosystem, as well as their application and commercial production are discussed.

11.2 Persistent Organic (POPs) and Inorganic Pollutants (PIPs)

POPs and PIPs are two classes of accumulative and toxic contaminants which have a perpetual presence in the environment because of their high resistance to being converted by biotic and abiotic transformation processes (Yong 2001; Duffus et al. 2017). Industrial activities generate pollution in different habitats, producing synthetic compounds which have toxic effects on living organisms (carcinogenic and mutagenic or teratogenic) (Doble and Kruthiventi 2005). Thus, for human health protection, the United Nations Environmental Pollution (UNEP) arranged a treaty to stop the production and utilization of POPs. Moreover, for more than 10 years, 20 POPs have been listed in a treaty created by the Stockholm Convention which was signed by over 151 countries (Fiedler 2003; Bharagava 2017). Polluted areas are rarely contaminated by one kind of pollutants. In general, there are different types of organic and inorganic contaminants or a mixture of both which require different remediation methods in a specific order. Remediation strategy may involve a chain of treatments in order to destroy all contaminants coexisting with various concentrations in the environment, without neglecting the interaction existing between contaminants and other chemical elements which can influence the reaction of others in the ecosystem (Yong 2001; Kennen and Niall 2015)

Physicochemical strategies for the remediation of polluted sites are less efficient and cost prohibitive and frequently originate sludge (Prasad 2018). They lead to the alteration of soil structure (Andrews et al. 2003) and generate secondary more toxic compounds (Luo and Tu 2018). However, bioremediation is one of the most promising, more efficient, sustainable, cost-effective and environmentally friendly methods (Fuentes et al. 2018). This method basically refers to the use of living microorganisms or plants to reduce the concentration or toxic effects of pollutants (Fulekar 2010). AM fungi are one of the effectual bioremediation techniques that have been mostly used in the remediation of heavy metals (HM) and polycyclic aromatic hydrocarbons (PAHs) in polluted soils (Bertold 2012).

Plants and their related microbial populations in the rhizosphere play an important role in removal and transformation of pollutants. However, plants are not able to degrade some man-made contaminants. Researchers are working on developing these bioremediation strategies to clean up most of the synthetic pollutants at low costs and without affecting the environment (Singh and Ward 2010; Amirad 2011)

11.3 Potential of AMF in Bioremediation of Organic and Inorganic Compounds

AM fungi are obligate root symbionts that are able to associate with most of the plant species in several habitats (Souza 2015). They belong to the phylum of *Glomeromycota*, which has 3 classes (*Glomeromycetes*, *Archaeosporomycetes*, and *Paraglomeromycetes*), 5 orders (*Diversisporales*, *Glomerales*, *Paraglomerales*,

Gigasporales, and *Archaeosporales*), 14 families, and 26 genera (Singh 2016). AMF have the ability to resist toxic elements; their cell walls possess proteins that are able to absorb and sequester toxic substances (Prasad 2018). Although significant researches have been carried out to understand physiological reaction of AMF to soil contaminants and their ability in bioremediation, the key aspects are still not clear (Azooz and Parvaiz 2016). The exploration of the potential of AMF in degrading POPs has not been attempted because of their difficult culture outside host plant and their limited extracellular hyphal network compared to ectomycorrhizal (EcM) associations. However, they have a significant role in POP degradation by colonizing roots and promoting plants to survive on extremely polluted soils by crude oil, oil shale wastes, coal wastes, lignite, and calcite mine spoils (Gadd 2001).

Glomalin is a glycoprotein produced by AMF that helps in the sequestration of toxic trace elements in polluted soils and reduces their viability (Singh 2017; Prasad 2018). The extra radical mycelium of AM fungi plays a key role in enhancing host plant development, aggregation of soil particles, and ensuring soil stability (Singh 2017).

Mycorrhiza-assisted remediation is one of the successful bioremediation technologies. This technology ameliorates soil structure and enhances nutrient acquisition by plants, and it is used as an effective tool in the bioremediation of soil from both organic and inorganic toxic substances (Prasad 2018) since it has been well known as a rapid and cost-effective method (Chibuike 2013). Furthermore, arbuscular mycorrhizal fungi (AMF) play an essential role in restoring damaged ecosystems by facilitating plants installation and development in soil contaminated by POPs as well as protecting them from the toxic effect of those compounds. AMF also increase soil bioremediation by stimulating the activity of soil microbial population and enhancing their structure (Lenoir et al. 2016).

Arbuscular mycorrhizal fungi (AMF) constitute the main biosystem responsible for regulating the comportment of plants in polluted soil (Singh and Ward 2010). The adapted strains of AMF could be selected and inoculated on the plants of polluted ecosystems in order to improve the bioremediation. The interaction between adapted AMF and rhizobacteria enhances synergically the effectiveness of phytoremediation (Bertold 2012). AMF have the ability to tolerate and adapt to different environmental stresses; they are capable of cleaning up heavy metals (HM) from soil using their hyphal network, improving the phytoextraction of HM or breaking off their propagation in the substrate, accumulating toxic metals in fungal hyphae, or chelating them by hyphae or sticking these metals with root surface or chitin of fungal cell wall (Shakeel and Yaseen 2015). Heavy metals resistance AM fungi can be a biotechnological tool to increase phytoremediation efficiency by inoculating them in plants (Upadhyaya et al. 2010). The contribution of AMF–plant partnership in phytoremediation of heavy metals mostly concerns Cd, Zn, Cu, and Pb (Pichardo 2012; Jiang et al. 2016; Fu et al. 2016; Watts-Williams et al. 2017; Cornejo et al. 2017).

Further, the remediation of other metals has also been attempted, such as arsenic (Huang et al. 2018) and iron-cyanide (Fe-CN) complexes polluted soils which has been indicated for the first time by (Sut et al. 2016) inoculating AMF with other microorganisms in HM multicontaminated soils increased plant biomass, produced the greater symbiotic AMF colonization and nodulation rates as improved bioremediation efficiency of HM (Azcón et al. 2010).

Further studies demonstrate that arbuscular mycorrhizal (AM) hyphae showed a great potential to absorb polycyclic aromatic hydrocarbons (PAHs) (Gao et al. 2010). The results of Ren et al. (2017) confirmed that the beneficial relation between rhizobia and AMF helps to enhance phytoremediation efficiency of PAHs in which up 97% of phenanthrene and 85–87% of pyrene have been dissipated by the tripartite symbiosis AMF (*Glomus mosseae*) in association with rhizobia (*Ensifer* sp.) and a legume plant (*Sesbania cannabina*). The highest degradation of PAHs (93.4%) was obtained in the association of AMF (*Glomus caledoniun L.*), earthworms (*Eisenia foetida*) and tall fescue (*Festuca arundinacea*), in which after 120 days, a decrease of the concentration of PAHs from an initial value of 620–41 mg kg(−1) was recorded (Lu and Lu 2015). In previous studies, the inoculation of leek (*Allium porrum* L. cv. *Musselburgh*) with AMF (with either *Glomus intraradices* or *G. versiforme)* and the addition of soil microorganism increased PAHs removal as improved plant development (Liu and Dalpé 2009). The study conducted by Małachowska-Jutsz and Kalka (2010), demonstrated that the inoculation of AMF alone showed less effect than the use of inoculated wheat with AMF. Moreover, plants co-inoculated with PGPR and AMF had a greater potential to remediate saline-alkali soil polluted with petroleum (Xun et al. 2015). The inoculation of arbuscular mycorrhizal fungi (AMF) coupled with nutrient solution showed the highest effectiveness on degradation of petroleum hydrocarbons (PHs) with a rate of 83.27% and better plant biomass (Hatami et al. 2018). AMF also ameliorate drought tolerance by (Zarik et al. 2016; Abdel-Salam et al. 2017; Zhang et al. 2018; Pavithra and Yapa 2018) improving the resistance of water stress (Ming and Hui 1999; Asrar et al. 2012; Birhane et al. 2012), enhancing tolerance to salinity (Chen et al. 2017; Hashem et al. 2018; Wang et al. 2018; Pollastri et al. 2018) and preventing soilborne diseases (Veresoglou and Rillig 2012).

11.4 Production and Application of AMF Inoculum

To be produced, mycorrhizal inoculants necessitate biotechnological competences, with juridical, ethical, educative, and commercial exigencies (Ahmad et al. 2011). AMF are obligate biotrophs; they can be cultivated by either germination of spores or extension of mycelium through roots and soil (Siddiqui et al. 2008). The trials of AMF inoculum production afar/outside of their host plant were low or not successful. However, the unique method for culturing these fungi is in collaboration with their host plant root, and the only suitable technique for producing big quantities of inoculum is the pot culture method (Singh 2016). Nowadays, the challenging objective for microbiologists is to develop a rentable technology to produce AM fungal inoculum with a large amount and excellent quality (Varma et al. 2017). Consequently, several techniques for mass multiplication of AMF inoculum comprising substrate-based (pot culture, inoculum rich pallets) and substrate-free (aeroponic, monoaxenic, hydroponic) production systems were developed, since each of them has its benefits and disadvantages (Elevitch 2004; Khan et al. 2010). Nevertheless, pot culture offers often doubtable information concerning the abundance, diversity and the activity of the indigenous AMF which should be known before using, and preventing the possibility of transferring weed seeds and pathogens (Elevitch 2004). Unlike soil-based techniques, soilless cultural system allows better control of physical and chemical features of the cultivated media and provides limitation of contamination with other microorganisms (Arora et al. 2004),

While aeroponic culture ensures spores' growth with less contamination and facilitates nutrients' uptake of colonized plants. Generally, cultured plants are transported to a controlled aeroponic chamber after being inoculated with viable and pure spores, to obtain successful results, and fine droplets (of 45 mm size) of nutrient solution are added on plants' roots at an appropriate adequate time (Ahmad et al. 2011). Previous studies have demonstrated that monoaxenic culture is the most interesting and developed cultivation system for AMF, in which the development of symbiosis is established in vitro on a specific medium using root-inducing transfer DNA of transformed roots of the associated plant (Khan et al. 2010). Although this method provides effective production of AMF inoculum which can be applied directly as an inoculum or starter inoculum for large-scale production (Siddiqui et al. 2008), it requires complex, costly material (Varma et al. 2017). Compared to traditional pot culture method, root-organ culture with transformed or non-transformed roots produces lower AMF spores which is considered to be its main limitation (Mukerji et al. 2002). For hydroponics (nutrient film technique), host plants' growth is carried out in an inclined tray in which fine droplets of nutrient solution are added. Thus, as for aeroponic culture, this method requires precolonization of seedlings in another media (Mukerji 2004). Both of these techniques produce great colonized root inocula with a low difference in sporulation, where aeroponic culture was quite better (Mukerji et al. 2002). The great convenience between these two techniques is the discharge from any substrate in the production of colonized roots and spores which allows higher production, efficiency and distribution of inoculum (Varma and Oelmüller 2007). Therefore, the most used system in commercial nurseries for rehabilitation programs is the aeroponic culture method, because the on-site inoculum production allows a time-optimal utilization of fresh mycorrhizal propagules (Arora et al. 2004). The growth period of cultures is 4–6 months, to ensure that all genera are sporulated, and then stocked as air-dried soil for more than 5 years but the viability of AMF isolates throughout the entire time of storing cannot be guaranteed (Arora et al. 2004).

The storing of AMF spores is generally performed at 4 °C in dried pot-culture soil or by cryopreservation of spores at −60 °C to −70 °C (Varma and Oelmüller 2007). It can be applied by using two principal methods which are AM inoculum at sowing time or at the time of transplanting (Mehrotra 2005). In order to prepare a specific zone to mycorrhiza-assisted remediation technology, some steps must be taken in consideration: recognition and evaluation of risks generated by the characterization of the polluted compounds and choosing the correspondent phytoremediation technology. However, the occurrence of suitable technology is based on the choice of the adequate host plant, examination and selection of effectual AMF stains and initiation of plant—fungi pairs collaboration. Finally surveying the success of the chosen technology by using bioindicators and biomonitoring for human health risk should be taken in consideration (Zaidi et al. 2017).

11.5 Successful Inoculation and Commercialization of AMF Inocula

Different methods are used for isolating spores from the rhizosphere of host plants growing in polluted sites like wet sieving and decanting techniques and/or sucrose density layers. (Siddiqui et al. 2008). Most of the inoculation assays (65%) are effectuated in greenhouse-grown plants, while 24% in open field conditions, where AMF inoculation was successful in both these experimental conditions (Ciancio et al. 2017). The production of commercial AMF inoculants is carried out in controlled conditions in greenhouse. It is composed of diverse fungal forms (residues of mycorrhizal root, hyphal fragments and spores) which are obtained from host plants used for sporulation like maize, onion, sorghum; substrates like sand, peat, expanded clay, perlite, vermiculite, zeolite, rock wool or by hydroponics which are frequently used for mass production in bags, beds or pots (Mukerji et al. 2002) (Reddy and Saravanan 2013). It is commercialized by numerous companies in the form of powder, tablets/pellets or granules, gel beads and balls (Declerck et al. 2005). The inoculum production may include single AM fungal specie or more. In order to obtain maximal benefits, AMF can be combined/mixed with other microorganisms like growth-promoting rhizobacteria (PGPRs) or biocontrol fungi which together ameliorate soil conditions for better plant development (Varma and Kharkwal 2009). The co-inoculant AMF-PGPR is known by its synergic and beneficial effect of enhancing the productivity and agricultural development, protecting the host plant from environmental stress and improving the remediation of polluted soils (Singh 2016). AMF-PGPR partnership could enhance plant growth by producing phytohormones, solubilizing inorganic phosphate, fixing the atmospheric nitrogen, as well as enhancing surface adsorption of nutrients (Pandey et al. 2012). Egamberdieva et al. (2017) demonstrated that different mechanisms are involved in this partnership such as the generation of plant growth regulators such as the ACC deaminase which is the main key hormone to reduce the environmental stress, as well as producing exopolysaccharides, increasing antioxidant defense system, and induction of systemic tolerance. Additionally, the combination of AMF with biocontrol agents, suppress soil borne pathogens, improve bioprotection and activate Plant defense response (Kong and Glick, 2017). Nevertheless, there are diverse obstacles that hamper the production of AMF inoculum: the obligatory requirement of suitable host plant for fungi propagation, the compulsory need of time for plant cultivation and sporulation of AMF, physical room for growing host plant and various mechanical operations for the collection and the preparation of inocula (Goss et al. 2017). The success of inoculation depends on species' compatibility with the biotope conditions and soil nature, inoculation time (Berruti et al. 2016) (Gupta and Sharma 2014) and other factors such as the host plant and fungi, density of AMF propagules, soil nutrients, the efficacy of AMF species and the competition between them and the microbial populations around (Gianinazzi et al. 2003). Other parameters can also affect AMF inoculum production such as lightning, temperature, size of pot, particle of the growth substrate and soil fertility (Varma and Oelmüller 2007).

The reintroduction of AM fungi in places that are characterized by low or inexistent AMF inoculum like eroded or mine soils requires the use of previously inoculated seedlings in nursery (Cardoso et al. 2017).

Commercial inoculant is generally used for enhancing sustainable crop production, but it can be used for non-agricultural reasons including rehabilitation of highly affected soils, degraded or bare area and environmental conservation (Goss et al. 2017).

The high-effective isolates could be identified and selected or developed to be a genetically ameliorated AMF strains adapted for several conditions or host plants (Bhalerao 2013).

11.6 Conclusion

The inoculation of pollutants resistant and/or degrading AMF strains can clean up the contaminated sites and being a starter for an ecological succession thus rehabilitation of perturbed ecosystems. Numerous studies have shown the ability of AMF inoculation in the degradation of some organic and inorganic pollutants. Furthermore, the inoculum could be produced by different methods (soil-based media and soilless media) and could be used for a large-scale application and even commercialized. Many factors can perturb the success of inoculation and bioremediation of the site. These obstacles must be taken into consideration, and more multidisciplinary advanced researches are needed to understand more about their aspects and their physiological response to soil pollutants which are still not fully comprehended. These information may help to evolve the usage of arbuscular mycorrhizal fungi as good candidates to assist phytoremediation and the restoration of the environment.

References

Abdel-Salam E, Abdulrahman A, Mohamed AE (2017) Inoculation with arbuscular mycorrhizal fungi alleviates harmful effects of drought stress on damask rose. Saudi J Biol Sci. https://doi.org/10.1016/j.sjbs.2017.10.015

Ahmad I, Ahmed F, Pichtel J (eds) (2011) Microbes and microbial technology: agricultural and environmental applications. Springer, New York

Amirad JC (2011) Les *risques chimiques environnementaux* – Méthodes d'évaluation et impacts sur les organismes. Tec & Doc Lavoisier, Paris

Andrews JE, Brimblecombe P, Jickells TD et al (2003) An Introduction to Environmental Chemistry (ed) blackwell sciences, 2nd edn, P, p 296

Arora DK, Bridge PD, Bhatnagar D (eds) (2004) Fungal biotechnology in agricultural, food, and environmental applications. Marcel Dekker, New York

Asrar AA, Abdel-Fattah GM, Elhindi KM (2012) Improving growth, flower yield, and water relations of snapdragon (Antirhinum Majus L.) plants grown under well-watered and water-stress conditions using arbuscular mycorrhizal fungi. Photosynthetica 50(2):305–316

Azcón R, Perálvarez MC, Roldán A et al (2010) Arbuscular mycorrhizal fungi, bacillus cereus, and Candida Parapsilosis from a multicontaminated soil alleviate metal toxicity in plants. Microb Ecol 59(4):668–677

Azooz MM, Parvaiz A (2016) Plant environment interaction: responses and approaches to mitigate stress. Wiley, Hoboken

Berruti A, Erica L, Raffaella B et al (2016) Arbuscular mycorrhizal fungi as natural biofertilizers: let's benefit from past successes. Front Microbiol 6(426). https://doi.org/10.3389/fmicb.2015.01559

Bertold H (2012) Fungal Associations. Springer, New York

Bhalerao SA (2013) Arbuscular mycorrhizal fungi: a potential biotechnological tool for phytoremediation of heavy metal contaminated soils. Int J Sci Nat 4:1–15

Bharagava RN (ed) (2017) Environmental pollutants and their bioremediation approaches. CRC Press, Taylor & Francis Group, USA, Boca Raton

Birhane E, Sterck FJ, Fetene M et al (2012) Arbuscular mycorrhizal fungi enhance photosynthesis, water use efficiency, and growth of frankincense seedlings under pulsed water availability conditions. Oecologia 169(4):895–904

Cardoso EJBN, Nogueira MA, Zangaro W (2017) Importance of mycorrhizae in tropical soils. In: De Azevedo JL, Quecine MC (eds) Diversity and benefits of microorganisms from the tropics. Springer International Publishing, Cham. pp 245–267. https://doi.org/10.1007/978-3-319-55804-2

Chen J, Zhang H, Zhang X et al (2017) Arbuscular mycorrhizal symbiosis alleviates salt stress in black locust through improved photosynthesis, water status, and K+/Na+ homeostasis. Front Plant Sci. https://doi.org/10.3389/fpls.2017.01739

Chibuike GU (2013) Use of mycorrhiza in soil remediation: a review. Sci Res Essays 8(35):679–1687

Ciancio A, Pieterse CMJ, Mercado-Blanco J (2017) Harnessing useful rhizosphere microorganisms for pathogen and pest biocontrol. Front Res Topics, Frontiers Media SA. https://doi.org/10.3389/978-2-88945-059-6

Cornejo P, Meier S, García S et al (2017) Contribution of inoculation with arbuscular mycorrhizal fungi to the bioremediation of a copper contaminated soil using oenothera picensis. J Soil Sci Plant Nutr 17(1):14–21

Declerck S, Strullu DG, Fortin JA (eds) (2005) In vitro culture of mycorrhizas. Soil biology, vol 4. Springer, Berlin, pp 49–71

Diane P (2016) Fungal Applications in Sustainable Environmental Biotechnology. Springer, Berlin/Heidelberg/New York

Doble M, Kruthiventi AK (2005) Biotreatment of industrial effluents. Elsevier Butter worth-Heinemann, Amsterdam, Boston

Duffus J, Douglas MT, Michael S et al (2017) Comprehensive glossary of terms used in toxicology. Royal Society of Chemistry, Cambridge, pp 001–004. https://doi.org/10.1039/9781782623724-FP001

Egamberdieva D, Wirth S, Abd-Allah EF (2017) Tripartite interaction among root-associated beneficial microbes under stress. In: Samina M (ed) Rhizotrophs: plant growth promotion to bioremediation. Springer Berlin Heidelberg, New York. pp 219–236

Elevitch CR (ed) (2004) The overstory book: cultivating connections with trees. Permanent Agriculture Resources, Holualoa

Fiedler H (2003) Persistent organic pollutants, vol 3. Springer, Berlin/Heidelberg, p 445

Fu Y, Bruce KE, Wu H, David P et al (2016) Arbuscular mycorrhizal fungi enhance the copper tolerance of Tagetes Patula through the sorption and barrier mechanisms of intraradical hyphae. Metallomics 8(1):61–70

Fuentes MS, Colin VL, Saez JM (2018) Strategies for bioremediation of organic and inorganic pollutants. CRC Press, Boca Raton. https://doi.org/10.1201/b22045

Fulekar MH (ed) (2010) Bioremediation technology: recent advances. Springer, Dordrecht

Gadd GM (2001) Fungi in bioremediation. Cambridge University Press, Cambridge

Gao Y, Cheng Z, Ling W et al (2010) Arbuscular mycorrhizal fungal hyphae contribute to the uptake of polycyclic aromatic hydrocarbons by plant roots. Bioresour Technol 101(18):6895–6901. https://doi.org/10.1016/j.biortech.2010.03.122

Gianinazzi S, Schiiepp H, Barea JM et al (2003) Mycorrhizal technology in agriculture: from genes to bioproducts. Mycorrhiza 13(1):53–54. https://doi.org/10.1007/s00572-002-0217-2

Giasson P, Karam A, Jaouich A (2008) Arbuscular mycorrhizae and alleviation of soil stresses on plant growth. In: Siddiqui ZA, Akhtar MS, Futai K (eds) Mycorrhizae: sustainable agriculture and forestry. Springer, Dordrecht, pp 99–134. https://doi.org/10.1007/978-1-4020-8770-7_4

Goss MJ, Carvalho M, Brito I (2017) Functional diversity of mycorrhiza and sustainable agriculture: management to overcome biotic and abiotic stresses. Elsevier/Academic, London

Gupta SK, Sharma M (2014) Approaches and trends in plant disease management. Kindle edn. Scientific Publishers, Technology & Engineering, Jodhpur

Hashem A, Alqarawi AA, Radhakrishnan R et al (2018) Arbuscular mycorrhizal fungi regulate the oxidative system, hormones and ionic equilibrium to trigger salt stress tolerance in Cucumis Sativus L. Saudi J Biol Sciences. https://doi.org/10.1016/j.sjbs.2018.03.009

Hatami E, Abbaspour A, Dorostkar V (2018) Phytoremediation of a petroleum-polluted soil by native plant species in Lorestan Province, Iran. Environ Sci Pollut Res. https://doi.org/10.1007/s11356-018-1297-7

Huang Z, Fei Z, Jianfeng H et al (2018) Prediction of the distribution of arbuscular mycorrhizal fungi in the metal(Loid)-contaminated soils by the arsenic concentration in the fronds of *Pteris Vittata* L. J Soil Sediment 18(7):2544–51. https://doi.org/10.1007/s11368-018-1945-z

Jiang Q, Zhuo F, Long S et al (2016) Can arbuscular mycorrhizal fungi reduce Cd uptake and alleviate Cd toxicity of Lonicera Japonica grown in Cd-added soils? Sci Rep. https://doi.org/10.1038/srep21805

Kennen K, Niall K (2015) Phyto: principles and resources for site remediation and landscape design. Taylor & Francis Group, London

Khan MS, Musarrat J, Zaidi A (eds) (2010) Microbes for legume improvement. Springer, Vienna

Kong Z, Glick BR (2017) The role of bacteria in phytoremediation. In: Toshiomi Y (ed) Applied bioengineering: innovations and future directions. Wiley-VCH Verlag GmbH & CO. KGaA, Germany, pp 327–344

Lenoir I, Sahraoui AL, Fontaine J (2016) Arbuscular mycorrhizal fungal-assisted phytoremediation of soil contaminated with persistent organic pollutants: a review. Eur J Soil Sci 67(5):624–640

Liu A, Dalpé Y (2009) Reduction in soil polycyclic aromatic hydrocarbons by arbuscular mycorrhizal leek plants. Int J Phytoremediation 11(1):39–52

Lu Y, Lu M (2015) Remediation of PAH-contaminated soil by the combination of tall fescue, arbuscular mycorrhizal fungus and epigeic earthworms. J Hazard Mater 285(3):535–541. https://doi.org/10.1016/j.jhazmat.2014.07.021

Luo Y, Tu C (eds) (2018) Twenty years of research and development on soil pollution and remediation in China. Springer, Singapore. https://doi.org/10.1007/978-981-10-6029-8

Małachowska-Jutsz A, Kalka J (2010) Influence of mycorrhizal fungi on remediation of soil contaminated by petroleum hydrocarbons. Fresenius Environ Bull 19(12):8

Mehrotra VS (2005) Mycorrhiza: role and applications. Allied Publishers Limited, New Delhi, pp 136–157

Ming T, Hui C (1999) Effects of arbuscular mycorrhizal fungi alkaline phosphatase activities on Hippophae Rhamnoides drought-resistance under water stress conditions. Trees 14(3):113–115

Mukerji KG (2004) Fruit and vegetable diseases. Disease management of fruits and vegetables, vol 1. Kluwer Academic Publishers, Dordrecht, p 554

Mukerji KG, Manoharachary C, Chamola BP (eds) (2002) Techniques in mycorrhizal studies. Springer, Dordrecht. https://doi.org/10.1007/978-94-017-3209-3

Pandey P, Bisht S, Sood A et al (2012) Consortium of plant growth-promoting bacteria: future perspective in agriculture. In: Maheshwari DK (ed) Bacteria in agrobiology: plant probiotics. SpringerVerlag, Berlin, Heidelberg, pp 185–200

Pavithra D, Yapa N (2018) Arbuscular mycorrhizal fungi inoculation enhances drought stress tolerance of plants. Groundw Sustain Dev. https://doi.org/10.1016/j.gsd.2018.03.005

Pichardo TS (2012) The potential effects of arbuscular mycorrhizae (AM) on the uptake of heavy metals by plants from contaminated soils. J Bioremed Biodegr. https://doi.org/10.4172/2155-6199.1000e124

Pollastri S, Savvides A, Pesando M et al (2018) Impact of two arbuscular mycorrhizal fungi on Arundo Donax L. response to salt stress. Planta 247(3):573–585

Prasad R (ed) (2018) Mycoremediation and environmental sustainability. Springer, New York

Reddy CA, Saravanan RS (2013) Polymicrobial multi-functional approach for enhancement of crop productivity. In: Advances in applied microbiology. Elsevier, San Diego. https://doi.org/10.1016/B978-0-12-407679-2.00003-X

Ren C, Kong C, Bian B et al (2017) Enhanced phytoremediation of soils contaminated with PAHs by arbuscular mycorrhiza and rhizobium. Int J Phytoremediation 19(9):789–797

Shakeel M, Yaseen T (2015) An insight into phytoremediation of heavy metals from soil assisted by ancient fungi from glomeromycota-arbuscular mycorrhizal fungi. Sci Technol Dev 34(4):215–220

Siddiqui ZA, Akhtar MS, Futai K (eds) (2008) Mycorrhizae: sustainable agriculture and forestry. Springer, Dordrecht

Singh HB (2016) Microbial inoculants in sustainable agricultural productivity. Research perspectives. Springer, Berlin/Heidelberg/New York

Singh VM (2017) Agriculturally important microbes for sustainable agriculture, applications in crop production and protection. Springer, Berlin/Heidelberg/New York

Singh A, Ward OP (eds) (2010) Applied bioremediation and phytoremediation. Springer, Heidelberg/Berlin

Souza T (2015) Handbook of arbuscular mycorrhizal fungi. Springer, Cham. https://doi.org/10.1007/978-3-319-24850-9

Sut M, Boldt-Burisch K, Raab T (2016) Possible evidence for contribution of arbuscular mycorrhizal fungi (AMF) in phytoremediation of iron–cyanide (Fe–CN) complexes. Ecotoxicology 25(6):1260–1269

Upadhyaya H, Panda SK, Bhattacharjee MK et al (2010) ROLE OF Arbuscular mycorrhiza in heavy metal tolerance in plants: prospects for phytoremediation. Phytol J 2(7):16–27

Varma A, Kharkwal AC (eds) (2009) Symbiotic fungi: principles and practice. Springer, Berlin/Heidelberg

Varma A, Oelmüller R (eds) (2007) Advanced techniques in soil microbiology. Soil biology. Springer, Berlin/Heidelberg/New York

Varma A, Prasad R, Tuteja N (2017) Mycorrhiza – nutrient uptake, biocontrol, ecorestoration. Springer, Berlin/Heidelberg/New York

Veresoglou SD, Rillig MC (2012) Suppression of fungal and nematode plant pathogens through arbuscular mycorrhizal fungi. Biol Lett 8(2):214–217

Wang Y, Wang M, Li Y et al (2018) Effects of arbuscular mycorrhizal fungi on growth and nitrogen uptake of Chrysanthemum Morifolium under salt stress. PLoS One 13(4):e0196408

Watts-Williams SJ, Tyerman SD, Cavagnaro TR (2017) The dual benefit of arbuscular mycorrhizal fungi under soil zinc deficiency and toxicity: linking plant physiology and gene expression. Plant Soil 420(1–2):375–388

Whitacre DM (ed) (2015) Reviews of environmental contamination and toxicology. Springer, Cham

Xun F, Xie B, Liu S et al (2015) Effect of Plant Growth-Promoting Bacteria (PGPR) and Arbuscular Mycorrhizal Fungi (AMF) inoculation on oats in saline-alkali soil contaminated by petroleum to enhance phytoremediation. Environ Sci Pollut Res 22(1):598–608

Yong RN (2001) Geoenvironmental engineering: contaminated soils, pollutant fate and mitigation. CRC Press, Boca Raton

Zaidi A, Khan MS, Musarrat J (eds) (2017) Microbes for legume improvement. Springer, Cham

Zarik L, Meddich A, Hijri M et al (2016) Use of Arbuscular Mycorrhizal fungi to improve the drought tolerance of Cupressus Atlantica G. C R Biol 339(5–6):185–196

Zhang T, Hu Y, Zhang K et al (2018) Arbuscular Mycorrhizal fungi improve plant growth of Ricinus Communis by altering photosynthetic properties and increasing pigments under drought and salt stress. Ind Crop Prod 117(7):13–19. https://doi.org/10.1016/j.indcrop.2018.02.087

Light-Emitting Diode for the Inactivation of Microorganisms on Fruits and Vegetables

12

Anbazhagi Muthukumar

Abstract

The light-emitting diodes (LEDs) for the inactivation of microorganisms on fruits and vegetables have been recognized as an innovative, non-thermal, as well as non-chemical treatment for disinfection process. However, with the advancement of technology, including gaining devices with better luminous output; low radiant heat emissions; high emissions of monochromatic light; long life expectancy; mechanical robustness; and a greater diversity of peak wavelengths, LEDs have shown more applications in the food industry, especially in the areas of microbiological quality. Therefore, the methods seem to be rapid, efficient, and reliable in reducing thermal damage of food and are suitable in cold-storage applications, as well as more effective in increasing the shelf life of food materials along with good preservation ability. Executing LEDs in the food industry has been showing enhanced nutritive quality of foods in the postharvest stage, an increased ripening rate of fruits, and deterrence in fungal infections. LEDs can also be used together with photosensitizers or photocatalysts to inactivate pathogenic bacteria on fruits and vegetables. Even though the technique is more environmentally friendly rather than traditional technologies, several challenges and limitations are identified, including the difficulty in acceptability of fruits and vegetables stored and processed under LED lighting, optimizing LED lighting regimens for postharvest storage, the stability of inactive microorganisms, spoilage of enzymes, etc. However, these methods can provide a non-thermal means of keeping food safe without the use of chemical sanitizers or additives and may prevent the development of bacterial resistance.

Keywords

Bacterial inactivation · Disinfection · Food preservation · Light-emitting diode

A. Muthukumar (✉)
Department of Environmental Science, School of Earth Science Systems, Central University of Kerala, Kasaragod, Kerala, India

P. K. Arora (ed.), *Microbial Technology for the Welfare of Society*,
Microorganisms for Sustainability 17,
https://doi.org/10.1007/978-981-13-8844-6_12

12.1 Introduction

Light-emitting diode (LED) is a semiconductor which can emit light within a narrow spectrum of wavelength when an electric current passes through it. LEDs were of low power and were used mainly as indicator lamps in the early stages of development. The new semiconductor materials got incoporated, crystal growth techniques as well as optics got improved, and thermal dissipation in junctions were implemented in the subsequent years results in the rapid development of the LED technology (Bourget 2008; Yeh and Chung 2009; Chang et al. 2012; Branas et al. 2013).

In addition, it can be easily implemented into existing systems without requiring any special disposal technique due to its relatively smaller size (Schubert 2018; Held 2016; Hamamoto et al. 2007; Mori et al. 2007). LED has several advantages over the traditional visible light source, such as its inability to produce monochromatic wavelengths. Moreover, it consumes an only lesser amount of energy and has high durability (D'Souza et al. 2015). Furthermore, LED can produce a phototoxic effect on microbial cells caused due to the photosensitizing metabolites present in the bacterial cells mainly in the range of 400–480 nm (blue) (Durantini 2006; Luksiene 2003). There are mainly two methods for inactivation of bacterial cells by LEDs: photodynamic inactivation (PDI) and photo-eradicating effect.

An antibacterial effect through a phenomenon known as photodynamic inactivation (PDI) in microorganisms on specific wavelengths of LEDs is predominantly due to the presence of endogenous photosensitizers such as porphyrins, flavins, cytochromes, and NADH, where it absorbs light and gets excited. The excitation of endogenous porphyrins takes place due to the absorption of photons and results in the release of reactive oxygen species (ROSs) which degrade cellular components such as lipids, proteins, and DNA to bring about a cytotoxic effect. The ROSs which include hydroxyl ions, peroxides, superoxide, and singlet oxygen are the main tool of eradication of the bacterial cell (Bumah et al. 2013). Moreover, the photo-eradicating effect of the blue light would cause the chemical modification and clevage of nucleic acid present in the microbial cell, which results in the inactivation of the microorganisms as well as the destruction of the structure and composition of the protein, cell membranes and the other cellular materials. Based on this mechanism, use of LEDs has received increased attention in optics, electronics, and medicine. Moreover, it is a novel technology for bacterial inactivation in clinical and food processing and preservation applications as well (Wu et al. 2007; ANSES 2013; Guffey and Wilborn 2006; Mori et al. 2007; Maclean et al. 2009).

12.2 Properties of LED

LED's are the solid-state lighting devices which is having the emission wavelengths of narrow bandwidths, high photoelectric efficiency and photon flux, low thermal output, compactness, portability, and can easily integrated into electronic systems. In addition, LED brings ease of manipulation of the spectral characteristics, radiant or luminous intensity and temporal settings of the light, and thus it makes the LED more attactive to users (Branas et al. 2013). Long life expectancies, robustness, compactness, and the cost-effectiveness of LEDs ensure the advances in three different aspects, namely, food production, postharvest storage, and food preservation (D'Souza et al. 2015). Table 12.1 shows the peculiar properties of LED while comparing with compact florescent lamps and HPS lamps (D'Souza et al. 2015).

A LED is a semiconductor diode incapable of producing monochromatic light, consisting of a narrow bandwidth of wavelengths called as electroluminescence property, and it occurs when an electron–hole interaction takes place, and it results in the emission of light of a distinct wavelength, which appears as distinct colors to the eye. The color of the emitted light depends on the band gap energy of the material of the semiconductor (Gupta and Jatothu 2013; Yeh and Chung 2009).

The color of the emitted light depends on the band gap energy of the material of the semiconductor.

Gallium arsenide is used for red and IR light; indium gallium aluminum phosphide for green, yellow, orange, and red lights; and gallium nitride (GaN) and silicon carbide for blue lights. Besides, LEDs emitting UV radiation are typically composed of aluminum gallium nitride or indium gallium nitride, with a wavelength as low as 210 nm (Shur and Gaska 2010). In addition, for producing white light, either a UV LED and a tricolor phosphor coating combination, or a blue LED with yellow phosphor is required (Park et al. 2014). Alternatively, white light can be produced by mixing the light from red, blue, and green LEDs (DenBaars et al. 2013).

Table 12.1 Comparing the properties of LEDs to two commonly used lighting technologies. (D'Souza et al. 2015)

Properties	LEDs	Compact fluorescent lamps	HPS lamps
Spectral composition	Monochromatic UV LEDs, IR LEDs, and white LEDs available	Broad spectrum, cannot be controlled, radiation in UVC and IR range present	Broad spectrum, cannot be controlled, radiation in UV and IR range present
Size and compactness	Chips are small and compact (2–5 cm) and can be assembled into different formations, shapes, and fixtures	Bulky	Bulky
Luminous efficiency	Color – mixed white LEDs: 100–150 lm/W by 2025	45–80 lm/W	65–150 lm/W
Photon efficiency	0.89–1.70 μmolJ^{-1}	0.95 μmolJ^{-1}	1.30–1.70 μmolJ^{-1}
Time to full light output after switching on	Almost instantly, with no restrike delay. High frequency pulsing and dimming possible	Approximately 3 min to full brightness	Approximately 10 min of warm-up time and up to 20 min restrike time delay
Life expectancy	50000 h	10000–17000 h	10000–17000 h
Durability	Not affected by mechanical force	Brittle components in bulb and fixtures	Brittle components in bulb and fixtures

The photon efficiency, which is the number of photons produced per unit of input electrical energy (μmol J^{-1}), and the electrical efficiency, which represents the percentage of output power in the form of light per unit of input electrical power, are essential for evaluating LED performance. While comparing LEDs with HPS lamps and fluorescent lamps, photon efficiency and electrical efficiency are similar and slightly higher, respectively (Nelson and Bugbee 2014; Pinho et al. 2012).

12.3 Types of LEDs

There are different types of LEDs such as ultraviolet light-emitting diode (UV LED), visible light-emitting diode (Vis-LED), and deep ultraviolet light-emitting diode (DUV LED).

The wavelength of UV light for food processing varies from 100 to 400 nm. This range may be further subdivided into UV-A (315–400 nm), UV-B (280–315 nm), and UV-C (200–280 nm) which are responsible for tanning in human skin, skin burning, and skin cancer. They also affect several physiological processes in plant tissues and damage microbial, DNA respectively (Lucht et al. 1998; Forney and Moraru 2009; Koutchma et al. 2015). A precise wavelength range of 240–400 nm UV LED has been developed with the advantages such as low cost, energy efficient, long life, easy control of emission, and no production of mercury wastes, for example, aluminum nitride (AIN), gallium nitride (GaN), and intermediate alloys (Nagasawa and Hirano 2018). Furthermore, the effects of UV LEDs on inactivation of *E. coli* K12, *E. coli* 0157:H7, and polyphenol oxidase (PPO) in clear as well as cloudy apple juice were investigated. A better inactivation effect on PPO was obtained when applied UV-A and UV-C rays combined together. In addition, UV LED is an effective technology for inactivation of pathogens in drinking water using a different wavelength of UV LEDs (Song et al. 2016; Akgün and Ünlütürk 2017). Waterborne bacterial viruses such as MS2, QβQβ, and φX174φX174 can be decontaminated by using deep ultraviolet light-emitting diodes (DUV-LEDs) at 280 nm and 255 nm (Aoyagi et al. 2011).

Blue monochromatic light-emitting diode (LED) has been used for the inactivation of *Escherichia coli* in milk, where wavelength has applied between 405 and 460 nm, and the results show a maximum microbial reduction at a higher temperature and lower wavelengths (Srimagal et al. 2016).

12.4 Advantages of LED

An important advantage of LEDs is the low emissions of radiant heat in the form of IR radiation, which reduces undesirable or detrimental effects of heat on food (Morrow 2008; Mitchell et al. 2012). However, a substantial amount of heat is produced at the p–n junction but it compromises the luminous efficacy of the LED. Therefore, the proper cooling system is necessary (United States Department of Energy 2012).

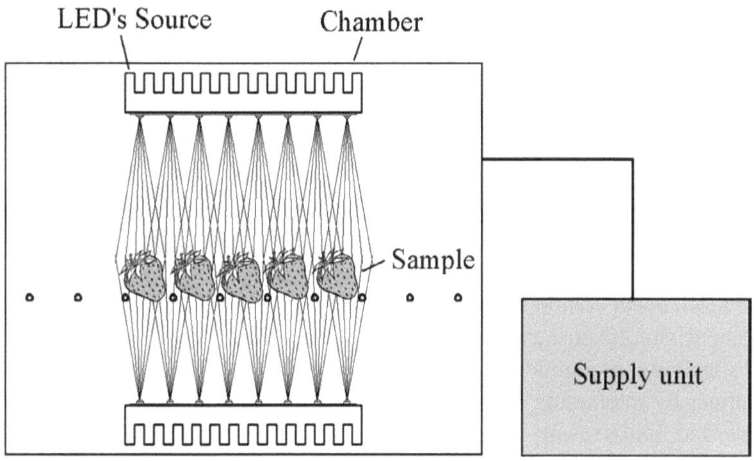

Fig 12.1 Schematic presentation of LED-based light source prototype. (Aponiene et al. 2015)

The compact size of LEDs allows better flexibility and long life expectancy, which is around 50000–100000 h compared with 15000 h for conventional lighting (Gupta and Jatothu 2013). Most conventional lighting lacks an instant on-off feature which is peculiar to LEDs, and this allows pulsed lighting periods in food safety applications (Yeh and Chung 2009; Branas et al. 2013). Figure 12.1 shows the schematic representation of LED-based light source prototype (Aponiene et al. 2015).

12.5 Role of LED for the Inactivation of Microorganisms

Preservation of raw fruits and vegetables, which are consumed widely as a part of staple food, has major importance to the food industry because they are highly susceptible to microbial contamination due to their high water and nutrient content. Therefore, instead of common preservation techniques, an efficient method is needed to preserve the food indeterminately without compromising the quality of food (Shital et al. 2017). Nowadays, LEDs are emerging as a novel technology in the field of food safety and preservation because the food manufacturing and processing industry need sterilization methods to eradicate pathogenic microorganisms from food materials. Fresh cut fruits and vegetables are at high risk of pathogenic microbial contamination, where several disease outbreaks have been reported due to foodborne pathogens brought by unhygienic conditions during the processing stage of fruits and vegetables. Use of LEDs for the eradication of pathogenic microorganisms in food industry is considered as one of the safe, efficacious, low-cost, and apparently harmless methods to produce fresh products (Bhagwat 2006; Frank et al. 2011; Nunes et al. 2013). Moreover, reduced heat generation and control over light quality enhance the nutrient content and antioxidant activity. Besides, the different wavelength can bring varieties of responses in plants. LED techniques potentially bring antibacterial effect on fruits and vegetables due to their peculiar phenomena called photodynamic inactivation, and LEDs also have many advantageous properties such as low radiant heat emissions, electrical luminosity, strong emission monochromatic light (D'Souza et al. 2015), high photon efficiency, long operational lifespan, etc. (Chen et al. 2016). These properties help LEDs find application in the areas of food safety and preservation.

The main objective of this article is to find applications of LEDs in the microbial safety of fruits and vegetables as well as the reported changes and side effects of LED irradiation on fruits and vegetables which are most relevant for the estimation of quality.

The mechanism behind the anti-bactericidal activity of LED was well explained by Luksienė and Zukauskas (2009), in which, the bacterial cells got exposed to light and results in the production of reactive oxygen species (ROS), it can bring a cytotoxic effect by interacting with adjacent intracellular components, such as DNA, proteins, and lipids, results in bacterial death and this mechanism is called photodynamic inactivation. Properties such as high selectivity, antimicrobial efficiency and non-thermal nature of photosensitization method would aids for the development of completely safe, non-thermal surface decontamination and food preservation technique in the near future (Luksiene et al. 2009).

12.5.1 Surface Sterilization

Luksiene et al. (2009) investigated the inactivation of foodborne pathogen Bacillus cereus by 5-aminolevulinic acid (ALA)-based photosensitization after adhesion on the surface of the packaging material. A remarkable reduction in the bacterial count on the surface of packaging material was observed after the incubation of *B. cereus* on the surface of food packaging material along with illumination of visible light at a 400 nm wavelength with illumination time 0–20 min. Moreover, the spores of *B. cereus* were also vulnerable to the visible light treatment. Inactivation of *B. cereus* is important because it can contaminate a wide range of foods, cereals, fresh vegetables, and berry and fruits, and additionally, this pathogen was frequently found in ready-to-eat (RTE) foods and sauces (Rosenquist et al. 2005). Processes such as washing, peeling and cutting brings RTE fresh-cut fruits foods to unhygienic environmental conditions would bring cross-contamination with pathogenic bacteria from raw foods or equipment, and similarly, susceptible to foodborne illness outbreaks since RTE foods had no further step to eliminate bacteria, if contaminated (Raybaudi-Massilia et al. 2013; Sim et al. 2013).

12.5.2 Photodynamic Inactivation Through Photosensitizers

Luksiene and Paskeviciute (2011) studied the possibilities to control microbial contamination of strawberries by chlorophyllin-based photosensitization, and the study also evaluated the photosensitization-induced effects on the key quality attributes of treated strawberries. The obtained results indicated that the decontamination of strawberries using photosensitization was 98% compared to the control sample. Naturally occurring yeasts and mesophiles were inhibited by 86 and 97%, respectively. The shelf life of treated strawberries was extended by 2 days. The total antioxidant activities of treated strawberries were increased by 19%, and no impacts were found on the number of phenols, anthocyanins, and surface color. Thus the study verifies that, deprived of any negative impact on antioxidant activity, phenols, anthocyanins, and surface color, photosensitization effectively helps for decontamination of microorganisms and which expands the shelf life of strawberries also (Luksiene and Paskeviciute 2011).

Study done by Kim (2016) demonstrated that LED of 405 ± 5 nm illumination on fresh-cut fruits such as papaya and mango at 4 and 10 °C inactivates the foodborne pathogens such as *E. coli* O157:H7, *L. monocytogenes*, and *Salmonella*, and the study also discloses that bacterial growth has inhibited or delayed by LED illumination even at room temperature without any significant differences in physicochemical and nutritional qualities between LED-illuminated and non-illuminated fruits. The extended study on the efficiency of LED illumination at 405±5 nm on inactivating foodborne pathogens in RTE foods at a refrigerated temperature would substantiate the same that the food chillier equipped with 405±5 nm LED can preserve RTE foods for long. Likewise, pathogenic microorganisms such as *Listeria monocytogenes* and *Salmonella* spp. on the rind of cantaloupe fruit can be eradicated using blue light-emitting diode having wavelength 405 and 460 nm in combination with sodium chlorophyllin and thus minimizing the risk of listeriosis and salmonellosis by consumption of fresh-cut cantaloupe (Sherrill et al. 2018).

The LED in combination with photosensitizers have emerged as a novel technology for microbial control on surface decontamination (Lukšiene 2005). At this point, surface sterilization of fruits and vegetables is exclusively needed in the food processing and manufacturing industry. Aihara et al. (2014) studied the inactivation of *E. coli* strain DH5α by spot inoculation onto the vegetable tissue, and treatments have been done under UVA LED irradiation, where possible effects of UVA LED on vegetables were quantified by using high-performance liquid chromatography (HPLC). The rate of bacterial inactivation by UVA LED after 10 min of treatment and the comparison to non-radiated vegetable tissues show that the bacterial survival rate grew slowly on the vegetable under irradiation, along with no changes in the nutritional content as well as vitamin C content. Therefore, UVA LED treatment is a highly safe, efficacious, low-cost, and apparently harmless surface sterilization method for vegetables.

Bacterial inactivation followed by the exposure to visible light ranging from 400 to 470 nm as well as blue LEDs is considered as recent development in LED technology (Shital et al. 2017). Another study conducted to determine the antibacterial activity at a wavelength above 405 nm such as 461, 521, and 642 nm showed a bactericidal activity on four major foodborne pathogens, such as *E. coli* O157:H7, *Listeria monocytogenes*, *Salmonella typhimurium*, and *Salmonella aureus*. On the other hand, influences of temperature variation also contributed to the efficiency of LED treatment, where the chilling of foodstuffs followed by LED illumination has evolved as novel food preservation technology (Ghate et al. 2013).

12.6 Postharvest Preservation

Fruits are prone to fungal infection, the main reason for postharvest loss. The studies on citrus fruits have revealed that the blue LED illumination at a moderate intensity was found to be an effective method in preventing fungal infection on citrus fruits, where illumination by deep UV on strawberries has also retained the anthocyanin and total soluble sugar levels in irradiated type than the strawberries stored in dark places (Shital et al. 2017).

Energy-efficient nature, small size, relatively cool surfaces, and long life expectancy encourage the use of LEDs in growth chambers, greenhouses, and postharvesting processes, where the desired aesthetic characteristics, enriched nutrition, and flavor quality can be obtained. Additionally, the LED illumination experiments on a petunia flower have brought an increased level of volatile compound 2-phenylethanol in the red and far-red light. Thus, the effect of LED at different wavelengths on accumulation of volatile compound related aroma and taste in fruit products has been revealed (Colquhoun et al. 2013). Furthermore, different spectral LEDs are existed to improve the nutritional quality of harvested vegetables like cabbage by increasing the accumulation of vitamin C, anthocyanin, and total phenolics (Lee et al. 2014 and Kanazawa et al. 2012). Furthermore, fungal infection in citrus fruit can be treated by exposure to a blue LED light, where mycelial growth, soft rot area, sporulation of *Penicillium digitatum*, *Penicillium italicum*, and *Phomopsis citri* on the surface of tangerines can be reduced when citrus fruit is exposed to blue light (Gupta, 2017).

12.6.1 Evaluation of LEDs in Postharvest Food Safety

Eradication of the pathogens on fruits and vegetables by LED illumination is very common nowadays, where the eradication of *Penicillium* species, which are responsible for the postharvest decay in citrus fruits, is possible by single-spectral blue LEDs (Liao et al. 2013 and D'Souza et al. 2015). Moreover, the fungal growth can be inhibited by suppressing sporulation and germination of fungi using blue light illumination (Suthaparan et al. 2010; Murdoch et al. 2013). Lesion development inhibition by red light induces expression of defense-related genes and also promotes synthesis of stilbene compound (Ahn et al. 2015). In addition, salicylic acid plays a vital role in plant disease resistance, where mutants of red/far-red light photoreceptors are known to be compromised in salicylic acid signaling stimulation and resistance to *Pseudomonas syringae* and the red LEDs induce salicylic acid content and expression of salicylic acid-regulated PR-1 and WRKY genes in pathogen-inoculated cucumber plants (Wang et al. 2010; Genoud et al. 2002). Furthermore, by reducing the expression of salicylic and jasmonic-responsive genes, the low red/far-red light ratio inhibits salicylic acid and jasmonic acid-mediated disease resistance in *Arabidopsis*, where the result shows the possible effect of LEDs on salicylic or jasmonic-mediated disease resistance (de Wit et al. 2013; Moreno et al. 2009; Hasan et al. 2017).

12.6.2 Effectiveness of LEDs in Postharvest Treatments

LED has been used as an interlighting system under greenhouse conditions because it generates less amount of heat as well as consumes less power; therefore, a significant amount of energy can be saved with its use. When compared with white fluorescent or solar light, the use of single-spectral blue or red LED has significant improvements in the quality and yield of vegetables and fruits such as cucumber, pepper, and strawberry (Table 12.2). Besides single-spectral light, use of mixed red: blue light can also increase the crop production (Choi et al. 2015; Dong et al. 2014; Hao et al. 2012; Sabzalian et al. 2014). Induced disease resistance in fruits and crops treated with different LED portrayed in Table 12.3 (Hasan et al. 2017).

Treatment with LEDs can assure the food safety by inactivating foodborne pathogens. In addition, the LEDs (especially blue LEDs) can be used as effective bactericides in cold storage practices. LEDs may offer an alternative to chemical sanitizers to satisfy the growing global demand for food microbiological safety (Ghate et al. 2013; Kumar et al. 2015). In conclusion, LEDs play a major role in the preservation of agricultural products and commodities, because LED can emit light within a very narrow wavelength spectrum and has the potential to inactivate microorganisms in foods. Studies done by Tamanna et al. 2014 investigated the efficiency of LED in the range of 400–480 nm and successfully proved that it can damage the DNA and results in an observed 5 log reduction in bacterial growth (Tamanna et al. 2014).

Table 12.2 Effect of LED lighting in postharvest preservation

Food	LED	Treatment time	Application
Broccoli	Red (660 nm)	Continuous	Delaying senescence in vegetables
Strawberries	Blue (470 nm)	Continuous	Accelerating secondary ripening processes
Tomatoes	Blue (440–450 nm)	Continuous	Delaying ripening
Lamb's lettuce	Warm white	16 cycles (5 h)	Enhancing/delaying loss of postharvest nutritional content
Broccoli	Red (660 nm)	Continuous	
Chinese bayberries	Blue (470 nm)	Continuous	
Cabbage	White, blue (436 nm)	Continuous	
	Green (524 nm) Red (665 nm)		

Adapted and modified from Shital et al. (2017)

Table 12.3 Induced disease resistance in fruits and crops treated with different light from LEDs. (Hasan et al. 2017)

LED light	Light intensity	Crops	Effect
Red	261–550 μW/cm^2	*Vicia faba*	Induces resistance against *Botrytis cinerea, Alternaria tenuissima*
	287 μW/cm^2	*Arabidopsis*	Induced resistance against *Meloidogyne javanica, Pseudomonas syringae*
	287 μW/cm^2	*Piper nigrum, Cucurbita, Solanum lycopersicum*	Induced resistance against *Phytophthora capsici*
	137 W/cm^2; 350 μmol m^{-2} s^{-1}	*Cucumis sativus*	Induced resistance against *Corynespora cassiicola*
	80 μmol m^{-2} s^{-1}	*Vitis vinifera*	Induced resistance against *B. cinerea*
Blue	200 μmol m^{-2} s^{-1}	*Lactuca sativa*	Induced resistance against gray mold by *B. cinerea*
	50–150 μ mol m^{-2} s^{-1}	*Solanum lycopersicum*	Induced resistance against gray mold disease by *B. cinerea*
Green	80 μ mol m^{-2} s^{-1}	*Fragaria ananassa*	*Glomerella cingulata*
	80 μ mol m^{-2} s^{-1}	*Cucumis sativus*	*Colletotrichum orbiculare, B. cinerea*

12.7 Conclusion

Traditionally, several types of antibacterial treatments exist in food manufacturing and processing industry, where most of them act by slowing down the microbial growth, and none of them can eliminate pathogens totally from the food matrix. On the other hand, freshly harvested fruits and vegetables are having high risk of microbial contamination, because of processing the food materials under less sanitary conditions. In this scenario, LED represents a great role in the eradication of microorganisms from food materials, because this decontamination technique works without any effects on nutritional and organoleptic characteristics of food matrix.

From the literatures understudied, it is clear that the side effects of LED irradiation on fruits and vegetables seem to be minimal, since there are no observed changes in fruits and vegetable tissues, where these criteria are considered to be the most relevant for estimation of fresh vegetable and fruit quality. The health of fruit and vegetable products after LED irradiation suggests that the treated tissues can retain their natural resistance to infection causing microorganisms. Thus, the LED technology is potentially useful for application in the food manufacturing and processing industry.

References

Ahn SY, Kim SA, Yun HK (2015) Inhibition of Botrytis cinerea and accumulation of stilbene compounds by light-emitting diodes of grapevine leaves and differential expression of defense-related genes. Eur J Plant Pathol 143(4):753–765

Aihara M, Lian X, Shimohata T et al (2014) Vegetable surface sterilization system using UVA light-emitting diodes. J Med Investig 61(3.4):285–290

Akgün MP, Ünlütürk S (2017) Effects of ultraviolet light emitting diodes (LEDs) on microbial and enzyme inactivation of apple juice. Int J Food Microbiol 260:65–74

ANSES (2013) French agency for food, environmental and occupational health and safety. LED e light-emitting diodes: health effects of lighting systems using light-emitting diodes (LEDs)

Aoyagi Y, Takeuchi M, Yoshida K (2011) Inactivation of bacterial viruses in water using deep ultraviolet semiconductor light-emitting diode. J Environ Eng 137(12):1215–1218

Aponiene K, Paskeviciute E, Reklaitis I et al (2015) Reduction of microbial contamination of fruits and vegetables by hypericin-based photosensitization: comparison with other emerging antimicrobial treatments. J Food Eng 144:29–35

Bhagwat AA (2006) Microbiological safety of fresh-cut produce: where are we now? In: Microbiology of fresh produce. American Society of Microbiology Press, Washington, DC, pp 121–165

Bourget CM (2008) An introduction to light-emitting diodes. Hortic Sci 43(7):1944–1946

Branas C, Azcondo FJ, Alonso JM (2013) Solid-state lighting: a system review. IEEE Ind Electron Mag 7(4):6–14

Bumah VV, Masson-Meyers DS, Cashin SE et al (2013) Wavelength and bacterial density influence the bactericidal effect of blue light on methicillin-resistant Staphylococcus aureus (MRSA). Photomed Laser Surg 31(11):547–553

Chang MH, Das D, Varde PV et al (2012) Light emitting diodes reliability review. Microelectron Reliab 52(5):762–782

Chen Z, Liu Z, Shen G et al (2016) Effect of chain flexibility of epoxy encapsulants on the performance and reliability of light-emitting diodes. Ind Eng Chem Res 55(28):7635–7645

Choi HG, Moon BY, Kang NJ (2015) Effects of LED light on the production of strawberry during cultivation in a plastic greenhouse and in a growth chamber. Sci Hortic 189:22–31

Colquhoun TA, Schwieterman ML, Gilbert JL et al (2013) Light modulation of volatile organic compounds from petunia flowers and select fruits. Postharvest Biol Technol 86:37–44

D'Souza C, Yuk HG, Khoo GH et al (2015) Application of light-emitting diodes in food production, postharvest preservation, and microbiological food safety. Compr Rev Food Sci Food Saf 14(6):719–740

de Wit M, Spoel SH, Sanchez-Perez GF et al (2013) Perception of low red: far-red ratio comprises both salicylic acid-and jasmonic acid-dependent pathogen defences in Arabidopsis. Plant J 75:90

DenBaars SP, Feezell D, Kelchner K et al (2013) Development of gallium-nitride-based light-emitting diodes (LEDs) and laser diodes for energy-efficient lighting and displays. Acta Mater 61(3):945–951

Dong C, Fu Y, Liu G et al (2014) Growth, photosynthetic characteristics, antioxidant capacity and biomass yield and quality of wheat (Triticum aestivum L.) exposed to LED light sources with different spectra combinations. J Agron Crop Sci 200(3):219–230

Durantini EN (2006) Photodynamic inactivation of bacteria. Current bioactive compounds. Bentham Sci Publ 2(2):127e142

Forney LJ, Moraru CI (2009) Ultraviolet light in food technology: principles and applications. CRC press, Boca Raton

Frank C, Werber D, Cramer JP et al (2011) Epidemic profile of Shiga-toxin–producing Escherichia coli O104: H4 outbreak in Germany. N Engl J Med 365(19):1771–1780

Genoud T, Buchala AJ, Chua NH et al (2002) Phytochrome signalling modulates the SA-perceptive pathway in Arabidopsis. Plant J 31(1):87–95

Ghate VS, Ng KS, Zhou W et al (2013) Antibacterial effect of light emitting diodes of visible wavelengths on selected foodborne pathogens at different illumination temperatures. Int J Food Microbiol 166(3):399–406

Ghate V, Leong AL, Kumar A, Bang WS, Zhou W, Yuk HG (2015) Enhancing the antibacterial effect of 461 and 521 nm light emitting diodes on selected foodborne pathogens in trypticase soy broth by acidic and alkaline pH conditions. Food Microbiol 48:49–57

Guffey JS, Wilborn J (2006) In vitro bactericidal effects of 405-nm and 470-nm blue light. Photomed Laser Ther 24(6):684–688

Gupta SD (ed) (2017) Light emitting diodes for agriculture: smart lighting. Springer, Singapore

Gupta SD, Jatothu B (2013) Fundamentals and applications of light-emitting diodes (LEDs) in in vitro plant growth and morphogenesis. Plant Biotechnol Rep 7(3):211–220

Hamamoto A, Mori M, Takahashi A et al (2007) New water disinfection system using UVA light-emitting diodes. J Appl Microbiol 103(6):2291–2298

Hao X, Little C, Khosla S (2012) LED inter-lighting in year-round greenhouse mini-cucumber production. Acta Hortic 956:335–340

Hasan MM, Bashir T, Ghosh R et al (2017) An overview of LEDs' effects on the production of bioactive compounds and crop quality. Molecules 22(9):1420

Held G (2016) Introduction to light emitting diode technology and applications. Auerbach Publications, New York

Kanazawa K, Hashimoto T, Yoshida S et al (2012) Short photoirradiation induces flavonoid synthesis and increases its production in postharvest vegetables. J Agric Food Chem 60(17):4359–4368

Kim MJ (2016) Photodynamic inactivation by 405±5 nm light emitting diode against foodborne pathogens on ready-to-eat foods and its antibactterial mechanism. Doctoral dissertation, National University of Singapore

Koutchma T, Forney LJ, Moraru CI (2015) Ultraviolet light in food technology: principles and applications. CRC Press, Boca Raton

Kumar A, Ghate V, Kim MJ et al (2015) Kinetics of bacterial inactivation by 405 nm and 520 nm light emitting diodes and the role of endogenous coproporphyrin on bacterial susceptibility. J Photochem Photobiol B Biol 149:37–44

Lee YJ, Ha JY, Oh JE et al (2014) The effect of LED irradiation on the quality of cabbage stored at a low temperature. Food Sci Biotechnol 23(4):1087–1093

Liao HL, Alferez F, Burns JK (2013) Assessment of blue light treatments on citrus postharvest diseases. Postharvest Biol Technol 81:81–88

Lucht L, Blank G, BOUSA J (1998) Recovery of foodborne microorganisms from potentially lethal radiation damage. J Food Prot 61(5):586–590

Luksiene Z (2003) Photodynamic therapy: mechanism of action and ways to improve the efficiency of treatment. Medicina (Kaunas) 39(12):1137–1150

Lukšiene Ž (2005) New approach to inactivation of harmful and pathogenic microorganisms by photosensitization. Food Technol Biotechnol 43(4):411–418

Luksiene Z, Paskeviciute E (2011) Novel approach to the microbial decontamination of strawberries: chlorophyllin-based photosensitization. J Appl Microbiol 110(5):1274–1283

Luksienė Ž, Zukauskas A (2009) Prospects of photosensitization in control of pathogenic and harmful micro-organisms. J Appl Microbiol 107:1415–1424

Luksiene Z, Buchovec I, Paskeviciute E (2009) Inactivation of food pathogen Bacillus cereus by photosensitization in vitro and on the surface of packaging material. J Appl Microbiol 107(6):2037–2046

Maclean M, MacGregor SJ, Anderson JG et al (2009) Inactivation of bacterial pathogens following exposure to light from a 405-nanometer light-emitting diode array. Appl Environ Microbiol 75(7):1932–1937

Mitchell CA, Both AJ, Bourget CM et al (2012) Horticultural science focus-LEDs: the future of greenhouse lighting! Chron Hortic Subscrip 52(1):6

Moreno JE, Tao Y, Chory J et al (2009) Ecological modulation of plant defense via phytochrome control of jasmonate sensitivity. Proc Natl Acad Sci 106(12):4935–4940

Mori M, Hamamoto A, Takahashi A et al (2007) Development of a new water sterilization device with a 365 nm UV-LED. Med Biol Eng Comput 45(12):1237–1241
Morrow RC (2008) LED lighting in horticulture. Hortic Sci 43(7):1947–1950
Murdoch LE, McKenzie K, Maclean M et al (2013) Lethal effects of high-intensity violet 405-nm light on Saccharomyces cerevisiae, Candida albicans, and on dormant and germinating spores of Aspergillus niger. Fungal Biol 117(7-8):519–527
Nagasawa Y, Hirano A (2018) A review of AlGaN-based deep-ultraviolet light-emitting diodes on sapphire. Appl Sci 8(8):1264
Nelson JA, Bugbee B (2014) Economic analysis of greenhouse lighting: light emitting diodes vs. high intensity discharge fixtures. PloS one 9(6):e99010
Nunes MM, de Alencar Mota ALA, Caldas ED (2013) Investigation of food and water microbiological conditions and foodborne disease outbreaks in the Federal District, Brazil. Food Control 34(1):235–240
Park JY, Lee JH, Raju GSR et al (2014) Synthesis and luminescent characteristics of yellow emitting GdSr2AlO5: Ce3+ phosphor for blue light based white LED. Ceram Int 40(4):5693–5698
Pinho P, Jokinen K, Halonen L (2012) Horticultural lighting–present and future challenges. Light Res Technol 44(4):427–437
Raybaudi-Massilia R, Calderón-Gabaldón MI, Mosqueda-Melgar J et al (2013) Inactivation of Salmonella enterica ser. Poona and Listeria monocytogenes on fresh-cut "Maradol" red papaya (Carica papaya L) treated with UV-C light and malic acid. J Verbr Lebensm 8(1-2):37–44
Rosenquist H, Smidt L, Andersen SR et al (2005) Occurrence and significance of Bacillus cereus and Bacillus thuringiensis in ready-to-eat food. FEMS Microbiol Lett 250(1):129–136
Sabzalian MR, Heydarizadeh P, Zahedi M et al (2014) High performance of vegetables, flowers, and medicinal plants in a red-blue LED incubator for indoor plant production. Agron Sustain Dev 34(4):879–886
Schubert EF (2018) Light-emitting diodes. E. Fred Schubert
Sherrill WJ, Min JK, Hyun GY (2018) Inactivation of Listeria monocytogenes and Salmonella spp. on cantaloupe rinds by blue light emitting diodes (LEDs). Food Microbiol 76:219–225
Shital P, Rahul CR, Pujari KH et al (2017) LEDs as Novel Chemical-Free Food Preservation Technology. Int J Soc Sci Dev 12(4):2454–6003
Shur MS, Gaska R (2010) Deep-ultraviolet light-emitting diodes. IEEE Trans Electron Devices 57:12–25
Sim HL, Hong YK, Yoon WB et al (2013) Behavior of Salmonella spp. and natural microbiota on fresh-cut dragon fruits at different storage temperatures. Int J Food Microbiol 160(3):239–244
Song K, Mohseni M, Taghipour F (2016) Application of ultraviolet light-emitting diodes (UV-LEDs) for water disinfection: A review. Water Res 94:341–349
Srimagal A, Ramesh T, Sahu JK (2016) Effect of light emitting diode treatment on inactivation of Escherichia coli in milk. LWT Food Sci Technol 71:378–385
Suthaparan A, Torre S, Stensvand A (2010) Specific light-emitting diodes can suppress sporulation of Podosphaera pannosa on greenhouse roses. Plant Dis 94(9):1105–1110
Tamanna R, Srimagal A, Vikram P et al (2014) Light Emitting Diode: a novel technique for preservation of agro and food commodities. In: 3rd international conference on agriculture & horticulture, Indian Institute of Crop Processing Technology, India, Agrotechnol 2014, 2:4
United States Department of Energy (2012) Using LEDs to their best advantage. Building technologies program: solid-state lighting technology fact sheet. Available from: http://apps1.eere.energy.gov/buildings/publications/pdfs/ssl/led_advantage.pdf. Accessed 13 Feb 2015
Wang H, Jiang YP, Yu HJ et al (2010) Light quality affects incidence of powdery mildew, expression of defence-related genes and associated metabolism in cucumber plants. Eur J Plant Pathol 127(1):125–135
Wu MC, Hou CY, Jiang CM (2007) A novel approach of LED light radiation improves the antioxidant activity of pea seedlings. Food Chem 101(4):1753–1758
Yeh N, Chung JP (2009) High-brightness LEDs—Energy efficient lighting sources and their potential in indoor plant cultivation. Renew Sust Energ Rev 13(8):2175–2180

Pseudomonas Species: Natural Scavenger of Aromatic Compounds from Industrial Effluents

13

Priyaragini and Rakesh Kumar

Abstract

An ever-increasing human demand has led to rapid industrialization, and almost all industrial effluents contain hazardous aromatic compounds, viz., dyes, azo dyes, phenolic dyes, other phenolic compounds, polyaromatic hydrocarbons, etc. There are a number of chemical and physical techniques to minimize the contents of harmful compounds discharged into the environment. However, their non-economical nature and production of secondary contaminants during the process has led to the demand for greener alternatives. Microbial degradation has been a great area of research for years. Out of numerous microbes, *Pseudomonas* species like *P. putida*, *P. aeruginosa*, and *P. fluorescens* are supposed to be the most reported one, for the degradation of aromatic compounds. Also metagenomics, an emerging molecular technique, can play an imperative role in biodegradation processes. Since, industrial effluents are a combination of compounds which in turn requires a consortium of microbes for their utilization. Metagenomic techniques can remodel any bacterial strain by combining the desired genes for the degradation of various classes of aromatic compounds from respective strains of *Pseudomonas.* This approach will modify the original genetic makeup of the bacterial strain which ultimately improves the microbial degradation method. In this book chapter, we are discussing a new paradigm in biodegradation involving metagenomics using *Pseudomonas* species.

Keywords

Industrial effluents · *Pseudomonas* · Biodegradation · Metagenomics · Aromatic compound

Priyaragini · R. Kumar (✉)
Department of Biotechnology, Central University of South Bihar, Gaya, India
e-mail: rakeshkr@cusb.ac.in

P. K. Arora (ed.), *Microbial Technology for the Welfare of Society*, Microorganisms for Sustainability 17,
https://doi.org/10.1007/978-981-13-8844-6_13

13.1 Introduction

In recent times, reckless industrialization has aggravated the environmental pollution, and it has now become an acute problem. The reason is that industrial effluents have negative impact on the environment. Each year, around 150 million tons of synthetic polymers are produced, and their non-reactive and stable nature after its use leads to limited degradation cycles. Plastic wastes and synthetic polymer wastes in the environment contain numerous types of aliphatic and aromatic compounds such as polyethylene, polypropylene, polystyrene (aromatic), polyvinyl chloride, etc. There are innumerable sources and causes which lead to discharge of aromatic compounds, hence polluting the environment. It has been estimated that aromatic species contribute about 10% to the total global anthropogenic non-methane organic carbon (NMOC) emissions.

In wastewater, aromatic pollutants play a major role because of their slightly soluble nature. A high number of aromatic compounds that are less soluble in water have been released into aquatic ecosystems leading to their accumulation in sediment (Fig. 13.1). Many kinds of organic compounds, such as polychlorinated biphenyls (PCBs), pesticides, herbicides, phenols, polycyclic aromatic hydrocarbons (PAHs), and aliphatic and heterocyclic compounds, are incorporated in the wastewater (discharged from industries) (Wu et al. 2003). These aromatic pollutants in water can harm the environment and also pose health risks for humans.

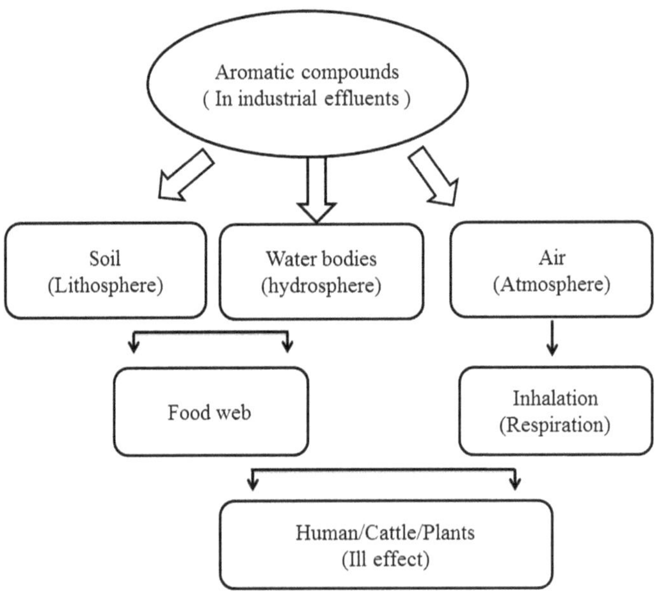

Fig. 13.1 Impact of aromatic compounds discharged from industrial effluents on environment

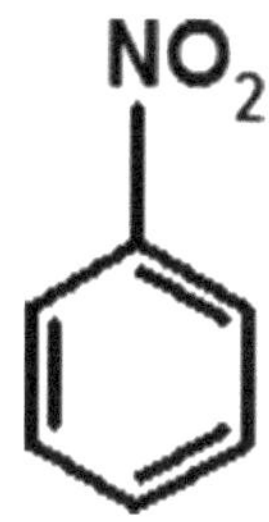

Fig. 13.2 Nitrobenzene

A few aromatic compounds discussed in details are given below:

(a) *Phenol:* Phenols are a class of chemical compounds consisting of a hydroxyl group (-OH) bonded directly to an aromatic hydrocarbon. There are various sources like textile industries, paper mills, oil refineries, and paint industries which lead to discharge and hence exposure of phenol and phenolic compounds to the environment (wastewater). Phenol and related compounds possess risks for living beings, as they have been reported to be highly toxic even at low concentration. It can be carcinogenic as well as teratogenic for humans (Santos and Linardi 2004). Phenol may also decrease the growth and reproductive capacity of aquatic organisms.
(b) *Nitrobenzene:* Nitrobenzene is an organic compound with the chemical formula $C_6H_5NO_2$ (Fig. 13.2). It is found all through the environment (air, water, soil) because of its enormous use in various industries, viz., in the production of fungicides, drugs, and dyes, and is mainly produced as a precursor of aniline. An adverse effect of this aromatic compound has also been reported. Human being may get exposed to this compound via food and/or water consumption or through air (respiration). Continuous exposure may lead to kidney or liver damage, impaired vision, anemia, and lung irritation and can also affect central nervous system (Ju and Parales 2010). Considering the ill effects, nitrobenzene and related compounds have been included in priority pollutants list by the US Environmental Protection (Tas and Pavlostathis 2014).
(c) *Polychlorinated biphenyls (PCBs):* This category of aromatic compounds includes biphenyl combined with 2–10 chlorine atoms. A long period of the accumulation of industrial PCBs and their inappropriate transfer have brought about pollution in numerous regions. Due to their gigantic and uncontrolled utilization in various industries, PCBs ended up as universal contaminants around the world (Brazova et al. 2012). Like other aromatic pollutants, PCBs are also carcinogenic; they can cause organ damage by accumulating in adipose tissues. Prolonged exposure may affect nervous, reproductive, and immune systems.
(d) *PAHs:* PAHs are toxic, mutagenic, and carcinogenic; and their toxicity ranges from moderate to high for aquatic life. PAHs in soil are non-toxic to terrestrial invertebrates at low concentration unlike at high concentration. PAHs can enter mammals via breathing, skin contact, and ingestion while plants can soak up

PAHs from soils (Dong et al. 2012). The absorption rate depends on the concentration of PAHs, water solubility, physicochemical state, and the type of soil (Inomata et al. 2012). The carcinogenic nature of some PAHs can cause significant damage to the well-being of humans. One of the most serious health effects caused due to breathing of PAHs is lung cancer (Kim et al. 2013). PAHs are not synthesized chemically for industrial purposes but are used commercially as intermediates in pharmaceuticals, agricultural and lubricating materials, thermosetting plastics, photographic products, and other chemical industries.

These are some examples where PAHs get produced as intermediates during several manufacturing processes (Abdel-Shafy and Mansour 2016):

(i) Acenaphthene is formed during manufacturing of dyes, pigments, plastics, pesticides, and pharmaceuticals.
(ii) Anthracene is formed during manufacturing of pigments and dyes and for wood preservations.
(iii) Fluoranthene is formed during production of chemicals used in agriculture, chemical dyes, and medicines.
(iv) Fluorene is formed during production of pharmaceuticals, pesticides, paints, dyes, and thermoset based plastics.
(v) Phenanthrene is formed during manufacturing of pesticides and resins.
(vi) Pyrene is formed during industrial production of pigments.

PAHs have long degradation periods, and recent studies have shown their high concentrations in soil, aquatic, and atmospheric environments. In winter season, pollution and its migration increase by the atmospheric influences on smog clouds, from air to soil, from air to water reserves, and from air to humans. That is the reason for high pollution index in the winter season all over the world especially in Beijing (China) and New Delhi (India).

It has been reported that the London soils contain more stable (4–6 ringed) PAHs originating from various industries, coal and oil burning, etc. The British Geological Survey reported the amount and distribution of PAH compounds including parent and alkylated forms in urban soils at 76 locations in Greater London. The study showed that fluoranthene and pyrene were generally the most abundant PAHs in the London soil (Vane et al. 2014). The content of PAH with 16 aromatic rings ranged from 4 to 67 mg/kg (dry soil weight) and an average PAH concentration of 18 mg/kg (dry soil weight). The total content of PAH with 33 aromatic rings ranged from 6 to 88 mg/kg.

Benzo[α]pyrene (BαP) is the most toxic PAH of the parent PAHs. It is widely considered as a key marker PAH for environmental assessments (Vane et al. 2014). The normal background concentration of BαP in the London urban sites has been reported as 6.9 mg/kg (dry soil weight). In the air, the background levels of some representative PAHs are reported to be 0.02–1.2 and 0.15–19.3 nanograms per cubic meter (ng/m^3) in rural and urban areas, respectively. Background levels of PAHs in drinking water range from 4 to 24 nanograms per liter (ng/L). The level of PAHs in the typical US diet is less than 2 parts of total PAHs parts per billion (ppb) of food,

Table 13.1 Physical property of less and heavily contaminated soil samples

	Components			
Texture or types of soil	Organic C (%)	PAH (mg/kg)	Priority pollutant PAHs (mg/kg)	Total petroleum hydrocarbons (mg/kg)
Less contaminated				
Silty clay loam	1.37	0.220	0.105	21
Silty clay	1.23	15.0	3.06	54
Loamy	0.99	37.0	1.46	3050
Heavily contaminated				
Silty clay loam	1.84	473	27.1	4650
Silty clay	5.65	86.0	10.5	2450
Loamy	3.34	320	18.2	27,500

Cheung and Kinkle (2001)

or less than 2 micrograms per kilogram (μg/kg) of food. Table 13.1 shows the presence of PAHs in less and heavily contaminated soil samples. The soil samples were collected from different locations and have been subjected to scrutiny to find out the amount of pollutants and infer their contamination level (Cheung and Kinkle 2001).

The identification of 16 PAHs as priority pollutants by the US Environmental Protection Agency (EPA) in 1976 has been a primary driver for the development of the analytical methods to determine PAHs. Some of these PAHs are considered to be possible or probable human carcinogens, and hence their distributions in the environment and possible exposure to humans have been of concern (Wise et al. 2015). The US EPA Toxic Release Inventory estimated that more than five million pounds of PAHs are among the toxicants released in the year 2000. Human contamination with PAHs is inevitable because its entry is mainly through skin absorption, inhalation, or food consumption. PAH-contaminated skin may be exposed to sunlight irradiation. It has been suggested that concomitant exposure to chemicals and light can cause toxic reactions and this process is termed as phototoxicity (Yan et al. 2004).

The Department of Health and Human Services (DHHS) and The International Agency for Research on Cancer (IARC) categorize aromatic compounds in various levels of probable human carcinogens. The federal government has set regulations to protect people from the possible health effects of eating, drinking, or breathing aromatic compounds. At present, there are numerous techniques for the degradation purpose, which includes physical, chemical, and biological methods. Physical methods that are currently being used are activated sludge (quick sorption), adsorption by low-cost adsorbents, powdered activated carbon (PAC) (Joaquín et al. 2009), and pyrolysed rice husk (PRH).

In chemical methods, photo-Fenton reaction has mainly been used for the removal of natural or synthetic aromatic compounds and solvent-impregnated resin (SIR) system for the removal of phenols and thiophenols (Cuypers et al. 2010). Other methods include ion exchange membranes (solid polymer electrolytes), H_2O_2/UV processes, O_3/UV processes, and Fenton processes (Fe^{2+}/H_2O_2). The foremost disadvantages of these techniques include expensive nature, and these methods are not completely effective (Klein and Lee 1978; Talley and Sleeper 1997).

Biological methods are among the emerging ways to deal with recalcitrant pollutants, mainly aromatic compounds. Various research strategies are being explored in order to find out the best option for the degradation of aromatic compounds present in effluents of various industries. A methodology more superior than the above-discussed conventional techniques is to completely degrade the aromatics or transform them into non-toxic substances. Bioremediation has been utilized at various sites around the world, including Europe, with varying degrees of success. Among all the biological methods, microbial remediation is a choice that offers the likelihood to degrade or render non-toxic contaminants utilizing microbial versatile metabolic activities.

13.2 *Pseudomonas* in Bioremediation Process

Microorganism-mediated removal and degradation or detoxification of hazards contaminants of soil, sediments, and water is known as bioremediation. Various bacterial, fungal, as well as algal species have been reported to be efficient in this process; but species of *Pseudomonas* have been well documented for their aromatic compound consumption. *Pseudomonas* species (family – *Pseudomonadaceae*) belongs to the class *Gammaproteobacteria*, Gram-negative, aerobic, and rod shaped. Almost all the members of this genus are free living and commonly found in soil and water.

For the treatment of effluents (aromatic compounds), a number of physical and chemical methods are used, but their high-cost input, energy-consuming nature, and the production of secondary pollutants (Shourian et al. 2009) require alternative methods of treatment. Microbial degradation has been proved to be a potential and promising option because of its eco-friendly as well as cost-effective characteristics nature (Liu et al. 2016). Various researches are being conducted on aromatic ring-degrading microbes in order to find the ideal way of treating industrial effluents. Among all the potent bacteria, *Pseudomonas* sp. has been reported in a number works as aromatic compound degraders (Banerjee and Ghoshal 2010; Shourian et al. 2009).

In this chapter, degradation of few aromatic compounds (commonly found in industrial effluents) by *Pseudomonas* sp. has been explained in a lucid manner. Majority of bacteria are capable of growing on two or more aromatic compounds. Among these aromatic ring degraders, *Pseudomonas putida* F1 is best characterized. This strain has been documented to consume (degrade) toluene, benzene, ethyl-benzene, and phenol as sole carbon sources. Mostly, single organisms degrade only a limited number of aromatic compounds. *Pseudomonas* sp. represents one of the most versatile bacteria involved in the degradation of aromatic compounds (Obayori et al. 2014).

13.2.1 Mechanism of Degradation

Degradation of most common aromatic compounds like benzene and phenol by *Pseudomonas* sp. has been reported by a number of researchers. Figure 13.3 shows the biodegradation pathways of benzene and phenol. In the first step, benzene and phenol are oxidized to catechol, and hydroxylase enzyme utilizes molecular oxygen and adds

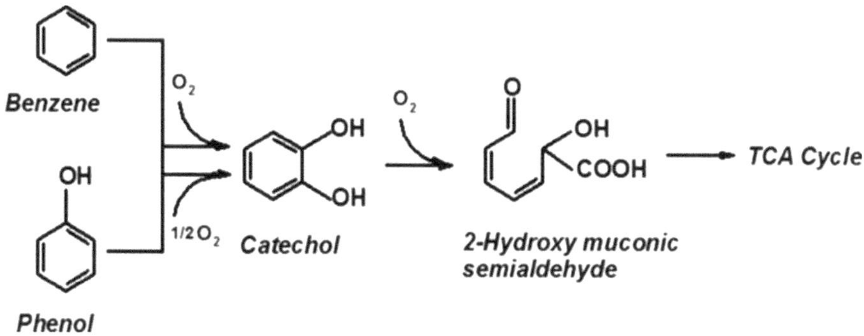

Fig. 13.3 Degradation pathway of Pseudomonas putida for benzene and phenol

hydroxyl group to the ring. The next step includes ring cleavage adjacent to, or in between, the two hydroxyl groups of catechol. Catechol can be oxidized either by ortho-cleavage pathway by catechol 1,2-dioxygenase or by meta-cleavage pathway to 2-hydroxymuconic semialdehyde by catechol 2,3-dioxygenase. Further degradation takes place through tricarboxylic acid cycle in the case of both the pathways (Powlowski and Shingler 1994; Harayama et al. 1992). Catechols are cleaved either by ortho-fission (intradiol, i.e., carbon bond between two hydroxyl groups) or by meta-fission (extradiol, i.e., between one of the hydroxyl groups and a non-hydroxylated carbon). Thus the ring is opened and subsequently degraded (Cerniglia 1984).

Hasan and Jabeen isolated and reported *Pseudomonas aeruginosa* as a versatile toxic aromatic compound degrader (Hasan and Jabeen 2015). *P. aeruginosa* along with *P. fluorescence* are found to utilize phenol, and the rate of degradation was reported to be 94.5% for *P. aeruginosa* and 69.4% for *P. fluorescence* (Afzal et al. 2007; Razika et al. 2010). Similarly, Ojumu et al. (2005) reported the complete removal of phenol (30 mg/l) within 60 h and 84 h using the same species of *P. aeruginosa* and *P. fluorescence*, respectively (Ojumu et al. 2005). In addition to this, both of the strains have been reported by Agarry et al. as potent phenol degraders (Agarry et al. 2008).

It has been stated by Kumar et al. that *P. putida* (MTCC1194) can degrade 20.59 mg/l and 9.66 mg/l of phenol and catechol completely (Kumar et al. 2005). However, the same pseudomonas strain can degrade 258 mg/l and 157 mg/l of phenol and catechol with a retarded rate of degradation resulting in substrate build-up (Bajaj et al. 2009).

Degradation ability of *Pseudomonas* sp. has also been reported through various researches. *Pseudomonas* along with *Sphingomonas* has been reported to possess the ability to utilize phenanthrene as a sole carbon and energy source. Bacterial PAH degradation involves mainly dioxygenase enzymes and partially monooxygenase-mediated reactions (Ghosal et al. 2016). Recently, Obayori et al. reported the biodegradation of fresh and used engine oils by *Pseudomonas aeruginosa* LP5 (Obayori et al. 2014). Authors isolated *P. aeruginosa* LP5 from the contaminated Nigerian soil which were able to degrade pyrene and consume fresh and

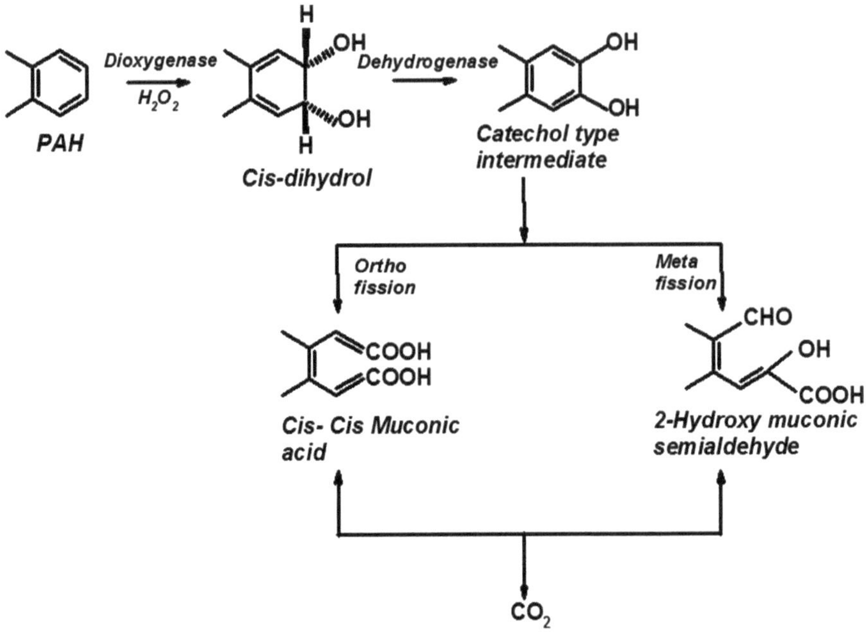

Fig. 13.4 Degradation mechanism of PAHs by bacterial species

used engine oil. Ma et al. in 2012 isolated PAHs (fluorene and phenanthrene)-degrading *Pseudomonas* spp. JM2 from sewage sludge of chemical plants. The isolated microbes were reported to be a high-performance strain, capable of consuming PAHs under harsh environmental conditions (Ma et al. 2012).

As described by Bamforth and Singleton, principal pathway for bacterial consumption of PAHs starts with oxidation of the benzene ring with the action of dioxygenase enzymes which leads to the formation of *cis*-dihydrodiols (Bamforth and Singleton 2005). These dihydrodiols then form dihydroxylated intermediates after dehydrogenation step, which can then be further metabolized via catechols to carbon dioxide and water (Fig. 13.4).

One of the best characterized species of *Pseudomonas* is *Pseudomonas putida* KT2440; it was reported to be a plasmid-free derivative of a toluene-degrading bacterium. *Pseudomonas* sp. possesses various non-specific enzymes; hence the metabolic pathway contains a great degree of convergences. This promotes the proficient consumption of a wide variety of aromatic compounds (growth substrates). The non-specificity of enzymes allows simultaneous utilization of similar compounds (substrates). *Pseudomonas* sp. is known to possess versatile metabolic systems, hence being able to degrade and consume various hazardous aromatic compounds as a carbon and energy source. It contains a number of enzymes, which helps in the metabolic process. Enzymes reported in *Pseudomonas* sp. responsible for aromatic ring degradation include Phenol hydroxylase, monooxygenase, dioxygenase, etc.

13.2.2 Factors Affecting the Degradation of Aromatic Compounds

There are a number of studies, showing the efficiency of microbes in aromatic compound degradation under ideal laboratory conditions. However, it is obvious that environmental conditions (soil pH, nutrient, temperature) vary from site to site. Major factors have been listed here.

(a) *Temperature:* Temperature has a substantial effect on the ability of the in situ microorganisms to degrade aromatic compounds. During every season of the year, most of the contaminated sites are at the optimum temperature. The solubility of PAHs increases with an increase in temperature, which in turn increases the bioavailability of the PAHs. On the contrary, oxygen solubility decreases with increasing temperature, which ultimately reduces the metabolic activity of aerobic microorganisms. Over a wide range of temperature, biodegradation of various aromatic compounds can happen. There are extensive numbers of studies that tend to focus on mesophilic temperatures rather than focusing on the efficiency of transformations at very low or high temperatures.
(b) *pH:* Many places which are contaminated with PAHs are not at the optimal pH for bioremediation. For an instance, gasworks sites that are no longer in use are often surrounded with substantial quantities of demolition waste such as concrete and brick. The pH of the native soil increases if this material leaches out; it ends up creating unfavorable conditions for microbial metabolism. Along with that, the oxidation and leaching of coal will also generate an acidic environment. This is caused by excessive release and oxidation of sulfides. The in situ microorganisms will not possess the ability to transform aromatic pollutants under acidic or alkaline conditions because the pH of contaminated sites is most of the times related to the pollutant. Hence, most commonly lime is added at these sites to adjust the pH level (Alexander 1995). The degradation of phenanthrene with *Burkholderia cocovenenans* in liquid culture has been examined at a wide range of pH values (pH 5.5–7.5).
(c) *Oxygen:* It is a well-known fact that the bioremediation of organic contaminants such as PAHs can be initiated under both aerobic and anaerobic conditions. Most of the existing literature has focused on the dynamics of aerobic metabolism of PAHs. During aerobic degradation, oxygen is essential to the action of mono- and dioxygenase enzymes, leading to the oxidation of the aromatic ring.
(d) *Nutrient availability:* Apart from readily degradable carbon source, microorganisms need mineral nutrients such as nitrogen, phosphate, and potassium (N, P, and K) for conducting cellular metabolism, which leads to successful growth. In contaminated sites due to the nature of the pollutants, organic carbon levels are often quite high. It can quickly deplete during microbial metabolism (Breedveld and Sparrevik 2000). Hence, it is common practice to provide contaminated land with nutrients; generally nitrogen and phosphates are added to kindle the in situ microbial community and which ultimately enhances bioremediation (Alexander 1994; Atagana et al. 2003).

(e) *Bioavailability*: Bioavailability can be characterized as the impact of physicochemical and microbiological factors on the rate and degree of biodegradation and is supposed to be among the most essential factors in bioremediation. Aromatic compounds like PAHs possess a low bioavailability and are considered as hydrophobic organic contaminants. These are synthetic compounds with low water solvency that are resilient to biological, chemical, and photolytic degradation. A few microorganisms can create surfactants (biosurfactants) that can improve the desorption of PAHs from the soil matrix (Makkar and Rockne 2003).

Considering the complexity of industrial effluents, single strain of microbes is not able to provide the desirable results. Use of microbial consortium can be an option to increase the level of degradation of aromatic compounds. Advances in the bioremediation techniques are also being considered, for example, bioaugmentation is in situ bioremediation practice which includes addition of cultured microbes into the contaminated sites to enhance the rate of degradation. Similarly, biostimulation technique can be used for advancement of industrial effluent removal by adding required or limiting nutrients to the site for upliftment of endogenous microbial growth.

13.3 Advances in *Pseudomonas* Sp.-Mediated Bioremediation

Different types of industries explore different types of raw materials and chemicals, so effluents coming out of these industries must be a combination of organic and inorganic compounds. Here as we are focusing on aromatic compounds, as discussed earlier, there are a variety of aromatic compounds in the industrial discharged water. Although *Pseudomonas* species has been found to be efficient in degradation of various aromatic compounds but incorporation of molecular approach may give an enhanced rate of degradation.

13.3.1 Metagenomics and Genetic Manipulation

Metagenomics as defined by Handelsman is the genomic analysis of microorganisms by direct extraction and cloning of DNA from an assemblage of microorganisms (Handelsman 2004). Metagenomics gained attention for the first time after the famous Diamond v. Chakrabarty: patent on oil-consuming bacteria case. He has got patent on a genetically engineered bacterium with unique ability to consume (degrade) various components of crude oil (Chakrabarty 1981). Rapid development in the sequencing technique leads to drastically dropped cost of sequencing; hence the size of sequencing ventures has been expanded. This has given the chance for the routine use of sequencing techniques in observing more and more microbes. Metagenomics applications have been routinely used to understand the ecology and diversity of microbes in a better way. Utilization of metagenomics in bioremediation and environmental monitoring is increasing with each passing day.

Metagenomics is a promptly growing field of research that has had a dramatic effect on the way we view and study the microbial world. By permitting the direct investigation of bacteria, viruses, and fungi irrespective of their culturability and taxonomic identities, metagenomics has changed microbiological theory and methods and has also challenged the classical concept of species (Techtmann and Hazen 2014). This new field of biology has proven to be rich and comprehensive and is making important contributions in many areas including ecology, biodiversity, bioremediation, bioprospection of natural products, and medicine.

It can also be considered as environmental genomics, as through this one can directly analyze a wide range of environmental microbes (uncultured microbes) on genomic level. Record of these genomic data depends on computational tools. A lot of articles are being published on this amusing as well as promising topic. This chapter addresses aspects of metagenomics and its potential application in bioremediation in a lucid manner. It has the power to boost and hence improve the current strategies of biodegradation of aromatic pollutants. Comprehensive understanding of the nature and mechanism of microbial communities, like how they adapt in environment comprising aromatic pollutants, could help to analyze the potential of the contaminated site (Chakraborty et al. 2014).

Hence, ultimately the level of bioremediation can be uplifted. In addition to that, uncultivable microbes can be analyzed by this approach and thus various aspects and application of a wide range of genes and enzymes can be studied. This could help researchers to plan and prepare consortia of desired strains of microbes for targeted use and also predict the degradation pathways. Basically, genetic fingerprinting technique covers the analysis of only limited microbial community. It cannot provide information beyond few selected genes that can be amplified and are well characterized. On the other hand, metagenomics techniques are based on the idea that the DNA recovered directly from natural (environmental) samples represents whole community genomes which could be sequenced and investigated similarly as that of an entire genome of pure bacterial culture. Hence, metagenomics-based studies are empowering ecologists to gather an exhaustive data of the entire community. In addition to this, metagenomics also helps in understanding the biochemical nature and roles of uncultured microorganisms.

Metagenomic libraries are developed by extracting complete DNA from natural sites. In the case of bioremediation, areas affected with industrial effluents must be considered for the purpose (Stenuit et al. 2008; Desai et al. 2010). Isolation of DNA is followed by cloning of random DNA fragments into suitable vector. These vectors are then transferred to bacterial hosts by the process of transformation. Screening of the transformed bacteria can be performed through sequencing or on the basis of physiological characteristics (Fig. 13.5). These metagenomic libraries are also being considered as a resource for new microbial enzymes with potential application in bioremediation (Riesenfeld et al. 2004).

There are various reports on the use of metagenomics, for example, metagenomic analysis was used by Hemme et al. to study the nature of adaptive bacterial communities in heavy metal-contaminated groundwater sites (Hemme et al. 2010). It was stated from this study that long-term exposure to heavy metals results in a

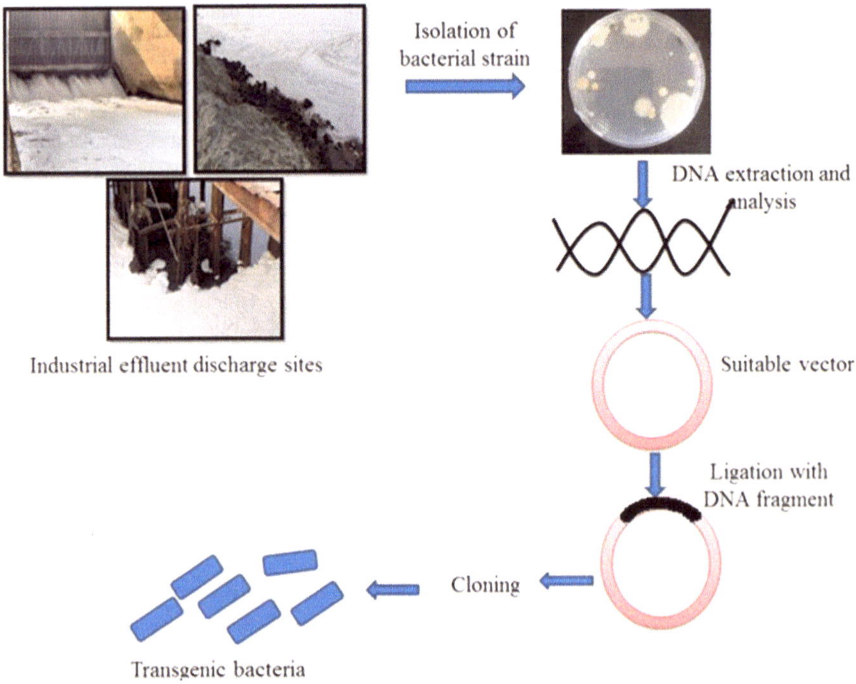

Fig. 13.5 Process of metagenomics (isolation of DNA, manipulation of DNA, and construction of library)

dramatic loss of groundwater microbial communities. It was concluded that lateral gene transfer could be the reason of adaptive and new mechanism by conferring metal-resistant genes. Hence, this technique is helping the microbial community to survive in harsh and stressed environment.

Another investigation by Abbai et al. utilized pyrosequencing-based metagenomics to describe the metagenome of a hydrocarbon-contaminated site, with respect to its composition and metabolic possibilities (Abbai et al. 2012). They also used the same technique and reported adaptation behavior of microbial consortium to hydrocarbon-contaminated environment (Abbai and Pillay 2013).

13.4 Conclusion

The constantly increasing problem of industrial effluent management can be solved by using diverse and versatile microbes, i.e., via bioremediation. Aromatic compounds are both substrates for microbial growth and also one of the major environmental threats. Only at lower concentration, aromatic compounds can act as substrates for microbial growth, while at higher concentration there is substrate inhibition which renders the growth of microbes. Microbes are well known for their adaptive nature; *Pseudomonas* species are among the mostly reported bacterial species for their aromatic ring degradation ability. Phenol, PAHs, and PCBs have been reported to be degraded by *Pseudomonas* species. They follow the classical method of degradation which starts with hydroxylation of the aromatic ring mediated by oxygenase enzyme and ultimately transforms hazardous aromatic compounds into carbon dioxide and other non-toxic simpler compounds which do not affect the environment. Because of their strong capabilities in biotransformation of xenobiotic pollutants, *Pseudomonas* species possess great potential to be used in different biotechnological applications in general and in bioremediation in particular. Hence the use of this invisible living organism can produce visible results in the field of biodegradation. Metagenomics approach can be utilized in case of *Pseudomonas* species that are unable to degrade industrial aromatic effluents.

References

Abbai NS, Govender A, Shaik R, Pillay B (2012) Pyrosequence analysis of unamplified and whole genome amplified DNA from hydrocarbon-contaminated groundwater. Mol Biotechnol 50(1):39–48

Abbai NS, Pillay B (2013) Analysis of hydrocarbon-contaminated groundwater metagenomes as revealed by high-throughput sequencing. Mol Biotechnol 54(3):900–912

Abdel-Shafy HI, Mansour MSM (2016) A review on polycyclic aromatic hydrocarbons: source, environmental impact, effect on human health and remediation. Egypt J Pet 25(1):107–123

Afzal M, Iqbal S, Rauf S, Khalid ZM (2007) Characteristics of phenol biodegradation in saline solutions by monocultures of *Pseudomonas aeruginosa* and *Pseudomonas pseudomallei*. J Hazard Mater 149:60–66

Agarry SE, Durojaiye AO, Yusuf RO, Aremu MO, Solomon BO, Mojeed O (2008) Biodegradation of phenol in refinery wastewater by pure cultures of *Pseudomonas aeruginosa* NCIB 950 and *Pseudomonas fluorescence* NCIB 3756. Int J Environ Pollut 32(1):3–11

Alexander M (1994) Biodegradation and bioremediation. Academic, San Diego

Alexander M (1995) How toxic are toxic chemicals in soil? Environ Sci Technol 29:2713–2717

Atagana HI, Haynes RJ, Wallis FM (2003) Optimization of soil physical and chemical conditions for the bioremediation of creosote-contaminated soil. Biodegradation 14:297–307

Bajaj M, Gallert C, Winter J (2009) Phenol degradation kinetics of an aerobic mixed culture. Biochem Eng J 46:205–209

Bamforth SM, Singleton I (2005) Bioremediation of polycyclic aromatic hydrocarbons: current knowledge and future directions. J Chem Technol Biotechnol 80:723–736

Banerjee A, Ghoshal AK (2010) Isolation and characterization of hyper phenol tolerant *Bacillus sp.* from oil refinery and exploration sites. J Hazard Mater 176:85–91

Brazova T, Hanzelova V, Miklisova D (2012) Bioaccumulation of six PCB indicator congeners in a heavily polluted water reservoir in eastern Slovakia: tissue specific distribution in fish and their parasites. Parasitol Res 111:779–786

Breedveld GD, Sparrevik M (2000) Nutrient limited biodegradation of PAHs in various soil strata at a creosote contaminated site. Biodegradation 11:391–399

Cerniglia CE (1984) Microbial transformations of aromatic hydrocarbons. In: Atlas RM (ed) Petroleum microbiology. Macmillan, New York, pp 99–128

Chakrabarty AM (1981) Microorganisms having multiple compatible degradative energy-generating plasmds and preparation thereof. US Patent 4259444

Chakraborty A, Gupta CKD, Bhadury P (2014) Application of molecular techniques for the assessment of microbial communities in contaminated sites. In: Das S (ed) Microbial Biodegradion and bioremediation. Elsevier, pp 85–113

Cheung PY, Kinkle BK (2001) Mycobacterium diversity and pyrene mineralization in petroleum-contaminated soils. Appl Environ Microbiol 67(5):2222–2229

Cuypers R, Sudhölter EJR, Zuilhof H (2010) Hydrogen bonding in phosphine oxide/phosphate-phenol complexes. ChemPhysChem 11(10):2230–2240

Desai C, Pathak H, Madamwar D (2010) Advances in molecular and "-omics" technologies to gauge microbial communities and bioremediation at xenobiotic/anthropogen contaminated sites. Bioresour Technol 101:15581569

Dong CD, Chen CF, Chen CW (2012) Determination of polycyclic aromatic hydrocarbons in industrial harbor sediments by GC-MS. Int J Environ Res Publ Health 9:2175–2188

Ghosal D, Ghosh S, Dutta TK, Ahn Y (2016) Current state of knowledge in microbial degradation of polycyclic aromatic hydrocarbons (PAHs): a review. Front Microbiol 7:1369–1395

Handelsman J (2004) Metagenomics: application of genomics to uncultured microorganisms. Microbiol Mol Biol Rev 68(4):669–685

Harayama S, Kok M, Neidle EL (1992) Functional and evolutionary relationships among diverse oxygenases. Annu Rev Microbiol 46:565–601

Hasan SA, Jabeen S (2015) Degradation kinetics and pathway of phenol by *Pseudomonas* and *Bacillus* species. Biotechnol Biotechnol Equip 29:45–53

Hemme CL, Deng Y, Gentry TJ, Fields MW, Wu L, Barua S (2010) Metagenomic insights into evolution of a heavy metal-contaminated groundwater microbial community. ISME J 4:660672

Inomata Y, Kajino M, Sato K, Ohara T, Kurokawa J, Ueda H, Tang N, Hayakawa K, Ohizumi T, Akimoto H (2012) Emission and atmospheric transport of particulate PAHs in Northeast Asia. Environ Sci Technol 46:4941–4949

Joaquín R, Domínguez-Vargas JA, Navarro-Rodríguez JBH, Eduardo M (2009) Removal of chlorophenols in aqueous solution by carbon black low-cost adsorbents. Equilibrium study and influence of operation conditions. J Hazard Mater 169:302–308

Ju KS, Parales RE (2010) Nitroaromatic compounds, from synthesis to biodegradation. Microbiol Mol Biol Rev 74:250–272

Kim KH, Jahan SA, Kabir E, Brown RJ (2013) A review of airborne polycyclic aromatic hydrocarbons (PAHs) and their human health effects. Environ Int 60:71–80

Klein JA, Lee DD (1978) Biological treatment of aqueous wastes from usual conversion processes. Biotechnol Bioeng 8:379–390

Kumar A, Kumar S, Kumar S (2005) Biodegradation kinetics of phenol and catechol using *Pseudomonas putida* MTCC 1194. Biochem Eng J 22:151–159

Liu J, Yu Y, Chang Y, Li B, Bian D, Yang W, Huo H, Huo M, Zhu S (2016) Enhancing quinoline and phenol removal by adding *Comamonastestosteroni* bdq06 in treatment of an accidental dye wastewater. Int Biodeterior Biodegrad 115:74–82

Ma J, Xu L, Jia L (2012) Degradation of polycyclic aromatic hydrocarbons by Pseudomonas sp. JM2 isolated from active sewage sludge of chemical plant. J Environ Sci 24(12):2141–2148

Makkar RS, Rockne KJ (2003) Comparison of synthetic surfactants and biosurfactants in enhancing biodegradation of polycyclic aromatic hydrocarbons. Environ Toxicol Chem 22:2280–2292

Obayori OS, Salam LB, Ogunwumi OS (2014) Biodegradation of fresh and used engine oils by *Pseudomonas aeruginosa* LP5. J Bioremed Biodegr 5:213–219

Ojumu T, Bello O, Solomon B (2005) Evaluation of microbial systems for bioremediation of petroleum refinery effluents in Nigeria. Afr J Biotechnol 4(1):31–35

Powlowski J, Shingler V (1994) Genetics and biochemistry of phenol degradation by *Pseudomonas* sp. CF600. Biodegradation 5(3):219–236

Razika B, Abbes B, Messaoud C, Soufi K (2010) Phenol and benzoic acid degradation by *Pseudomonas aeruginosa*. Int J Water Res Environ 2(9):788–791

Riesenfeld CS, Schloss PD, Handelsman J (2004) Metagenomics: genomic analysis of microbial communities. Annu Rev Genet 38:525–552

Santos VL, Linardi VR (2004) Biodegradation of phenol by a filamentous fungi isolated from industrial effluents-identification and degradation potential. Process Biochem 39:1001–1006

Shourian M, Noghabi KA, Zahiri HS, Bagheri T, Karballaei G, Mollaei M, Rad I, Ahadi S, Raheb J, Abbasi H (2009) Efficient phenol degradation by a newly characterized *Pseudomonas* sp. SA01 isolated from pharmaceutical wastewaters. Desalination 246:577–594
Stenuit B, Eyers L, Schuler L, Agathos SN, George I (2008) Emerging high throughput approaches to analyze bioremediation of sites contaminated with hazardous and/or recalcitrant wastes. Biotechnol Adv 26:561–575
Talley JW, Sleeper PM (1997) Roadblock to the implementation of biotreatment strategies. Bajpai RK, Zappi ME (eds) Ann N Y Acad Sci 829:17–29
Tas DO, Pavlostathis SG (2014) Occurrence, toxicity, and biotransformation of pentachloronitrobenzene and chloroanilines. Crit Rev Environ Sci Technol 44(5):473–518
Techtmann SM, Hazen TC (2014) Metagenomic applications in environmental monitoring and bioremediation. J Ind Microbiol Biotechnol 43(10):1345–1354
Vane CH, Kim AW, Beriro DJ, Cave MR, Knights K, Moss-Hayes V, Nathanail PC (2014) Polycyclic aromatic hydrocarbons (PAH) and polychlorinated biphenyls (PCB) in urban soils of Greater London, UK. Appl Geochem 51:303–314
Wise SA, Sander LC, Schantz MM (2015) Analytical methods for determination of polycyclic aromatic hydrocarbons (PAHs)—a historical perspective on the 16 U.S. EPA priority pollutant PAHs. Polycycl Aromat Compd 35(2–4):187–247
Wu WE, Ge HG, Zhang KF (2003) Wastewater biological treatment technology. Chemical Industry Press, Beijing. (In Chinese)
Yan J, Wang L, Fu PP, Yu H (2004) Photomutagenicity of 16 polycyclic aromatic hydrocarbons from the US EPA priority pollutant list. Mutat Res 557(1):99–108

Bioremediation of Persistent Toxic Substances: From Conventional to New Approaches in Using Microorganisms and Plants

14

Mihaela Rosca, Raluca-Maria Hlihor, and Maria Gavrilescu

Abstract

Bioremediation – the use of microorganisms and/or plants to detoxify different environmental compartments polluted with organic and inorganic compounds – received an increased attention across scientific media for its eco-friendliness, efficiency, and low cost. After the Stockholm Convention, persistent toxic substances (PTS) have also received an increased attention due to their persistence and bioaccumulative and toxic characteristics. Different bioremediation options started to be used to degrade and/or remove PTS. One of the major breakthroughs of our century is the use of bioremediation of genetically modified microorganisms (GMMOs) and plant-microbe interactions, including plants engineered by transferring different genes from organisms such as other plants, animals, bacteria, or fungi, meaning the so-called transgenic plants. Accordingly, this chapter discusses major in situ and ex situ technologies used for PTS removal by

M. Rosca
"Cristofor Simionescu" Faculty of Chemical Engineering and Environmental Protection, Department of Environmental Engineering and Management, "Gheorghe Asachi" Technical University of Iasi, Iasi, Romania

R.-M. Hlihor
"Cristofor Simionescu" Faculty of Chemical Engineering and Environmental Protection, Department of Environmental Engineering and Management, "Gheorghe Asachi" Technical University of Iasi, Iasi, Romania

Faculty of Horticulture, Department of Horticultural Technologies, "Ion Ionescu de la Brad" University of Agricultural Sciences and Veterinary Medicine of Iasi, Iasi, Romania

M. Gavrilescu (✉)
"Cristofor Simionescu" Faculty of Chemical Engineering and Environmental Protection, Department of Environmental Engineering and Management, "Gheorghe Asachi" Technical University of Iasi, Iasi, Romania

Academy of Romanian Scientists, Bucharest, Romania
e-mail: mgav@tuiasi.ro

P. K. Arora (ed.), *Microbial Technology for the Welfare of Society*, Microorganisms for Sustainability 17,
https://doi.org/10.1007/978-981-13-8844-6_14

bioremediation, their limitations and advantages in practice, and future perspectives by applying genetically engineered alternatives to improve process efficiency and for successful application at a large scale.

Keywords
Persistance toxic substances · Bioremediation · Biodegradation · Genetically modified microorganisms

14.1 A Short Overview of Persistent Toxic Substances

Human and environmental health is nowadays among major objectives when it comes to releasing of toxic substances in the environment. Ever since the book *Silent Spring* written by Rachel Carson raised the attention of pesticides toxicity, generations of scientists, environmentalists, and officials, under the assessment of different regulatory frameworks, have attempted to put into public attention a major breakthrough into the subject. In this respect, a document for decision-makers listing an initial set of 12 persistent organic pollutants (POPs) has been agreed within the Stockholm Convention (2001), which was subsequently extended. POPs were classified as being persistent, toxic, and bioaccumulative substances.

Significant advances have increased the public awareness on toxic substances. For example, by implementing "The Regionally Based Assessment of Persistent Toxic Substances (RBA PTS)" Project proposed by Global Environment Facility (GEF), the sources, behavior, and effects of persistent toxic substances (PTS) were evaluated in 12 regions of the globe (RBA PST 2003). In view of supporting the advancement of knowledge on PTS assessment, some of the Arctic indigenous peoples' organizations – Permanent Participants of the Arctic Council and the Secretariat of the Arctic Monitoring and Assessment Programme (AMAP) – initiated another project with the support of Global Environmental Facility (GEF), that is, "Persistent Toxic Substances, Food Security and Indigenous Peoples of the Russian North" Project. The overall goal of the project was to reduce PTS contamination in the Arctic region (AMAP 2004).

Depending on the context, authors, audience, and date of publication, there are actually different ways to define persistent and toxic chemicals. On a broader view, the *group of persistent toxic substances* (PTS) includes the Stockholm POPs (e.g., dichlorodiphenyltrichloroethane, polychlorinated biphenyls, dioxin/furan, etc.) and metallic compounds (containing mercury, cadmium, lead, and the others) (Wong et al. 2012). Another preferred term is *persistent, bioaccumulative, and toxic* (PBT) chemicals (Matthies et al. 2016; Moermond et al. 2011). Aside the general recognized designation of POPs, other classifications provide terms such as emerging contaminants of concern (ECC), organic wastewater contaminants (OWC), pharmaceuticals and personal care products (PPCP), and hypothalamic-pituitary-ovarian (HPO) axis (Wong et al. 2012).

Considering the environmental and human health points of view, all terms agree with the characteristics of these chemicals as of being persistent, toxic, bioaccumulative, and susceptible to long-distance transport through different environmental compartments. PTS can be even transported to regions where they have never been used. The screening criteria of PBT under REACH (Registration, Evaluation, Authorization and Restriction of Chemicals) have been included in the work of Moermond et al. (2011) and focus on clearly defining the terms persistence, bioaccumulation, and toxicity by considering the type of data and criterion in the PBT assessment and all criteria as shown in Annex XIII of REACH.

14.2 Sources of PTS

The release of PTS in the environment occurs either intentionally or unintentionally. Once in the environment, their long-range transport by atmospheric, oceanic, and riverine pathways makes them available in different regions across the globe. An interesting view into different pathways into the Arctic has been released by the Commission for Environmental Cooperation (CEC 2018). As provided by the Persistent Toxic Substances, Food Security and Indigenous Peoples of the Russian North Report (AMAP 2004), there are multiple sources of PTS, divided into three categories as shown in Fig. 14.1. Some examples in literature demonstrate that the occurrence in the environment of PTS involves various, but specific, sources.

Coal combustion in industrial areas can generate contamination and distribution of PTS (e.g., HCHs, DDTs, PCBs, and PAHs) in an agricultural soil. Data analysis in Lishui County, China, performed by Hu et al. (2009) revealed that PAHs had as source coal combustion in the neighbor industrial areas. However, soil analyses addressing HCHs and DDTs showed that these chemicals had no sources in the concerned area; therefore, they were transported from longer distances (Hu et al. 2009). Another source of PTS is represented by discharges to freshwater. In a comprehensive assessment based on PTS, Barakat (2004) identified potential sources and locations of contamination with pesticides in the Egyptian Mediterranean. The authors identified discharges to freshwater as localized inputs from different coastal areas such as Nile estuaries and sewage discharges from industrialized cities such as Alexandria. Minh et al. (2016) pointed out on the global attention of toxic chemicals release from e-waste recycling in different regions of Asia. Informal and

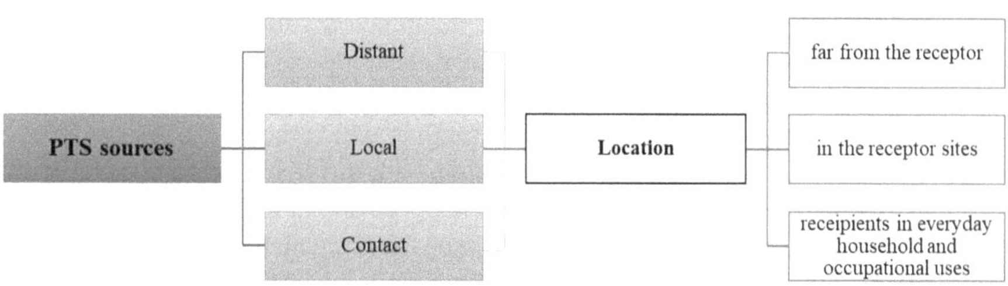

Fig. 14.1 Sources of persistent toxic substances

uncontrolled waste treatment and recycling facilities were identified as hotspots for PTS. Lammel et al. (2015) identified 26 PTS near-ground air and 55 PTS in surface seawater, around the Aegean Sea. Interestingly, the authors found increasing concentrations of PCB in seawater even if emissions have been declining since 1970s.

14.3 Environmental Bioremediation

The large variety of persistent toxic substances found in the environment call for the application of adequate technologies for the removal or degradation of these hazardous pollutants. In a common sense, the potential remediation technologies have been grouped as ex situ and in situ, and, depending on the nature of the process involved in decontamination, these are physical, chemical, or biological technologies. The main physical remediation techniques are "soil washing, vitrification, encapsulation of contaminated areas by impermeable vertical and horizontal layers, electrokinesis, and permeable barrier systems" (Bradl and Xenidis 2005; Gavrilescu et al. 2015). The cleanup of contaminated components by chemical treatments involves the use of chemical agents to achieve extraction, degradation, or immobilization of the pollutant. Chemical oxidation, chemical immobilization, dehalogenation, ion exchange, and stabilization/solidification are among the most representative technologies (Bradl and Xenidis 2005; Gavrilescu and Chisti 2005; Gavrilescu et al. 2009; Hamby 2000).

By definition, bioremediation is "the process used to degrade, remove, immobilize, or detoxify the soils and water contaminated with various pollutants through the action of microorganisms (bacteria, yeast, fungi) and plants" (Mosa et al. 2016; Ayangbenro and Babalola 2017). Compared to physical and chemical alternatives, bioremediation has a number of advantages but also disadvantages. Application of biological processes for the remediation of environmental compartments requires a moderate level of investment and operating costs and is considered safe and eco-friendly for the environment.

In this context, our major attention is focused, in this chapter, on bioremediation, as part of environmental biotechnology, applied for the removal of PTS – among other pollutants, in different environmental media with a special emphasis on microorganisms and plant-microorganisms interactions, including those with improved performance through gene modification.

The microorganisms possess the ability to degrade a large number of contaminants and do not generate waste, while the degradation compounds are less toxic than the parent substance (Iosob et al. 2016; Abatenh et al. 2017; Apostol and Gavrilescu 2016; Smaranda et al. 2016). The disadvantages of bioremediation technologies are especially related to the time frame in which the processes are able to degrade or remove some persistent compounds with different toxic characteristics. Bioremediation is considered a long-term process (Iosob et al. 2016; Abatenh et al. 2017; Malschi et al. 2018; Pavel and Gavrilescu 2008), which cannot be applied to any type of pollutant (e.g., highly chlorinated compounds have very low biodegradability), while, in some cases, at the end of the process, degradation compounds

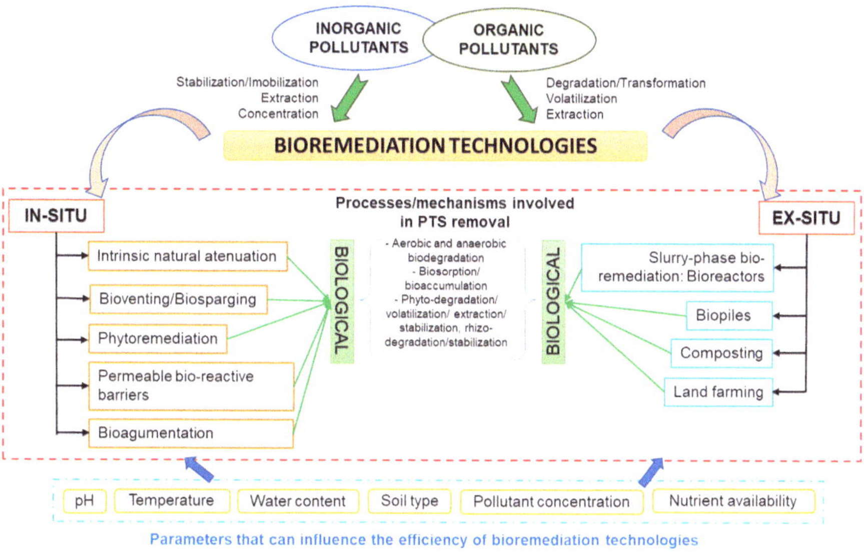

Fig. 14.2 Bioremediation technologies applied for PTS removal from soil

with high toxicity could be released in the environment (e.g., the transformation of some PAHs to less degradable PAHs with carcinogenic effects) (Iosob et al. 2016). The efficiency of biological processes is strongly influenced by many factors and parameters such as temperature, pH, water content, nutrient availability, soil type, aeration, the chemical nature and concentration of pollutants, and the availability of indigenous microorganisms' strains in soil (Abatenh et al. 2017; Apostol and Gavrilescu 2016).

The ex situ technologies include bioreactors, land farming, anaerobic digestion, composting, and biopile, while the well-known in situ bioremediation technologies are the intrinsic natural attenuation, air sparging, bio-venting (aerobic), bio-slurping, enhanced bio-degradation, and phytoremediation (Fig. 14.2) (Apostol and Gavrilescu 2016; Cozma et al. 2016). The in situ biotechnologies are considered advantageous because these processes can be applied without soil excavation or water pumping (operations considered expensive and resource-consuming). Other advantages of in situ technologies consist in the possibility of their application on extended polluted surfaces and for removing lower concentrations of contaminants in soil. The main disadvantages of these technologies are related to (i) the time frame of bioremediation, which is long, (ii) the difficulty in controlling the parameters that could affect the biological processes, and (iii) the usage in treating a smaller range of contaminants compared to ex situ biotechnologies (Malschi et al. 2018; Abraham et al. 2016; Cojocaru et al. 2016).

Bioventing-Biosparging technology was applied by Eslami and Joodat (2018) to biodegrade the ethylbenzene and pyrene found in a construction site located in Garmdareh, Iran. The results showed that the efficiency of the process varied between 60% and 40%, decreasing with the increase in the depth of the pollutants in soil. The highest efficiency (60%) for ethylbenzene biodegradation was reached after 40 days and for pyrene after 50 days (55%).

Permeable Bio-reactive Barriers represent a sustainable groundwater remediation technique in which both organic and inorganic toxic substances can be removed (Obiri-Nyarko et al. 2014). At the laboratory scale, this technique combined with enricher reactor technique was used by Kasi et al. (2010) to biodegrade benzene found in groundwater at a concentration of 10 mg/L. The results of this study showed that more than 90% of benzene was removed in the first 72 h.

Another in situ biotechnology successfully applied is ***phytoremediation***, based on the ability of plants to retain or degrade toxic substances. Often the phytoremediation process is stimulated by the presence of some types of microorganisms in the rhizosphere, such as plant growth-promoting bacteria (PGPB) (Ojuederie and Babalola 2017) and mycorrhizal fungi (Cabral et al. 2015; Turnau et al. 2006). The ability of plants such as *Loliumperenne*, *Festucapratensis*, *Stipacapillata*, *Agrostis alba*, *Cynodon dactylon*, *Agrostis tenuis*, and *Luzula campestris* to retain Cu^{2+}, Zn^{2+}, Sn^{2+}, Pb^{2+}, Co^{2+}, Ni^{2+}, Mn^{2+}, and Cr^{3+} ions was studied by Elekes (2014). Plants naturally grown on an industrial platform of Targoviste city, Romania, proved their ability in phytoremediation of heavy metal-contaminated soil. The results showed that *Loliumperenne* had the capacity to remove 921.67 mg/kg Zn^{2+}, 201.23 mg/kg Pb^{2+}, 114.19 mg/kg Cr^{3+}, and 61.95 mg/kg Cu^{2+} (Elekes 2014; Roşca et al. 2017a). Another plant species able to remove the heavy metal ions from soil is *Populus nigra* L., which has the capacity to transport and accumulate ions in leaves and bark (Roşca et al. 2017a; Barbeş and Bărbulescu 2017).

Composting is considered a biotechnology procedure usually applied to convert organic materials into humic substances, but it can also be applied for bioremediation of polluted soils and sludges. During this process, the toxic organic substances are biodegraded by microorganisms, while the bioavailability of metals is reduced (Hua et al. 2008). Composting was applied by Sasek et al. (2003) for the biodegradation of 12 PAH substances. The biodegradation efficiency of individual PAH after 54 days was in the range of 17–71%, while after other 100 days, the process efficiency increased to 28–98%. Similar results regarding the degradation of PAH were presented by Hua et al. (2008). Results on the degradation efficiency of organic substances such as explosive TNT, herbicide dicamba, 2-chlorophenol, and 2,4-dichlorophenol during composting process were presented by EPA (1998).

Biopiles is an ex situ remediation biotechnology, similarly to bioventing. It can be applied for heavy metal removal or for biodegradation of organic toxic substances (Pavel and Gavrilescu 2008; EPA 1994). This technology was applied, for example, in a research project implemented by Kao et al. (2009) for the degradation of BTEX,

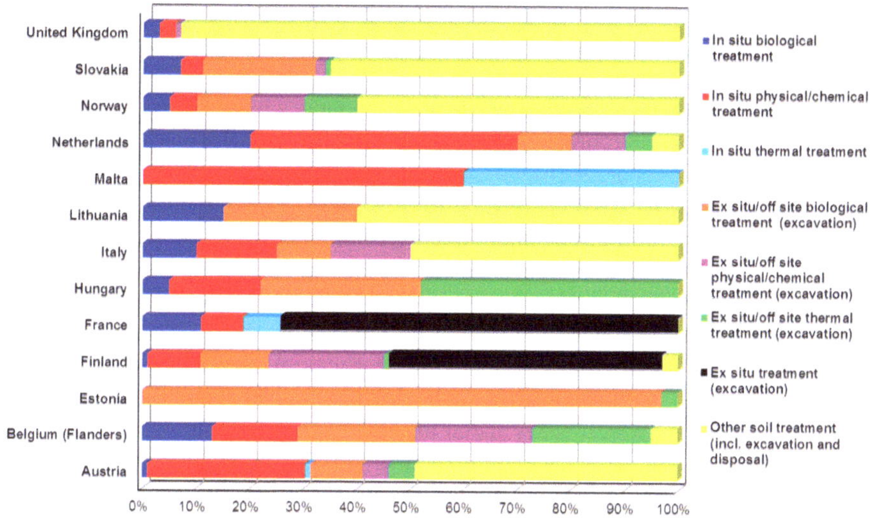

Fig. 14.3 Most frequently applied remediation techniques for contaminated soils. (Adapted after EEA (2014b))

TPH, and PAH substances. The results revealed that after 20 months of operation, the concentrations of BTEX were reduced with 90–99.9%, TPH with 65–90%, and PAH with 50–75% .

According to the European Environment Agency reports, in situ biological treatments in Europe represent approximately 7% of the remediation techniques applied for contaminated soils, Netherlands being the country that uses most frequently this type of remediation (approximately 20% of applied techniques) compared to other European countries (EEA 2014a) (Fig. 14.3). Looking at ex situ biotechnologies at the European level, their average usage is approximately 19% of the remediation techniques applied, the highest application being in Estonia (97%) followed by Hungary (30%) (EEA 2014a). The classes of pollutants whose presence in the solid matrix causes the worst effects are heavy metals, polycyclic aromatic hydrocarbons, chlorinated hydrocarbons, and aromatic hydrocarbons, which represent 34.8%, 10.9%, 8.3%, and 10.2%, respectively, of contaminants that affect soils, sludges, and sediments in Europe (EEA 2014b).

14.3.1 Potential of Microorganism Species for PTS Removal

14.3.1.1 Removal of Heavy Metals as Inorganic PTS

Metals are inorganic substances present in all the environmental components, with direct and/or indirect involvement in the microorganism growth and metabolism. The interaction between microorganism cells and metals depends on many factors, such as type of metal ion, microorganism strain, environmental factors, the structural components of the cell wall and metabolic activity, metal speciation, solubility, mobility, bioavailability, and toxicity. Some metals are considered essential microelements for life in small quantities (e.g., Na, K, Cu, Zn, Co, Ca, Mg, Mn, and Fe) compared to the other metals whose presence induce toxic effects on the metabolic functions of microorganisms (Ayangbenro and Babalola 2017; Naik and Dubey 2013; Hlihor et al. 2016).

Although heavy metals ions (e.g., Cd^{2+}, Pb^{2+}, Ni^{2+}, As^{3+}, Cr^{3+}, Cr^{6+}) pose toxic effects, they can be removed by microorganisms through a series of mechanisms, as indicated in Fig. 14.4. Microorganisms can immobilize metal ions on the surface of the cell wall or inside the cell and can convert some ions to less toxic species, e.g., Cr^{6+} is reduced at Cr^{3+} (Naik and Dubey 2013; Chen and Hao 1998; Hlihor et al. 2017), or through a series of enzymatic transformations, they can even volatilize some metal ions (possible only for Hg^{2+} and Se^{5+}) (Ayangbenro and Babalola 2017). The mechanisms by which heavy metal ions are immobilized by microorganisms are a response of their resistance to the toxic effects caused by heavy metals (Naik and Dubey 2013).

To maintain the vital functions of microorganisms' cells and control metals toxicity levels inside the cell, the *ATPases proteins* transport them outside of the cell membrane, positively influencing the degree of microorganisms' resistance to heavy metals (Naik and Dubey 2013; Argüello et al. 2007). These transporter proteins prevent overaccumulation of heavy metal ions inside the cell, due to the presence of genes such as *copA*, *zntA*, or *cadA* in their structures, shown to be responsible for binding of Cu^{2+}, Zn^{2+}, or Cd^{2+} ions (Takahashi et al. 2012; Dlugonski 2016). Other important proteins involved in the export of heavy metal ions outside of the cell wall are resistance-nodulation-cell division proteins (RND protein) which are closely related to genes such as MFP (periplasmatic adapter protein family), OMFs (outer membrane factors), and CDF (cation diffusion facilitators). Examples of MFP and OMFs genes are *CzcCBA*, *CnrCBA*, *NccCBA*, *CzrCBA*, *CusCFBA*, *SilCFBA*, etc. (Dlugonski 2016).

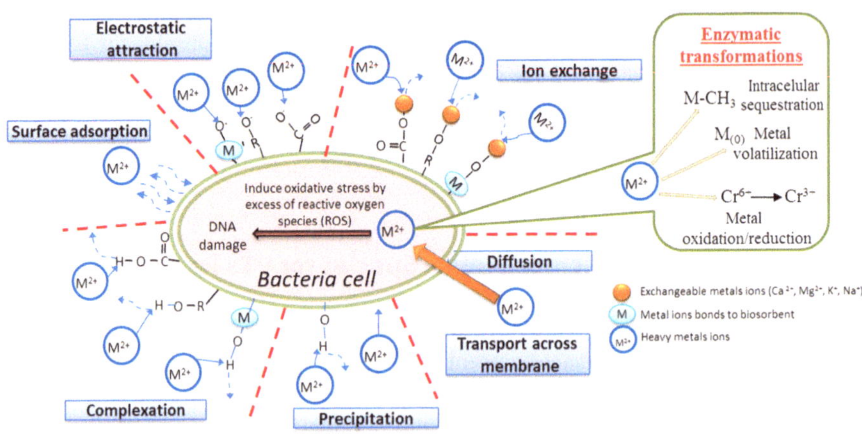

Fig. 14.4 Mechanisms of heavy metal ions removal by bacterial cell. (Roşca et al. 2017b)

The sequestration/bioaccumulation of toxic metals inside the cell is also due to the presence in the microorganisms structure of some binding proteins called *metallothioneins* (MTs), whose main role is to protect the bacterial metabolic processes catalyzed by enzymes (Naik and Dubey 2013). MTs are the products of mRNA translation, with low molecular weight and a large number of cysteine residue and metal-binding proteins (Cobbett and Goldsbrough 2002) with the role in detoxification of non-essential and excess of essential metals and/or heavy metals (Carpene et al. 2007).

Considering extracellular sequestration, the *extracellular high molecular weight biopolymers* which are secreted by bacterial cells and known as exopolysaccharides (EPS) are involved in pollutant removal. This category comprises polysaccharides, proteins, nucleic acids, humic substances, lipids, and other constituents with low molecular weight. Their structure contains functional groups such as hydroxyl, carboxyl, amides, and phosphoryl, which have been shown to be directly involved in the removal of heavy metal ions. The extracellular sequestration process is called biosorption and is mediated by several mechanisms such as ion exchange, chelation, adsorption, and diffusion through cell walls and membranes (Naik and Dubey 2013; Quintelas et al. 2013).

The multitude of studies in the field of environmental bioremediation led to the identification of a large variety of microorganisms' strains capable in removing heavy metals ions. Species such as *Bacillus megaterium*, *Rhodotorula* sp., *Arthrobacter viscosus*, *Saccharomyces cerevisiae* were used as dead biomass to remove Cd(II), Pb(II), and Cr(VI) ions from aqueous solutions (Hlihor et al. 2011, 2015, 2017; Roşca et al. 2018). To study the bioaccumulation process of metal ions, Hlihor et al. (2017) used *Arthrobacter viscosus* for Cr(VI); da Costa and Duta (2001) studied the bioaccumulation of Cu^{2+}, Zn^{2+}, Cd^{2+}, and Pb^{2+} by using *Bacillus* sp. like *Bacillus cereus*, *Bacillus sphaericus*, and *Bacillus subtilis*; and Sharma and Fulekar (2009) evaluated the bioaccumulation of Cu^{2+} ions by using *Citrobacter freundi*.

14.3.1.2 Removal of Various Organic Pollutants as PTS

According to Stockholm Convention, 33 chemical substances with persistent characteristics were identified up to now, of which 22 substances are forbidden to be used or produced and 2 substances are restricted for production and use, while for the rest of the substances several measures for reduction of unintentional release into the environment are necessary to be taken (Stockholm Convention 2008). Chemicals included in the list of substances are pesticides, substances used in various industrial processes, or represent the degradation compounds of other organic toxic substances. The most well-known pesticides with persistent, toxic, and bioaccumulative characteristics, forbidden to be used or produced, are aldrin, chlordane, chlordecone, lindane, mirex, heptachlor, toxaphene, and DDT, with half-lives longer than 5 years (Stockholm Convention 2008). Chemical substances belonging to the industrial chemicals category are polychlorinated biphenyls (PCBs), polychlorinated naphthalenes, hexachlorobutadiene, hexabromocyclododecane (HBCDD), tetrabromodiphenyl ether and pentabromodiphenyl ether, hexabromodiphenyl ether and heptabromodiphenyl ether, hentachlorobenzene, hexabromobiphenyl, short-chain chlorinated paraffins, and decabromodiphenyl ether. The unintentionally released chemicals included in the POP list are hexachlorobenzene, PCBs, polychlorinated dibenzo-p-dioxins, polychlorinated naphthalenes, pentachlorobenzene, polychlorinated dibenzofurans, and hexachlorobutadiene (Stockholm Convention 2008).

For the biodegradation of PCBs by microorganisms, four specific enzymes, *biphenyl dioxygenase (BphA), dihydrodiol dehydrogenase (BphB), 2,3-dihydroxybiphenyl dioxygenase (BphC),* and *2-hydroxyl-6-oxo-6-phenylhexa-2,4-dienoic acid hydrolase (BphD),* are sequentially involved in their oxidative degradation, being transformed by microorganisms into chlorobenzoates and 2-hydroxypenta-2,4-dienoate which are less toxic than PCBs (Taguchi et al. 2007; Weiland-Bräuer et al. 2017; Zhang et al. 2015). Several species of bacteria were identified as being PCB-degrading bacteria, e.g., *Pseudomonas* sp., *Shigella* sp., *Subtercola* sp., *Chitinophaga* sp., *Janthinobacterium* sp. (Weiland-Bräuer et al. 2017), *Anabaena* PD-1 (Zhang et al. 2015), *Alcaligenes xylosoxidans, Pseudomonas stutzeri, Ochrobactrum anthropic, Pseudomonas veronii (Murínová* et al. 2014), and others. The PCB substances can be also biodegraded by fungal microorganisms such as *Trametes versicolor*, *Phanerochaete chrysosporium*, and *Lentinus edodes* with an effectiveness of more than 30% for concentrations above 600 mg/L (Ruiz-Aguilar et al. 2002).

Verma et al. (2014) showed that in the biodegradation of lindane by microorganisms, the *lin* genes encoding found in the genome of different microorganisms strains are involved. The degradation of lindane can be done in an upper degradation pathway, when dehydrochlorinase (LinA), haloalkane dehalogenase (LinB), and dehydrogenase (LinC) enzymes can transform lindane into pentachlorocyclohexane, tetrachloro-1,4-cyclohexadiene, trichloro-2,5-cyclohexadiene-1-o, and dichloro-2,5-cyclohexadiene-1,4-diol, dichlorohydroquinone, and in a lower degradation pathway when reductive dechlorinase (LinD), ring cleavage oxygenase (LinE), maleylacetatereductase (LinF), an acyl-CoA transferase (LinG, H), a thiolase (LinJ), and transcription factors (LinI and LinR) enzymes transform the products resulted from upper degradation into chlorohydroquinone and hydroquinone (Verma et al. 2014; Saez et al. 2017).

The bacterial strains able to degrade lindane were mostly from genera *Sphingobium* spp., *Kocuria* sp., S*taphylococcus* sp. (Kumar et al. 2016), *Escherichia coli*, *Bacillus* sp., *Pseudomonas* sp., and *Azotobacter* sp. (Kumar and Pannu 2018). Also, yeasts species such as *F. poae* and *F. solani (Sagar and Singh* 2011), *Rhodotorula* sp. VITJzN03 (Abdul Salam et al. 2013), and *Candida* sp. VITJzN04 (Salam and Das 2014) and fungi species such as *Phanerochaete chrysosporium*, *Trametes hirsutus (Singh and Kuhad* 1999), and *Conidiobolus* 03–1-56 (Nagpal et al. 2008) have the capacity to degrade lindane in different concentrations.

On the other side, dieldrin can be degraded by microorganisms such as *Mucor racemosus (Kataoka* et al. 2010), *Clostridium*, *Pseudomonas* sp., *Micrococcus* sp.,

Arthrobacter sp., *Bacillus* sp., and *Trichodermaviride (Matsumoto* et al. 2009*)*. Other species of microorganisms capable to degrade POPs are *Pseudoalteromonas* sp., *Psychrobacter* sp., *Rhodococcus erythropolis* S-7, *Hydrogenophaga taeno-spiralis* IA3, and *Burkholderia xenovorans* LB400 (Michaud et al. 2007; Bajaj and Singh 2015).

14.3.2 Genetically Modified Microorganisms Applied in Bioremediation

The enzymatic activity of microorganisms plays an important role in the bioremediation process, enzymes being considered biological catalysts (Karigar and Rao 2011) which can enhance the rate of some reactions by lowering the activation energy of molecules (Sharma et al. 2018). The main enzymes found in microorganisms structure directly involved in pollutants degradation pathways (Sharma et al. 2018) or in their retention are *oxidoreductases, laccases, hydrolases, peroxidases, dehalo-genases*, and *transferases (Karigar and Rao* 2011*; Sharma* et al. 2018*)*. Even if the degradation of pollutants by indigenous microorganisms is a slow process, and therefore the feasibility of bioremediation in large-scale applications decreases, this process is considered an effective, safe, and less expensive technology (Karigar and Rao 2011). In the last few years, genetically engineered microorganisms stated to be considered an appropriate alternative capable in removing pollutants from contaminated environments at a high speed, enhancing efficiency of various bioremediation technologies.

Genetically modified microorganisms (GMMOs) were defined by World Health Organization as "microorganisms in which the genetic material (DNA) has been altered in a way that does not occur naturally by mating and/or natural recombination" (World Health Organization 2014). Researchers found suitable GMMOs applications in different fields of activity including environmental bioremediation.

The development of GMMOs for bioremediation purposes can be done by a wide range of techniques, as suggested in Fig. 14.5: *modification of enzyme specificity, designing of new metabolic pathways and their regulation, introduction of marker gene for identification of recombinant in contaminated environment, and construction of biosensors for detection of specific chemical compounds (Wasilkowski* et al. 2012*)*. In bioremediation, the GMMOs are used (i) to increase the decontamination speed of polluted sites and water, (ii) to enhance the ability of retaining or degrading contaminants, (iii) to improve microorganism resistance in various environmental conditions, and (iv) to reduce the amount of cell biomass needed for bioremediation (Abatenh et al. 2017).

Despite their ability in improving process efficiency, the use of GMMOs in bioremediation is still limited to academic research, due to possible ethic, conflicting, or other issues that might occur. If GMMOs are used in bioremediation of a contaminated site, it would be difficult to distinguish between GMMOs specific

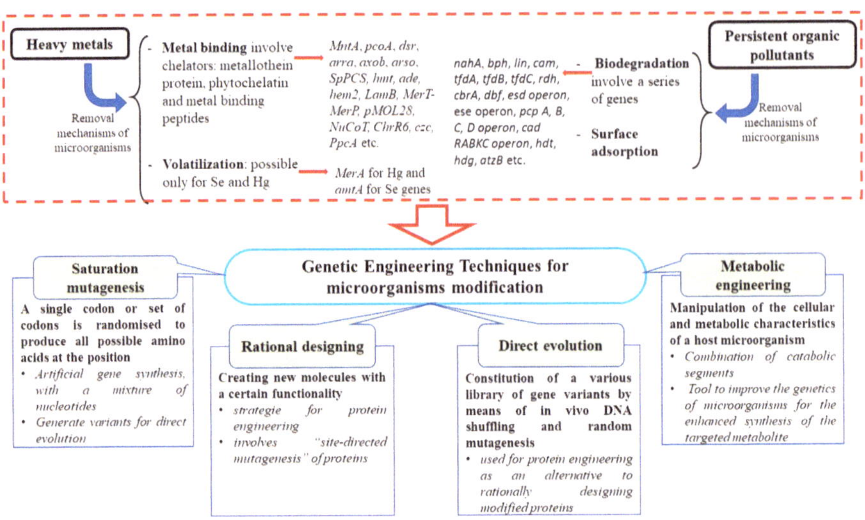

Fig. 14.5 Genes involved in PTS removal by microorganisms and genetic engineering techniques for the development of GMMOs

degradation and biodegradation, due to the presence of indigenous microbial consortia. Also, the consideration of GMMOs used in bioremediation depends on their effectiveness compared with their counterparts present in nature, their influence on indigenous microorganisms, and their resistance in nature and their containment. Until these issues are clarified, the future use of GMMOs in bioremediation will remain uncertain (Kumar et al. 2013; Jafari 2013).

Screening of literature showed that there are numerous publications on the application of techniques involving microbial genetic modification aiming to develop species with high capacities for desired pollutant removal. Some research groups have altered or transferred genes from other species in order to improve the ability of some microorganisms to remove heavy metal ions from soil or water (Brim et al. 2003; Gonzalez et al. 2005). Other examples focusing on the efficiency of GMMOs for the biodegradation of PTS are presented in Table 14.1.

14.3.3 Plant-Microbe Interactions

The most abundant organisms on Earth, microorganisms play nowadays a vital role in bioremediation. In the last decade, bioremediation – with its branches phytoremediation (the use of plants to remove/degrade pollutants from environmental compartments) and rhizoremediation (the use of plants-microorganisms interactions to remediate contaminated sites) – has started to play a central role in environmental cleanup. This technology has become more and more attractive in the scientific literature because it offers better remediation potential than the conventional

Table 14.1 Removal of persistent toxic substances by GMMOs

Host strain	Pollutant	Genes/enzymes/plasmids involved	Donor strain	Initial concentration	Degradation bioaccumulation/ biosorption capacity/ efficiency	References
Organic substances						
S. japonicum UT26	Lindane Parathion	Transformation of pPNCO33 using the electroporation method	*E. coli* and *Pseudomonas*	10 mg/kg soil 100 mg/kg soil	100% in 15 days	Cao et al. (2013)
Anabaena sp.	Lindane	BamHI-EcoRI fragment from plasmid pIMA2, operon fcbABC and plasmid pRL623	*P. paucimobilis*, *A. globiformis*, and *E. coli*	5 μg/mL	100% in 72 h	Kuritz and Wolk (1995)
P. fluorescens strains *F113pcb* and *F113 L:1180*	PCB mixture	bph gene transfer	*Burkholderia* sp.	$C_0 = 1$ mg/L	30–100%	Rein et al. (2007)
P. putida and *Sphingomonas* sp.	Carbofuran methyl parathion	Recombinant plasmid pBBR-mpd	*E. coli*	C = 0.2 g/L	100% in 24 h	Liu et al. (2006)
C. necator JMS34	PCB	Incorporation of *bph* genes into genome, plasmid pT5K012 transfer	*E. coli* SM10(λpir) and *Burkholderia xenovorans*	In soil C = 10 mg/kg	50% in 1 week	Saavedra et al. (2010)
S. paucimobilis UT26XEGM	Lindane	Insertion of *mpd*, *ecoRIR*, and *gef* genes into chromosome	*E. coli*	In soil 10 mg/kg soil	100% in 24 h in soil	Lan et al. (2014)
				In water 5 μg/mL	100% in 28 h	

(continued)

Table 14.1 (continued)

Host strain	Pollutant	Genes/enzymes/plasmids involved	Donor strain	Initial concentration	Degradation bioaccumulation/ biosorption capacity/ efficiency	References
Inorganic substances						
E. coli	Cd(II)	pCLG2 and pGPMT plasmids	–	60 mg/L	63.26 mg/g	Deng et al. (2007)
	Cd(II)	Recombinant methalothionein I of mice – pMt-Thio	Liver tissues from adult mice	100 mg/L	37%	Almaguer-Cantú et al. (2011)
	Pb(II)				93%	
	Cd(II)	Human MT (hMT-1A)	Gene sequence of human MT, hMT-1A	0.2 mM	6.36 mg/g	Ma et al. (2011)
	As(III)			100 μM	7.59 mg/g	
A. eutrophus	Cr(VI)	pMOL28 and pMOL30 plasmids	–	10–100 mgCr/L	93.8–48.7% in 72 h	Srivastava et al. (2010)
S. cerevisiae	Cd(II)	CP2, HP3	*E. coli* TG1	10–200 μM	29.1–30.1 nmol/g	Vinopal et al. (2007)
	Zn(II)				27.1–29.8 nmol/g	
P. chrysogenum	Cu(II)	PEI modification	–	100 mg/L	60.4 mg/g	Deng and Ting (2005)
	Pb(II)				76.3 mg/g	
	Ni(II)				30.2 mg/g	

technologies and advantages such as lower costs and outstanding efficiency. The association of plants and microorganisms used for environmental clean-up has become one of the most evolved processes in biotechnology (Marihal and Jagadeesh 2013; Nie et al. 2011; Prabha et al. 2017). Microorganisms have the capacity to create symbiotic relationship with plants and especially with their roots which enhances the bioavailability of different types of pollutants (Kumar et al. 2017).

Recent advances in biotechnology propose multiple remediation schemes, some involving the use of *transgenic plants, low-input "designer" plants or nanotechnology*. These strategies enhance the processes benefits from both environmental and economic ways (Abhilash et al. 2012; Krämer and Chardonnens 2001). To improve the ability of plants to degrade/mineralize/remove organic and inorganic pollutants, *transgenic plants* started to be engineered by transferring different genes from organisms such as other plants, animals, bacteria, or fungi (Truua et al. 2015). This remediation strategy is focused on the use of genetically engineered plants to develop new characteristics/abilities. Although controversial from a commercial point of view due to their association with genetically modified organisms (GMO), their new abilities were remarkably proved for research purposes (Dixit et al. 2015; Hussain et al. 2018; Ibañez et al. 2016; Van Aken and Doty 2010). Table 14.2 emphasizes studies regarding the removal of persistent toxic substances by transgenic plants. Manipulation of desired plant species with microbial genes able to degrade/remove organics and inorganics from different environmental compartments has become an attractive strategy in bioremediation due to its eco-friendliness. Application of genetically engineered plants, namely, *"designer" plants*, is a stratagem which plays an important role in maximizing the benefits of phytoremediation (Dixit et al. 2015).

Nanotechnology has also been used to boost bioremediation which started to present an enhanced potential. This innovative technology, sometimes called nanoremediation, acts as a carrier system for microorganisms (Abhilash et al. 2012; Barh et al. 2015).

14.4 Conclusions

Increased human activities leading to constantly increased pollutants in our environment have led to the development of new and innovative technologies for their removal. From the applied technologies for PTS degradation/removal, biotechnology – with its branch, environmental bioremediation – proved to be a rational strategy due to high efficiency and low costs. The concern remains on the time frame in which bioremediation occurs and disposal of pollutant-laden plants and microorganisms. In this sense, new approaches on gene transfer to improve the capacity of microorganisms and plants in bioremediation, hence creating a lower amount of biomass and improving tolerance to toxic pollutants were developed. GMMOs received a wide attention for bioremediation purposes. Also, the use of transgenic plants was found appropriate for PTS removal. Studies revealed many challenges in

Table 14.2 Removal of persistent toxic substances by transgenic plants

Target plant	Pollutant	Gene(s)	Donor strain	Removal/degradation efficiency/uptake	References
Organic substances					
Arabidopsis thaliana	PAHs – naphthalene, phenanthrene, and pyrene	*nidA*, *nidB*, *NahAa*, and *NahAb genes*	*Mycobacterium vanbaalenii*	90–100%	Peng et al. (2014a)
			Pseudomonas putida G7		
Arabidopsis thaliana	Phenanthrene	*NahAa*, *NahAb*, *NahAc*, and *NahAd genes*	*Pseudomonas putida*	23.5–42.7%	Peng et al. (2014b)
Nicotianatabacum	Lindane	*Cytochrome P4502E1*	*Human* and *Agrobacterium tumefaciens*	81.7–90.6%	Singh et al. (2011)
Arabidopsis thaliana	Lindane	*HCH dehydrochlorinase (LinA) protein*	*Sphingobiumjaponicum UT26*	100%	Dick (2014)
Alfalfa plants	PCB	*BphC gene*	*Agrobacterium tumefaciens* EHA105	98.8%	Wang et al. (2015)
Oryza sativa cv.	Atrazine Metolachlor Simazine	*Human CYP1A1*, *CYP2B6*, and *CYP2C19 genes*	*Agrobacterium*	42.6%, 18.3%, and 30.7%, respectively	Kawahigashi et al. (2006)
Alfalfa, *Arabidopsis thaliana*, and *tobacco*	Atrazine	*Modified bacterial atzA gene*	*Agrobacterium*	No data	Wang et al. (2005)
Inorganic substances					
Nicotiana tabacum L.	Cd(II)	*HisCUP*, *HisGUS*	*Agrobacterium tumefaciens*	33.2 ± 2.7 mgCd/kg dry mass	Pavlíková et al. (2004)
	Zn(II)			104.3 ± 9.4 mgZn/kg dry mass	
	Ni(II)			6.5 ± 2.4 mgNi/kg dry mass	
	Cd(II)	*ScMTII gene*	*Agrobacterium tumefaciens*	730.44 mg Cd/kg dry mass in the above-ground parts and 2965.00 mg Cd/kg dry mass in the roots	Daghan et al. (2010)

Target plant	Pollutant	Gene(s)	Donor strain	Removal/degradation efficiency/uptake	References
P. tomentosa	Cd(II)	*P1B-ATPase gene*	*Populus tomentosa* Carr	114.3 mg/kg dry biomass	Wang et al. (2018)
Populus tremula P. alba	Cd(II)	*transgenic lines ggs11* and *ggs28*	*E. coli*	63.1 ± 3.1 μmol/cm^2/s	He et al. (2015)
M. sativa L.	Cd(II) Trichloroethylene	*expressing human CYP2E1* and *glutathione S-transferase*	*Agrobacterium tumefaciens* LBA4404	No data	Zhang and Liu (2011)

using genes transfer for bioremediation purposes, while public acceptance still remains to debate.

Acknowledgments This work was supported by a grant of the Romanian National Authority for Scientific Research, CNCS – UEFISCDI, project number PN-III-P4-ID-PCE-2016-0683, Contract no. 65/2017.

References

Abatenh E, Gizaw B, Tsegaye Z et al (2017) Application of microorganisms in bioremediation-review. J Environ 1(1):02–09

Abdul Salam J, Lakshmi V, Das D, Das N (2013) Biodegradation of lindane using a novel yeast strain, Rhodotorula sp. VITJzN03 isolated from agricultural soil. World J Microbiol Biotechnol 29:475–487

Abhilash PC, Powell JR, Singh HB, Singh BK (2012) Plant–microbe interactions: novel applications for exploitation in multipurpose remediation technologies. Trends Biotechnol 30(8):416–420

Abraham J, Gea T, Vicent T (2016) Impact of biostimulation and bioaugmentation as bioremediation systems on diesel contaminated soils. Environ Eng Manag J 15:1743–1753

Almaguer-Cantú V, Morales-Ramos LH, Balderas-Rentería I (2011) Biosorption of lead (II) and cadmium (II) using Escherichia coli genetically engineered with the mice metallothionein I. Water Sci Technol 63(8):1607–1613

AMAP (2004) Persistent toxic substances, food security and indigenous peoples of the Russian North final report. Arctic Monitoring and Assessment Programme, AMAP Secretariat, Oslo, Norway

Apostol LC, Gavrilescu M (2016) Conventional and emerging technologies for environmental bioremediation – current state, constraints, developments and perspectives. In: Gavrilescu M (ed) Biosorption and bioaccumulation: principles and applications in environmental bioremediation. Politehnium Publishing House, Iasi

Argüello JM, Eren E, Gonzalez-Guerrero M (2007) The structure and function of heavy metal transportP1B-ATPases. Biometals 20(3–4):233–248

Ayangbenro AS, Babalola OO (2017) A New Strategy for Heavy Metal Polluted Environments: A Review of Microbial Biosorbents. Int J Environ Res Public Health 14(1):94

Bajaj S, Singh DK (2015) Biodegradation of persistent organic pollutants in soil, water and pristine sites by cold-adapted microorganisms: Mini review. Int Biodeterior Biodegrad 100:98–105

Barakat AO (2004) Assessment of persistent toxic substances in the environment of Egypt. Environ Int 30(3):309–322

Barbeş L, Bărbulescu A (2017) Monitoring and statistical assessement of heavy metals in soil and leaves of Populus Nigra L. Environ Eng Manag J 16:187–196

Barh A, Singh S, Chandra D, Pankaj KPR, Chandra S, Singh NK (2015) Enhanced bioremediation techniques for agricultural soils. Int J Curr Res Acad Rev 3(7):166–173

Bradl H, Xenidis A (2005) Chapter 3: Remediation techniques. In: Bradl HB (ed) Interface science and technology, vol 6. Elsevier Ltd.

Brim H, Venkateswaran A, Kostandarithes HM, Fredrickson JK, Daly MJ (2003) Engineering Deinococcusgeothermalis for bioremediation of hightemperature radioactive waste environments. Appl Environ Microbiol 69:4575–4582

Cabral L, Soares CR, Giachini AJ, Siqueira JO (2015) Arbuscular mycorrhizal fungi in phytoremediation of contaminated areas by trace elements: mechanisms and major benefits of their applications. World J Microbiol Biotechnol 31(11):1655–1664

Cao X, Yang C, Liu R, Li Q, Zhang W, Liu J, Song C, Qiao C, Mulchandani A (2013) Simultaneous degradation of organophosphate and organochlorine pesticides by Sphingobiumjaponicum UT26 with surface-displayed organophosphorus hydrolase. Biodegradation 24(2):295–303

Carpene E, Andreani G, Isani G (2007) Metallothionein functions and structural characteristics. J Trace Elem Med Biol 21(Suppl 1):35–39

CEC (2018) Persistent bioaccumulative toxic substances, Commission for Environmental Cooperation. http://www3.cec.org/islandora/en/item/992-north-american-mosaic-overview-key-environmental-issues-en.pdf. Accessed 5 Nov 2018

Chen JM, Hao OJ (1998) Microbial chromium (VI) reduction. Crit Rev Environ Sci Technol 28(3):219–251

Cobbett C, Goldsbrough P (2002) Phytochelatins and metallothioneins: roles in heavy metal detoxification and homeostasis. Annu Rev Plant Biol 53(1):159–182

Cojocaru P, CotiuscaZauca D, Biali G (2016) Decontamination of soils polluted with petroleum products by bioremediation. Environ Eng Manag J 15:1419–1425

Cozma P, Hlihor RM, Apostol LC, Gavrilescu M (2016) Biosystems and bioreactors for biosorption and bioaccumulation. In: Gavrilescu M (ed) Biosorption and Bioaccumulation: Pricimples and Applications in Environmental Bioremediation. Politehnium Publishing House, Iasi

da Costa ACA, Duta FP (2001) Bioaccumulation of copper, zinc, cadmium and lead by Bacillus sp., Bacillus cereus, Bacillus sphaericus and Bacillus subtilis. Braz J Microbiol 32(1):1. https://doi.org/10.1590/S1517-83822001000100

Daghan H, Arslan M, Uygur V, Koleli N, Eren A (2010) The cadmium phytoextraction efficiency of ScMTII gene bearing transgenic tobacco plant. Biotechnol Biotechnol Equip 24(3):1974–1978

Deng S, Ting YP (2005) Characterization of PEI-modified biomass and biosorption of Cu(II), Pb(II) and Ni(II). Water Res 39(10):2167–2177

Deng X, Yi XE, Liu G (2007) Cadmium removal from aqueous solution by gene-modified *Escherichia coli* JM109. J Hazard Mater 139(2):340–344

Dick J (2014) Phytoremediation of Lindane in Transgenic Arabidopsis thaliana Expressing a Bacterial HCH-dehydrochlorinase (LinA) Protein. Doctoral dissertation. University of Sheffield

Dixit R, Malaviya WD, Pandiyan K, Singh UB, Sahu A, Shukla R, Singh BP, Rai JP, Kumar Sharma P, Lade H, Paul D (2015) Bioremediation of heavy metals from soil and aquatic environment: an overview of principles and criteria of fundamental processes. Sustainability 7:2189–2212

Dlugonski J (2016) Microbial biodegradation. Academic, Caister

EEA (2014a) Most frequently applied remediation techniques for contaminated soil, European Environment Agency. https://www.eea.europa.eu/data-and-maps/daviz/dominant-remediation-technologies-for-contaminated-1#tab-chart_1_filters=%7B"rowFilters"%3A%7B%7D%3B"columnFilters"%3A%7B%7D%3B"sortFilter"%3A%5B"country_region"%5D%7D. Accessed 8 Nov 2018

EEA (2014b) Contaminants affecting soil and groundwater in Europe. European Environment Agency. https://www.eea.europa.eu/data-and-maps/daviz/overview-of-contaminants-affecting-soil#tab-chart_3. Accessed 10 Nov 2018

Elekes CC (2014) Chapter 10: Eco-technological solutions for the remediation of polluted soil and heavy metal recovery. In: Environmental risk assessment of soil contamination. InTechEditure, pp 309–335

EPA (1994) How to evaluate alternative cleanup technologies for underground storage tank sites. A guide for corrective action plan reviewers. https://www.epa.gov/sites/production/files/2014-03/documents/tum_ch4.pdf. Accessed 24 Oct 2017

EPA (1998) An analysis of composting as an environmental remediation technology, United States Environmental Protection Agency EPA530-R-98-008 April 1998. https://www.epa.gov/sites/production/files/2015-09/documents/analpt_all.pdf. Accessed on 23 Oct 2018

Eslami E, Joodat SHS (2018) Bioremediation of oil and heavy metal contaminated soil in construction sites: a case study of using bioventing-biosparging and phytoextraction techniques. Geophysics. https://arxiv.org/pdf/1806.03717.pdf

Gavrilescu M, Chisti Y (2005) Biotechnology-a sustainable alternative for chemical industry. Biotechnol Adv 23(7–8):471–499
Gavrilescu M, Pavel LV, Cretescu I (2009) Characterization and remediation of soils contaminated with uranium. J Hazard Mater 163(2–3):475–510
Gavrilescu M, Demnerová K, Aamand J, Agathos S, Fava F (2015) Emerging pollutants in the environment: present and future challenges in biomonitoring, ecological risks and bioremediation. New Biotechnol 32(1):147–156
Gonzalez CF, Ackerley DF, Lynch SV, Matin A (2005) ChrR, a soluble Quinone reductases of Pseudomonas putida that defends against H2O2. J Biol Chem 280(24):22590–22595
Hamby DM (2000) Site remediation techniques supporting environmental restoration activities: a review. University of Michigan press, Michigan
He J, Li H, Ma C, Zhang Y, Polle A, Rennenberg H et al (2015) Overexpression of bacterial g-glutamyl cysteine synthetase mediates changes in cadmium influx, allocation and detoxification in poplar. New Phytol 205:240–254
Hlihor RM, Diaconu M, Gavrilescu M (2011) Influence of pH on Cr(VI) reduction and removal by Saccharomyces cerevisiae. Bull Polytechnic Inst Iasi Sect Chem Chem Eng LVII(LXI), pp 147–153
Hlihor RM, Diaconu M, Leon F, Curteanu S, Tavares T, Gavrilescu M (2015) Experimental analysis and mathematical prediction of Cd(II) removal by biosorption using support vector machines and genetic algorithms. New Biotechnol 32(3):358–368
Hlihor RM, Apostol LC, Smaranda C, Simion IM, Fortuna EM, Diaconu M, Bulgariu L, Gavrilescu M (2016) Target compounds and pollutants, biosorbents and bioaccumulators. In: Gavrilescu M (ed) Biosorption and bioaccumulation: principles and applications in environmental bioremediation. Politehnium Publishing House, Iasi
Hlihor RM, Figueiredo H, Tavares T, Gavrilescu M (2017) Biosorption potential of dead and living Arthrobacter viscosus biomass in the removal of Cr(VI): batch and column studies. Process Saf Environ 108:44–56
Hu G-J, Chen S-L, Zhao Y-G, Sun C, Li J, Wang H (2009) Persistent toxic substances in agricultural soils of Lishui County, Jiangsu Province, China. Bull Environ Contam Toxicol 82(1):48–54
Hua L, Wu W, Liu Y et al (2008) Effect of composting on polycyclic aromatic hydrocarbons removal in sewage sludge. Water Air Soil Pollut 193:259–267
Hussain I, Aleti G, Naidu R, Puschenreiter M, Mahmood Q, Mahmudur Rahman M, Wang F, Shaheen S, Hussain J, Reichenauer STG (2018) Microbe and plant assisted-remediation of organic xenobiotics and its enhancement by genetically modified organisms and recombinant technology: a review. Sci Total Environ 628–629:1582–1599
Ibañez S, Talano M, Ontanon O, Suman J, Medina MI, Macek T, Agostini E (2016) Transgenic plants and hairy roots: exploiting the potential of plant species to remediate contaminants. New Biotechnol 33(5. Pt B):625–635
Iosob G-A, Prisecaru M, Stoica I, Călin M, Cristea TO (2016) Biological remediation of soil polluted with oil products: an overview of available technologies. http://pubs.ub.ro/dwnl.php?id=SCSB201602V25S01A0014. Accessed 23 Oct 2018
Jafari M (2013) Bioremediation and genetically modified organisms. In: Goltapeh EM et al (eds) Fungi as bioremediators. Springer, Berlin/Heidelberg
Kao CM, Chien HY, Surampalli RY, Sung WP (2009) Application of biopile system for the remediation of petroleum-hydrocarbon contaminated soils. In: World environmental and water resources congress. Kansas City, Missouri, pp 17–21
Karigar CS, Rao SS (2011) Role of microbial enzymes in the bioremediation of pollutants: a review. Enzyme Res 2011:1. https://doi.org/10.4061/2011/805187
Kasi M, McEvoy J, Padmanabhan G, Khan E (2010) In situ groundwater remediation using enricher reactor – permeable reactive biobarrier. In: Proceedings of the Water Environment Federation. WEFTEC 2010: Session 1 through Session, vol 10, pp 300–317
Kataoka R, Takagi K, Kamei I, Kiyota H, Sato Y (2010) Biodegradation of dieldrin by a soil fungus isolated from a soil with annual endosulfan applications. Environ Sci Technol 44(16):6343–6349

Kawahigashi H, Hirose S, Ohkawa H, Ohkawa Y (2006) Phytoremediation of the herbicides atrazine and metolachlor by transgenic rice plants expressing human CYP1A1, CYP2B6, and CYP2C19. J Agric Food Chem 54:2985–2991

Krämer U, Chardonnens AN (2001) The use of transgenic plants in the bioremediation of soils contaminated with trace elements. Appl Microbiol Biotechnol 55:661–672

Kumar D, Pannu R (2018) Perspectives of lindane (γ-hexachlorocyclohexane) biodegradation from the environment: a review. Bioresour Bioprocess 5:29

Kumar S, Dagar VK, Khasa YP, Kuhad RC (2013) Genetically modified microorganisms (GMOs) for bioremediation. In: Kuhad R, Singh A (eds) Biotechnology for environmental management and resource recovery. Springer, New Delhi

Kumar D, Kumar A, Sharma J (2016) Degradation study of lindane by novel strains Kocuria sp. DAB-1Y and Staphylococcus sp. DAB-1W. Bioresour Bioprocess 3(1):53

Kumar SS, Kadier A, Malyan SK, Ahmad A, Bishnoi NR (2017) Phytoremediation and rhizoremediation: uptake, mobilization and sequestration of heavy metals by plants. In: Singh D, Singh H, Prabha R (eds) Plant-microbe interactions in agro-ecological perspectives. Springer, Singapore

Kuritz T, Wolk CP (1995) Use of filamentous cyanobacteria for biodegradation of organic pollutants. Appl Environ Microbiol 61(3):1169

Lammel G, Audy O, Besis A, Efstathiou EK, Kohoutek J et al (2015) Air and seawater pollution and air–sea gas exchange of persistent toxic substances in the Aegean Sea: spatial trends of PAHs, PCBs, OCPs and PBDEs. Environ Sci Pollut Res Int 22(15):11301–11313

Lan WS, Lu TK, Qin ZF, Shi XJ, Wang JJ, Hu YF, Chen B, Zhu YH, Liu Z (2014) Genetically modified microorganism Spingomonaspaucimobilis UT26 for simultaneously degradation of methyl-parathion and c-hexachlorocyclohexane. Ecotoxicology 23(5):840–850

Liu Z, Hong Q, Xu JH, Jun W, Li SP (2006) Construction of a genetically engineered microorganism for degrading organophosphate and carbamate pesticides. Int Biodeterior Biodegrad 58:65–69

Ma Y, Lin J, Zhang C, Ren Y, Lin J (2011) Cd(II) and As(III) bioaccumulation by recombinant Escherichia coli expressing oligomeric human metallothioneins. J Hazard Mater 185(2–3):1605–1608

Malschi D, Muntean L, Oprea I, Roba C, Popita G, Stefanescu L, Malschi Florian B, Rinba E (2018) Research on wastewaters bioremediation with aquatic species for constructed wetlands. Environ Eng Manag J 17:1753–1764

Marihal AK, Jagadeesh KS (2013) Plant–microbe interaction: a potential tool for enhanced bioremediation. In: Arora N (ed) Plant microbe symbiosis: fundamentals and advances. Springer, New Delhi

Matsumoto E, Kawanaka Y, Yun SJ, Oyaizu H (2009) Bioremediation of the organochlorine pesticides, dieldrin and endrin, and their occurrence in the environment. Appl Microbiol Biotechnol 84(2):205–216. https://doi.org/10.1007/s00253-009-2094-5

Matthies M, Solomon K, Vighi M, Gilman A, Tarazona JV (2016) The origin and evolution of assessment criteria for persistent, bioaccumulative and toxic (PBT) chemicals and persistent organic pollutants (POPs). Environ Sci Process Impacts 18(9):1114–1128

Michaud L, Di Marco G, Bruni V, Lo Giudice A (2007) Biodegradative potential and characterization of psychrotolerant polychlorinated biphenyl-degrading marine bacteria isolated from a coastal station in the Terra Nova Bay (Ross Sea, Antarctica). Mar Pollut Bull 54:1754–1761

Minh NH, Anh DH, Tri TM, Anh HQ, Ngoc Mai PT, Nam VD, Viet PH, Minh TB (2016) Persistent toxic substances in Vietnam: a review of environmental contamination and human exposure. In: Persistent organic chemicals in the environment: status and trends in the Pacific Basin countries I contamination status. ACS Symposium Series, vol 1243, pp 55–83

Moermond CTA, Janssen MPM, de Knecht JA, Montforts MHMM, Peijnenburg WJGM, Zweers PGPC, Sijm DTHM (2011) PBT assessment using the revised Annex XIII of REACH: A comparison with other regulatory frameworks, Integr Environ Assess Manag 8(2):359–371

Mosa KA, Saadoun I, Kumar K, Helmy M, Dhankher OP (2016) Potential biotechnological strategies for the cleanup of heavy metals and metalloids. Front Plant Sci 7:303. https://doi.org/10.3389/fpls.2016.00303

Murínová S, Dercová K, Dudášová H (2014) Degradation of polychlorinated biphenyls (PCBs) by four bacterial isolates obtained from the PCB-contaminated soil and PCB-contaminated sediment. Int Biodeterior Biodegrad 91:52–59

Nagpal V, Srinivasan MC, Paknikar KM (2008) Biodegradation of γ-hexachlorocyclohexane (Lindane) by a non-white rot fungus conidiobolus 03-1-56 isolated from litter. Indian J Microbiol 48:134–141

Naik MM, Dubey SK (2013) Lead resistant bacteria: lead resistance mechanisms, their applications in lead bioremediation and biomonitoring. Ecotoxicol Environ Saf 98:1–7

Nie M, Wang Y, Yu J, Xiao M, Jiang L et al (2011) Understanding plant-microbe interactions for phytoremediation of petroleum-polluted soil. PLoS One 6(3):e17961. https://doi.org/10.1371/journal.pone.0017961

Obiri-Nyarko F, Grajales-Mesa SJ, Malina G (2014) An overview of permeable reactive barriers for in situ sustainable groundwater remediation. Chemosphere 111:243–259

Ojuederie OB, Babalola OO (2017) Microbial and plant-assisted bioremediation of heavy metal polluted environments: a review. Int J Environ Res Public Health 14(12). https://doi.org/10.3390/ijerph14121504

Pavel LV, Gavrilescu M (2008) Overview of ex situ decontamination techniques for soil cleanup. Environ Eng Manag J 7:815–834

Pavlíková D, Macek T, Macková M, Surá M, Száková J, Tlustoš P (2004) The evaluation of cadmium, zinc and nickel accumulation ability of transgenic tobacco bearing different transgenes. Plant Soil Environ 50(12):513–517

Peng R, Fu X, Tian Y, Zhao W, Zhu B, Xu J et al (2014a) Metabolic engineering of Arabidopsis for remediation of different polycyclic aromatic hydrocarbons using a hybrid bacterial dioxygenase complex. Metab Eng 26:100–110

Peng R-H, Fu X-Y, Zhao W, Tian Y-S, Zhu B, Han H-J et al (2014b) Phytoremediation of phenanthrene by transgenic plants transformed with a naphthalene dioxygenase system from Pseudomonas. Environ SciTechnol 48:12824–12832

Prabha R, Singh DP, Verma MK (2017) Microbial interactions and perspectives for bioremediation of pesticides in the soils. In: Singh D, Singh H, Prabha R (eds) Plant-microbe interactions in agro-ecological perspectives. Springer, Singapore

Quintelas C, Pereira R, Kaplan E, Tavares T (2013) Removal of Ni(II) from aqueous solutions by an Arthrobacter viscosus biofilm supported on zeolite: from laboratory to pilot scale. Bioresour Technol 142:368–374

RBA PST (2003) Regionally based assessment of persistent toxic substances, Global report 2003. UNEP Chemicals, Châtelaine

Rein A, Fernqvist MM, Mayer P, Trapp S, Bittens M, Karlson UG (2007) Degradation of PCB congeners by bacterial strains. Appl Microbiol Biotechnol 77:469–481

Roşca M, Hlihor RM, Comăniţă ED, Cozma P, Simion IM, Gavrilescu M (2017a) The potential of some indigenous microorganisms and plants for the removal of heavy metals from soil. Sci Papers Hortic Ser 60:65–72

Roşca M, Hlihor RM, Diaconu M, Tavares T, Gavrilescu M (2017b) Efficiency of living Bacillus megaterium in cadmium removal from aqueous solutions. Paper presented at the 9th international conference on environmental engineering and management – circular economy and environmental sustainability, Bologna, Italy, 6–9 September 2017

Roşca M, Hlihor RM, Cozma P, Dragoi EN, Diaconu M, Silva B, Tavares T, Gavrilescu M (2018) Comparison of Rhodotorula sp. and Bacillus megaterium in the removal of cadmium ions from liquid effluents. Green Processes Synth 7:74–88

Ruiz-Aguilar GML, Fernandez-Sanchez JM, Rodrıguez-Vazquez R, Poggi-Varaldo H (2002) Degradation by white-rot fungi of high concentrations of PCB extracted from a contaminated soil. Adv Environ Res 6(4):559–568

Saavedra JM, Acevedo F, González M, Seeger M (2010) Mineralization of PCBs by the genetically modified strain Cupriavidusnecator JMS34 and its application for bioremediation of PCBs in soil. Appl Microbiol Biotechnol 87(4):1543–1554
Saez JM, Alvarez A, Fuentes MS, Amoroso MJ, Benimeli CS (2017) An overview on microbial degradation of lindane. In: Singh S (ed) Microbe-induced degradation of pesticides. Environmental science and engineering. Springer, Cham
Sagar V, Singh DP (2011) Biodegradation of lindane pesticide by non white-rots soil fungus Fusarium sp. World J Microbiol Biotechnol 27(8):1747–1754
Salam JA, Das N (2014) Lindane degradation by Candida VITJzN04, a newly isolated yeast strain from contaminated soil: kinetic study, enzyme analysis and biodegradation pathway. World J Microbiol Biotechnol 30(4):1301–1313
Sasek V, Bhatt M, Cajthaml T, Malachová K, Lednická D (2003) Compost-mediated removal of polycyclic aromatic hydrocarbons from contaminated soil. Arch Environ Contam Toxicol 44(3):336–342
Sharma J, Fulekar MH (2009) Potential of Citrobacterfreundii for bioaccumulation of heavy metal – copper. Biol Med 1(3):7–14
Sharma B, Dangi AK, Shukla P (2018) Contemporary enzyme based technologies for bioremediation: a review. J Environ Manag 210:10–22
Singh BK, Kuhad RC (1999) Biodegradation of lindane (g-hexachlorocyclohexane) by the white-rot fungus Trameteshirsutus. Lett Appl Microbiol 28(3):238–241
Singh S, Sherkhane PD, Kale SP, Eapen S (2011) Expression of a human cytochrome P4502E1 in Nicotianatabacum enhances tolerance and remediation of g-hexachlorocyclohexane. New Biotechnol 28(4):423–429
Smaranda C, Hlihor RM, Apostol LC, Bulgariu L, Diaconu M, Fortuna ME, Gavrilescu M (2016) Biosorption and bioaccumulation: principles and potential applications in bioremediation. In: Gavrilescu M (ed) Biosorption and Bioaccumulation: Principles and Applications in Environmental Bioremediation. Politehnium Publishing House, Iasi
Srivastava NK, Jha MK, Mall ID, Singh D (2010) Application of genetic engineering for chromium removal from industrial wastewater. Int J Environ Ecol Eng 4(12):433–438
Stockholm Convention (2001) Stockholm convention on persistent organic pollutants. http://chm.pops.int/Portals/0/Repository/convention_text/UNEP-POPS-COP-CONVTEXT-FULL.English.PDF. Accessed 15 Oct 2018
Stockholm Convention (2008) All POPs listed in the Stockholm Convention. http://chm.pops.int/TheConvention/ThePOPs/AllPOPs/tabid/2509/Default.aspx. Accessed 12 Oct 2018
Taguchi K, Motoyama M, Iida T, Kudo T (2007) Polychlorinated biphenyl/biphenyl degrading gene clusters in Rhodococcus sp. K37, HA99, and TA431 are different from well-known bph gene clusters of rhodococci. Biosci Biotechnol Biochem 71(5):1136–1144
Takahashi R, Bashir K, Ishimaru Y, Nishizawa NK (2012) Nakanishi H The role of heavy-metal ATPases, HMAs, in zinc and cadmium transport in rice. Plant Signal Behav 7(12):1605–1607
Truua J, Truu M, Espenberg M, Nõlvak H, Juhanson J (2015) Phytoremediation and plant-assisted bioremediation in soil and treatment wetlands: a review. Open Biotechnol J 9. https://doi.org/10.2174/1874070720150430E009
Turnau K et al (2006) Role of mycorrhizal fungi in phytoremediation and toxicity monitoring of heavy metal rich industrial wastes in Southern Poland. In: Twardowska I, Allen HE, Häggblom MM, Stefaniak S (eds) Soil and water pollution monitoring, protection and remediation, NATO science series, vol 69. Springer, Dordrecht
Van Aken B, Doty SL (2010) Transgenic plants and associated bacteria for phytoremediation of chlorinated compounds. Biotechnol Genet Eng Rev 26:43–64
Verma H, Kumar R, Oldach P, Sangwan N, Khurana JP, Gilbert JA, Lal R (2014) Comparative genomic analysis of nine Sphingobium strains: insights into their evolution and hexachlorocyclohexane (HCH) degradation pathways. BMC Genomics 15:1014. https://doi.org/10.1186/1471-2164-15-1014

Vinopal S, Ruml T, Kotrba P (2007) Biosorption of Cd2+ and Zn2+ by cell surface-engineered Saccharomyces cerevisiae. Int Biodeterior Biodegrad 60:96–102

Wang L, Samac DA, Shapir N, Wackett LP, Vance CP, Olszewski NE, Sadowsky MJ (2005) Biodegradation of atrazine in transgenic plants expressing a modified bacterial atrazine chlorohydrolase (atzA) gene. Plant Biotechnol J 3(5):475–486. https://doi.org/10.1111/j.1467-7652.2005.00138.x

Wang Y, Ren H, Pan H, Liu J, Zhang L (2015) Enhanced tolerance and remediation to mixed contaminates of PCBs and 2,4-DCP by transgenic alfalfa plants expressing the 2,3-dihydroxybiphenyl-1,2-dioxygenase. J Hazard Mater 286:269–275. https://doi.org/10.1016/j.jhazmat.2014.12.049

Wang X, Zhi J, Liu X, Zhang H, Liu H, Xu J (2018) ITransgenic tobacco plants expressing a P1B-ATPase gene from Populustomentosa Carr (PtoHMA5) demonstrate improved cadmium transport. Int J Biol Macromol 113:655–661

Wasilkowski D, Swędzioł Ż, Mrozik A (2012) The applicability of genetically modified microorganisms in bioremediation of contaminated environments. Chemik 66(8):817–826

Weiland-Bräuer N, Fischer MA, Schramm KW, Schmitz RA (2017) Polychlorinated Biphenyl (PCB)-degrading potential of microbes present in a cryoconite of Jamtalferner Glacier. Front Microbiol 8:1105. https://doi.org/10.3389/fmicb.2017.01105

Wong MH, Armour M-A, Naidu R, Man M (2012) Persistent toxic substances: sources, fates and effects. Rev Environ Health 27(4):207–213

World Health Organization (2014) Frequently asked questions on genetically modified foods. http://www.who.int/foodsafety/areas_work/food-technology/faq-genetically-modified-food/en/. Accessed 5 Nov 2018

Zhang Y, Liu J (2011) Transgenic alfalfa plants co-expressing glutathione S-transferase (GST) and human CYP2E1 show enhanced resistance to mixed contaminates of heavy metals and organic pollutants. J Hazard Mater 189(1–2):357–362. https://doi.org/10.1016/j.jhazmat.2011.02.042

Zhang H, Jiang X, Lu L, Xiao W (2015) Biodegradation of polychlorinated biphenyls (PCBs) by the novel identified cyanobacterium Anabaena PD-1. PLoS One 10(7):e0131450. https://doi.org/10.1371/journal.pone.0131450.eCollection

Next-Generation Sequencing and Its Application: Empowering in Public Health Beyond Reality

15

Nidhi Gupta and Vijay K. Verma

Abstract

Next-generation sequencing has the ability to revolutionize almost all fields of biological science. It has drastically reduced the cost of sequencing. This allows us to study the whole genome or part of the genome to understand how the cellular functions are governed by the genetic code. The data obtained in huge quantity from sequencing upon analysis gives an insight to understand the mechanism of pathogen biology, virulence, and phenomenon of bacterial resistance, which helps in investigating the outbreak. This ultimately helps in the development of therapies for public health welfare against human pathogen and diagnostic reagents for the screening. This chapter includes the basic of Sanger's method of DNA sequencing and next-generation sequencing, different available platforms for sequencing with their advantages, and limitations and their chemistry with an overview of downstream data analysis. Furthermore, the breadth of applications of high-throughput NGS technology for human health has been discussed.

Keywords

Pyrosequencing · Roche 454 · ABISOLiD · Nonopore

N. Gupta
Department of Biochemistry, Central University of Rajasthan, Kishangarh, Rajasthan, India

V. K. Verma (✉)
Department of Microbiology, Central University of Rajasthan, Kishangarh, Rajasthan, India
e-mail: vijayverma@curaj.ac.in

P. K. Arora (ed.), *Microbial Technology for the Welfare of Society*,
Microorganisms for Sustainability 17,
https://doi.org/10.1007/978-981-13-8844-6_15

15.1 Introduction and History of Sequencing

Earlier, Sanger's method of DNA sequencing was the only easy and popular method to determine the sequence of DNA molecule. On top of everything, this dideoxy chain termination method was diligently used for more than three decades from its discovery. Since the start of the twenty-first century, high-throughput sequencing technology has made an impact on the genomics research because it enabled genome-wide sequencing and screening far easier, inexpensive, and reproducible with lesser need of manpower (Metzker 2010). Using NGS, the genome of the bacteria can be sequenced in a single run. Sequencing of the DNA allows us to provide the basic information, i.e., the sequence/order of the nucleotides. After recognizing that this could be a magic tool to understand the gene sequences and location of the gene regulatory molecules, a UK-based team led by Dr. Frederick Sanger initiated work on DNA sequencing in 1972. He finally developed the "dideoxy chain termination" method for DNA sequencing and also published his work in 1977 (Sanger et al. 1977a). The method was based upon the base-specific termination of the growing chain. DNA polymerase adds the nucleotides on the chain, but upon the incorporation of dideoxynucleotide, the chain terminates due to its inability of the phosphodiester bond formation between newly coming nucleotide and dideoxynucleotide. In the same year 1977, a US team led by Maxam and Gilbert also published a chemical-based sequencing method in which the sequencing of DNA was dependent on chemical cleavage protocol (Maxam and Gilbert 1977). This method was based upon the use of harmful chemicals such as dimethyl sulfate (DMS) and hydrazine and hence was not that popular among molecular biologists. In contrast, the Sanger's method was accepted widely because of its easy protocol and use of lesser harmful reagents (Obenrader 2003). However, both Fredrick Sanger and Walter Gilbert were awarded with the Nobel Prize in chemistry in 1980 for their DNA sequencing method discovery. Using Sanger's method, bacteriophage ΦX174 genome of 5386 bp was sequenced, and it was the beginning of the full DNA genome sequencing (Sanger et al. 1977b).

Sanger's method was the method of choice and used for the genome sequencing of a number of organisms. *Haemophilus influenzae* was the first bacterium of genome size 1,830,140 bp sequenced using an approach, namely, shotgun, in 1995 (Fleischmann et al. 1995). Soon after this, *Saccharomyces cerevisiae* was the first eukaryotic genome of size 12,156,677 bp sequenced in 1996 (Goffeau et al. 1996). Major breakthrough happened when human genome (14.8 billion bp) was sequenced and got published in 2001. The human genome was sequenced using two different approaches by two different independent teams. Using shotgun approach (where the genome got fragmented randomly), Dr. Craig Venter's team from a company called Celera Genomics published its dataset in 2001 (Venter et al. 2001), whereas Francis Collins group from the National Human Genome Research Institute (NHGRI, NIH) used the BAC contigs (bactigs) approach for the mapping of sequence.

The major limitation of the Sanger's method was that it could sequence very less number of DNA sequences in one go and also the cost per base was very high. Apparently, the high rise in the complex diseases and their relation with the

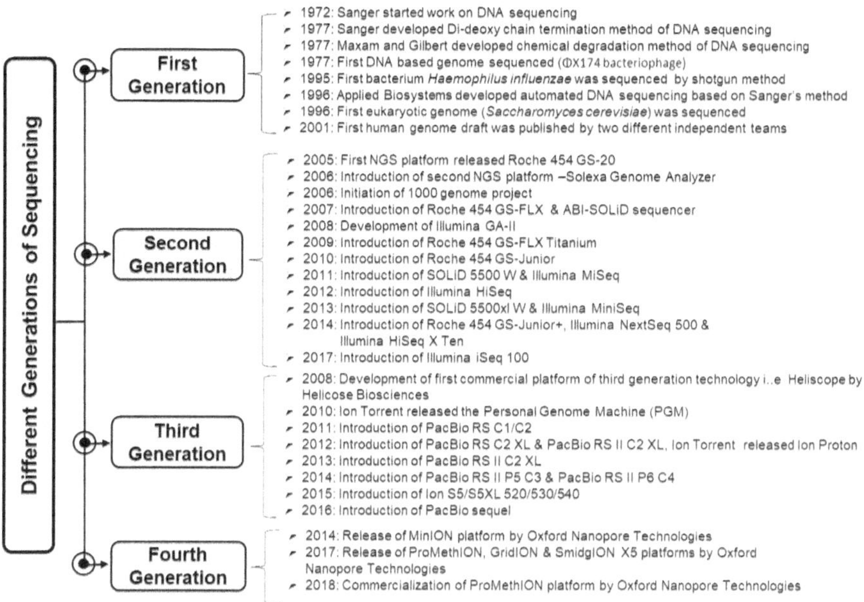

Fig. 15.1 Sequencing events, developments, and introduction of different generations' sequencing platforms in chronological order

mutations/changes in the genome demand a far-flung knowledge of the genome sequence. Thus, it is required to sequence the genome of huge number of individuals as well as other organisms for the diagnosis and treatment in short period of time with low cost. This prompted the need of high-throughput sequencing technologies, which can provide information at substantially lower cost. Further, the development of high-throughput next-generation sequencing technology has proven that it can generate enormous data (millions of sequences) cost effectively and rapidly.

Here we have shown a timeline of sequencing events and introduction of platforms of different generations' sequencing technologies (Fig. 15.1).

15.2 Different Generations of Sequencing

It is very prominent that for almost four decades, the Sanger's method was considered the gold standard because there was not a considerable development in DNA sequencing techniques in those years. In the last almost 15 years (from 2005), there are massive changes in the generation of sequencing technology. In short, changes in generation means the change in chemistry and platform. The widely accepted method of sequencing, i.e., Sanger's method of DNA sequencing, covered the first-generation sequencing along with Maxam-Gilbert method. Both methods were capable of sequencing about 1 kilobase length DNA fragment at one shot. In the process of analyzing longer sequences, researchers used the "shotgun technique" where overlapping sequences were cloned and sequenced separately followed by assembly into contigs (Anderson 1981). Sanger's method is one of the best methods for years to come for the sequencing of gene cloned in heterologous system because of its precision, robustness, and ease of use.

Scientists presented a next-generation sequencing technology, which comes under second-generation sequencing technology that includes Roche 454, Illumina Solexa, and ABI-SOLiD and has transformed the field of omics. This technology was able to produce enormous amount of data at very economic cost and expeditiously. Moreover, this technology is very rapid than traditional method that whole genome of small organisms can be sequenced in a single day. However, in recent years, Illumina platforms have contributed very much to the second-generation sequencers and hence are considered to be one of the best platform providers.

The third-generation sequencing that is also known as next next-generation sequencing refers to those technologies which do not depend on the PCR amplification of DNA molecule. Thus, the problem related to biasness through PCR amplification and dephasing was ruled out. Platforms of this generation, which include Helicos and PacBio, are capable of sequencing single molecule.

There is a platform called Ion Torrent that has been kept between second and third generation because it is based upon first "post light sequencing" technology. This does not require fluorescence or luminescence. Nanopore sequencers offered by Oxford Nanopore Technologies (ONT), namely, GridION and MinION, lie under fourth generation. These platforms are based on different chemistry from third-generation sequencers.

Note that Roche 454 platform was commercially introduced first, but nowadays it is not available in the market. This shows that the changes made in this field in the past 12–15 years are very rapid.

15.3 Comparison of First-Generation and Second-Generation Sequencing Principle

First generation includes two separate methods, namely, Sanger and Maxam Gilbert method. Both methods were equally accepted in the beginning, but Sanger's method was extensively used for the routine sequencing purposes, which was based on the chain termination method. In the chain termination method, upon incorporation of the dideoxynucleotides (ddNTPs), the growing chain terminates. Fragments of varied length (length varied by single nucleotide) of the DNA molecule then run on the traditional slab gels and pattern of bands obtained for the sequence determination. Subsequently, radiolabeling is replaced with fluorescently labeled (automated method of sequencing) ddNTPs, and the laser light at different wavelengths does the sequence determination (Smith et al. 1986). This method can generate a maximum read length that ranges from 800 to 1000 bp. In this method, only one fragment can be sequenced in one capillary, which means the output of one run is the length of the sequenced fragment.

Conversely, the principle of second-generation sequencers or next-generation sequencers is based on clonal amplification of DNA molecule where billions of different DNA fragments get sequenced at the same time in parallel fashion and generate enormous data. For sequencing of the whole genome of an organism, random fragmentation is done at particular size range and then fragments ligated to the oligonucleotide adaptors, which are platform specific, followed by independent parallel sequencing. Parallel analysis increases the sequencing speed. NGS offers the capability to produce massive volume of data from a single run at a very low cost in a very short duration without the need of fragment cloning strategy, which is generally used in conventional method. As we can see, there is a vast difference in the cost per genome sequencing in 2001 ($100M) as compared to the cost in 2017 (<$1K) (data from NHGRI genome sequencing program). Complete draft sequence of the human genome with the help of automated Sanger's method was published in 2001, which was the outcome of 13 years of rigorous efforts of international project of $2.7 billion. On the contrary, using NGS platform, the whole human genome can be sequenced in a week for a few thousand dollars (Gullapalli et al. 2012).

15.4 Second-Generation/Next-Generation Sequencing Technologies

NGS and high-throughput sequencing generally denote to technologies that permit the millions of sequencing reactions in parallel on the same solid surface which may be beads or glass slide. This does not require the physical separation of reaction in different well or in lane/tube but spatially separated. Hence, thousands of million different reactions ensue simultaneously, because of which there is a dramatic decrease in the labor input as compared to other conventional methods and the huge reduction in cost per se. The path involves several commercial NGS platforms that are based on different technologies but typically follow a general pattern or steps. General steps involved in the DNA sequencing using NGS are (i) library preparation (random fragmentation of genome, ligation with appropriate adaptors), (ii) amplification of library, and (iii) sequencing using different approaches. The basic steps involved are presented in a flowchart in Fig. 15.2. The generated results differ with respect to read length, quality of the data, and quantity of the data based upon the platform used. Classification of different sequencing technologies based upon the type of technology, chemistry, detection system used, and method of amplification used in different generations of sequencing platforms is described in Fig. 15.3. Here in the following section, we will discuss about the current sequencing technologies, their principle, and their advantages and limitation.

15.4.1 Pyrosequencing Technology

Nyren's group pioneered this sequencing by synthesis (SBS) approach technology in 1993, where DNA sequencing is based upon detection of released pyrophosphate

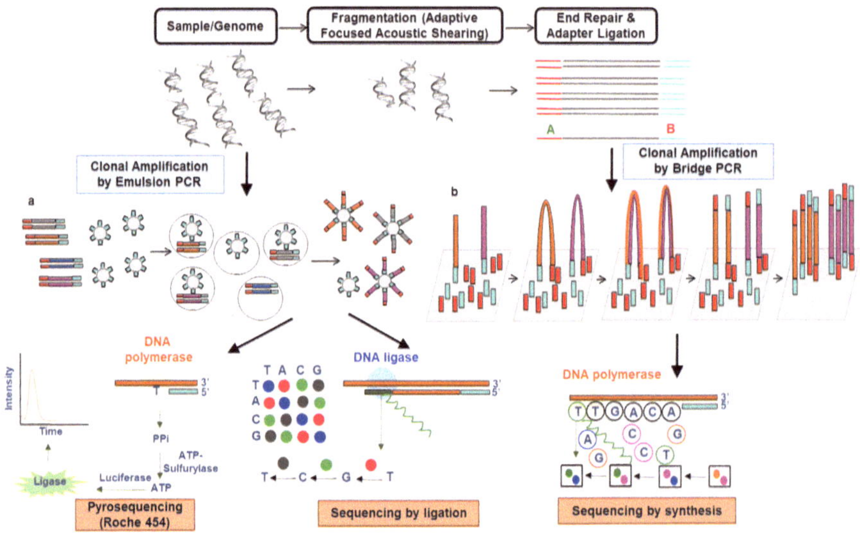

Fig. 15.2 Schematic representation of the basic steps involved in DNA sequencing using different NGS platforms

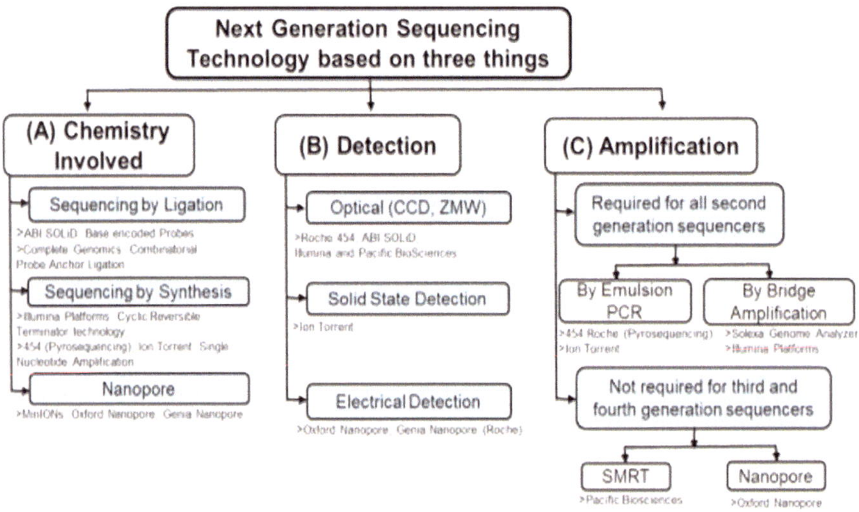

Fig. 15.3 Classification of the next generation sequencing (NGS) on the basis of types of technology, chemistry, detection system and clonal application

(PPi) molecule during addition of nucleotide by DNA polymerase (Ronaghi et al. 1996). The speed of the reaction is very fast as it just takes 3–4 s at room temperature to complete the reaction from addition of nucleotide to chemiluminescent detection. Later, 454 Life Sciences (a USA-based biotechnology company which was later acquired by Roche) took over this technology and commercialized it with some modifications. Pyrosequencing uses the enzyme luciferase from *Photinus pyralis* (American firefly) and recombinant ATP sulfurylase from *Saccharomyces cerevisiae* (Karamohamed et al. 1999). Two different approaches are available in pyrosequencing: (i) solid phase (Ronaghi et al. 1996) where DNA gets immobilized (based on three-enzyme cascade method with washing step) and (ii) liquid phase (Ronaghi et al. 1998) in which a nucleotide-degrading enzyme, pyrase, has been added (based on four-enzyme cascade method without washing step) which excludes the requirement of solid support because of which reaction can be performed in a single tube.

15.4.2 Roche 454 (GS FLX plus)

In 2007, 454 Life Sciences, a pyrosequencing technology-based sequencer, was taken over by Roche and later known as Roche 454. Steps involved in pyrosequencing are fragmentation of the nucleic acid to be sequenced followed by synthesis of template strand with the assistance of polymerase enzyme. On incorporation of new nucleotide by polymerase, the pyrophosphate molecule is released. This pyrophosphate molecule converts the ADP to ATP in the presence of the enzyme ATP sulfurylase. ATP molecule supplies the energy for the oxidation of luciferin, which subsequently emits light and recorded by charge-coupled device (CCD). The identity of nucleotides added (all four dNTPs added in the reaction separately) into the reaction for polymerization is known to help in the determination of template sequence.

Pyrosequencing-based platform uses a massively parallel system for sequencing in picolitre volumes in microfluidic format. In brief, the methodology of sequencing is the fragmentation of DNA (~800 bp) using spray method (nebulizer), adaptors ligation to the fragmented DNA, library preparation followed by attachment of library to beads (DNA capture beads, which allows attachment of one fragment per bead). The beads make individual compartments, which are usually referred to as microreactors/microvesicles. Clonal amplification takes place in those compartments by emulsion PCR; subsequently, emulsion gets broken and beads attached clonally amplified DNA becomes enriched in microreacters/microvesicles (Margulies et al. 2005). All clonally amplified DNA-bound beads are individually loaded on picotitre plate (PTP; imprinted on the fiber-optic slide), which contains approx. 3.4×10^9 wells of ~55 μm in depth and 75 picolitre calculated size. The slide/plate comprising the picolitre-sized wells mounted in a flow cell, which forms the channel for the flow of the sequencing reagents above the wells. The base of the plate is connected with the imaging device called as CCD that captures the emitted light and provides the results in the form of flowgrams. Genome sequencer GS FLX produces nearly 450 MB data from a single run whereas new genome sequencer (GS FLX plus) can produce 700 MB data from a single run in ~10 h with an approximate cost of 5K to 7K USD.

Advantage This is a fast (700 MB data in a day), accurate (~99.9% after filter) and reliable technology for high throughput real-time sequencing. The Roche 454 is an upgraded platform, which can give read length>700 bp. Further, the technology does not require labeled nucleotides and primers followed by gel electrophoresis and suitable for *de novo* sequencing as well as confirmatory sequencing (Ronaghi 2001). The technology provides flexibility in primer designing since it starts sequencing immediately downstream of primer sequence without keeping gap of 30–40 bp.

Limitation The platform sure has some limitation. The main constraint is the problem in sequencing same nucleotide repeat (>8 bp), that is, homopolymer sequencing (Mardis 2008). Another, it is relatively costlier than other NGS technologies.

15.5 Reversible Terminator Technology

This technology also depends on the sequencing by synthesis (SBS) strategy. Dr. JingyueJu was the first person who described reversible terminator sequencing technology (Li et al. 2003). The basic difference between traditional sequencing and this is that the traditional sequencing uses ddNTPs to irreversibly terminate the extension of the primer whereas reversible sequencing technology employs modified analogue of nucleotide to terminate the extension of primer reversibly (Guo et al. 2010). In the past decade, numerous reversible terminators have identified based upon reversible blocking groups and can be categorized into two types. One category belongs to 3'-O-blocked reversible terminators and other is 3' unblocked reversible terminator. Illumina Solexa commercializes this technology because of its comprehensive acceptance in second-generation sequencers (Bentley et al. 2008).

15.5.1 Illumina Solexa

As stated above, this is the popular NGS platform of second generation. David Klenerman and Shankar Balasubramanian gave the idea for only one DNA molecule sequencing attached to microsphere with the foundation of Solexa in 1998. The "Solexa Genome Analyzer" system, which came in 2006 and later, it was acquired by Illumina for the sequencing of clonally amplified DNA (Voelkerding et al. 2009).

A flow cell, which is used by Illumina Solexa, is made up of optical transparent slide having eight lanes on its surface. Oligonucleotide anchors are immobilized to the surface of the flow cell. In brief, the methodology of the sequencing is the fragmentation of template DNA, end repairing of fragments (blunting and 5' end phosphorylation). Adenylation of 3' ends by the addition of single 'A' nucleotide to facilitate ligation with the oligonucleotide adapters carrying a 'T' overhang at 3'. Since ligated adapters are complementary to the flow cell, anchors thus get hybridized. The DNA template attached to the anchors of the flow cell relies for the cluster generation by "bridge amplification" in contrast to emulsion PCR (Adessi et al. 2000). Further, DNA fragment makes an arc and hybridizes with its distal end to neighboring anchor oligonucleotide to its complementary part. Because of clonal amplification, each template generates thousands of copies (clusters) of same template DNA and subsequently millions of separate (unique) clusters are generated on the single flow cell followed by addition of DNA polymerase, and four different fluorescent-labeled reversible terminators aid in sequencing the millions of clusters in parallel fashion onto the flow cell. Polymerization terminates because of the incorporation of fluorescent-labeled reversible terminators (ddNTPs) and incorporated nucleotide is identified by the fluorescence captured (Guo et al. 2008). By the enzymatic cleavage, the fluorescence label gets cleaved, which permits the incorporation of next nucleotide (www.illumina.com). Recently, several technical improvements are happening rapidly in this sequencing technology including library preparation method, which involves fragmentation of the DNA in acceptable sizes by Covaris (Adaptive Focused Acoustic wave) sonication method and improved adapter ligation efficiency, etc.

Illumina platforms dominate in high throughput sequencing market. Currently, Illumina is producing a series of platforms (MiSeq, HiSeq series and NextSeq series). Different platforms are augmented for throughput and turnaround time of the run. Most recognized platforms are MiSeq and series of HiSeq platforms. The MiSeq is a personal tabletop sequencer marketed in 2011, where a run can be completed in as fast as in four hours for the targeted bacterial sequencing. On the contrary, HiSeq 2500 is applicable for high throughput sequencing like 1 TB data from a single run in 5–6 days. New model of HiSeq platform, that is, HiSeq 2500, can also be run in fast mode. However, run will not be cost effective and will sequence 30X human genome in approximately 27 h. In the beginning of 2014, Illumina launched another two NextSeq 500 and HiSeq X Ten. NextSeq 500 is similar to MiSeq made for individual labs. HiSeq X Ten platform works as whole genome sequencer at population scale. Presently, Illumina only supports sample of human for whole genome sequencing on HiSeq X Ten. MiniSeq platform from Illumina also came into the market in 2016. Most recently released platforms from Illumina are HiSeq 3000 and HiSeq 4000 based on the patterned flow cell technology. Their data output and run time lie between HiSeq X Ten and HiSeq 2500 (Reuter et al. 2015).

The latest machine from Illumina launched at the end of 2017 is HiSeq 100. This is the smallest and most inexpensive sequencer in the portfolio of Illumina with a maximum data output of 1.2 GB and 4 million reads per run, with run time ranging from 9 to 17.5 h (www.illumina.com/iseq).

Advantage First and foremost, the technology provides high throughput data in a very short period of time with very low amount of sample per run (Buermans and den Dunnen 2014). The newer platforms of Illumina like HiSeq 2500, HiSeq 2000 and MiSeq generate more data (up to 600 GB) at low cost per base. Using the platform based on reversible terminator technology, 1 TB data can be generated in a single day. Another big advantage is the longer read lengths, that is, 300 bp paired-end sequencing in Illumina (MiSeq) platform is now possible, which was earlier 25 bp single-end reads by Solexa. Besides this, Illumina platforms provide 99.9% accuracy of the sequencing data (Morey et al. 2013). Because of the presence of blocking group, the addition of only one nucleotide per cycle facilitates the sequencing of homopolymeric regions efficiently (Mardis 2013).

Limitation Major limitation is guanine and cytosine (GC) biasness, which gets introduced during bridge amplification (Mardis 2013). Another concern is dephasing, which means the different copies of DNA in a cluster get out of sync (inconsistent). In other words, inappropriate deblocking of nucleotide results in varying length fragments in a cluster. This decreases the accuracy in base calling at 3' end of the DNA fragments, especially in invert repeat sequence (Nakamura et al. 2011).

15.5.2 Sequencing by Ligation Technology

This is a DNA sequencing technique, which determines the DNA sequence by utilizing the mismatch sensitivity of DNA ligase enzyme (Ho et al. 2011). Applied Biosystems, USA marketed this technology in 2008. The platforms of this technology rely upon oligonucleotide probes of variable lengths, labeled with different fluorescence tag liable to the nucleotide to be sequenced.

15.5.3 ABI-SOLiD

The expansion of SOLiD is small oligonucleotide ligation and detection system. The technology was invented in 2005 by George Church. Later in 2008, it was further upgraded and marketed by Applied Biosystems (Voelkerding et al. 2009), which is now acquired by Life Technologies. The sequencing reaction can be divided into five broad steps: (1) preparation of DNA library (2) clonal amplification in microreactors by emulsion PCR (3) attachment of the beads (4) sequencing and (5) resetting of primer. In brief, the methodology for sequencing using this technology includes fragmentation of DNA, attachment of fragments to the beads and clonal amplification of fragments attached to the beads by emulsion PCR. Following this, adapter sequences in amplified fragments hybridized to the specific primers. This facilitates the ligation of octamer (eight base pair) interrogation probe (fluorescently labeled) by offering 5' -PO_4 group in place of 3' –OH group. The first two bases of interrogation probes are specific while the other six are degenerate. Set of four different fluorescent-tagged probes (interrogation probe), having one out of sixteen possible combinations (e.g., AC, AT, AC, AG, CG, TC, GT and TT) of two base (specific) at the end compete to ligate with the primer. After ligation, fluorescence is imaged that is equivalent to the interrogation probe ligated. For another round of cycle, 5' –PO_4 group regenerated by the deletion of fluorescence label of the attached interrogation probe. Further, steps of the previous cycle are repeated after injection of set of four different fluorescent-tagged probes. Generally after seven cycles of ligation, template is reset to the n-1 position of complementary primer for another round of ligation. This procedure is repeated every time with a consecutive offset like n-1, n-2 and so on of a new primer. A single run on SOLiD 5500 platform takes approx. 6–7 days to complete and produce 120–240 GB data with 75 bases read length while SOLiD 4 platform generates 100 GB data.

Advantage This technology offers the highest accuracy of ~99.99% (Voelkerding et al. 2009) because each nucleotide sequenced two times, thus, there is pretty less chances of miscall from two contiguous colors.

Limitation One of the limitations is time taken in one run is so long (6–7 days) along with production of less data as compared to Illumina platforms. Another limitation is that if we compare with other second-generation methods (sequencing by synthesis), the analysis of the data is complex in this technology, which impedes the marketing of the platform. *De novo* sequencing is another limitation in using this technology.

15.6 Third-Generation Sequencing Technologies

Second-generation technology platforms were the most widely used platforms. But the major limitation of them was the occurrence of biasness because of PCR amplification step. On the contrary, third-generation sequencing technologies do not require amplification step and are capable of sequencing single DNA molecule in real time. These platforms have the capability to provide single run at very low cost as well as made the preparation of sample easier. Further, third-generation platforms produce generally longer read of about some kilobase length, which resolve the problem of assembling the reads.

Dr. Stephen Quake's team established the first single-molecule sequencing technology (SMT) (Braslavsky et al. 2003), which was further commercialized by Helicos Biosciences.

15.6.1 Single-Molecule Real-Time Sequencing

Nanofluidics Incorporation pioneered single-molecule real-time sequencing (SMRT). This technology for sequencing is based upon two key inventions: phospholinked nucleotides and zero-mode waveguides (ZMW). The key feature of ZMW is that it only permits light to illuminate the bottom of a well where the immobilized template and DNA polymerase are present.

15.6.1.1 Pacific Biosciences (PacBio)

The PacBio depends on the SMRT technology, which was commercialized in 2010 by Pacific Biosciences (Roberts et al. 2013). This allows us to know about the synthesis of DNA in real time. This is possible because of the presence of zero-mode waveguide (ZMW) holes where DNA synthesis takes place.

Methodology in brief, single stranded adapters (hairpin adapters) ligated to the fragmented DNA, which is then known as capped template. Here to increase the accuracy, a strand displacing DNA polymerase is used to sequence the same template several times (Travers et al. 2010). DNA polymerase and template immobilize at the base of the ZMW where DNA synthesis takes place (Levene et al. 2003). Approximately 75K ZMWs/SMRT cells are present allowing 75K single-molecule reactions in parallel fashion. There is concept of physics, which does not allow the laser light (600 nm) to pass completely through the ZMWs (because of zepto-litre holes) and it decays exponentially after entering into ZMW. Thus, the laser light only illuminates the 30 nm of the hole. Phospholinked nucleotides of all four types subsequently pass over the ZMWs. The nucleotides get excited and fluoresce when reaching the base because laser cannot penetrate up within the hole. Therefore, nucleotides cannot be fluoresced till they are present on the surface of the hole. Hence, polymerization reaction takes place continuously; fluorescent signals can be detected in real time so the sequence can be read (Eid et al. 2009). RS II platform of Pacific Biosciences, which is commercially available, was released in 2010, whereas PacBio Sequel is the latest platform released by Pacific BioSciences in autumn 2015 in collaboration with Roche Diagnostics for the development of clinical grade sequencer for diagnostics. RS II platform is the foremost platform that offers sequencing read length >20 Kb and PacBio Sequel is analogous to its former. The new platform PacBio sequel generates almost seven times (~365,000) more number of reads than RS II (55,000). The driving force for the long read length is single-molecule real-time technology in combination with zero-mode waveguides.

Advantage The reaction can be monitored in real time is the biggest advantage per se, which permits to gather the data related to base composition or sequence of the DNA template as well as the enzyme kinetics. Difference in enzyme kinetics provides us the clue about different modifications present in the DNA like methylation (6-methyladenine, 5-methylcytosine) (Flusberg et al. 2010; Fang et al. 2012). With the help of the identification of this modification sites genomewide, the approach can also be used to identify the potential modification present in different genetic diseases. Additional advantage includes longer reads and unbiased data. Moreover, using the SMRT approach, not only DNA, ribosome can also be observed at single-molecule resolution (Uemura et al. 2010). *De novo* sequencing can be easily performed because of the longer read length. Hence, short read allows error in the assembly of fragments and formation of scaffold in repeat and GC-rich regions (Bahassi el and Stambrook 2014).

Limitation The cost per base sequencing being relatively high and the lower throughput are disadvantages, which limit its use in maximum genomewide studies. Higher error rate (insertion and deletion) is the main limitation of the technology (Mardis 2011). Along with this, the data generated per run is very less as compared to second-generation platforms.

15.6.2 True Single-Molecule Real-Time Sequencing

This technique also provides us the sequence of single molecule of template DNA. Hence, evade the requirement of clonal amplification and library preparation (Harris et al. 2008). It is also known as single-molecule fluorescent sequencing. In this method, virtual terminators were employed for the fluorescent detection of nucleotide. The virtual terminators were introduced in 2009 (Bowers et al. 2009) for the third-generation sequencing technology. Their working methodology is similar to the second-generation reversible terminators. However, different fluorescent dyes and different blocking groups express different features rendering to structure and nucleotide binding region of them. Generally, virtual terminators are made up of free 3' –OH group, which interacts with DNA polymerase and fluorescent molecule bounded with the linker group (Korlach et al. 2010).

15.6.2.1 Helicos Biosciences (HeliScope)

This is the first commercial platform of third generation, which has again revolutionized the DNA resequencing technology in 2008. Helicos Biosciences Corporation, Cambridge MA, USA has launched the HeliScope platform and produces 3×10^7 reads/channel in a channel slide format (Metzker 2010). In this method, the template to be sequenced is fragmented and polyadenylated at the 3' end by terminal transferase. The flow cell, which is coated with the oligo-dT containing primers gets hybridized to poly-A tail present in the fragments. To avoid Poly-A tail sequencing present in the fragments, virtual terminators of nucleotides other than dTTP are added in the initial sequencing step. The principle/chemistry of the sequencing is same as for sequencing by reversible terminator. Likewise, in the cyclic extension manner template, molecules get sequenced and imaged using CCD camera after cleavage of blocking group and fluorescent dye (Thompson and Steinmann 2010). Approximately 35 GB data can be generated from a single run on this platform.

Advantage It does not require clonal amplification step, which resulted in reduction of biasness. So, it is a beneficial substitute for applications that are mostly affected by PCR biasness like RNASeq. This is the first sequencing method able to sequence every single nucleotide of each DNA molecule from the fossils, which provide the information regarding DNA damage (Krause et al. 2010). Furthermore, this requires very less concentration of template molecule.

Limitation Despite the advantages, there are some limitations as well, like shorter reads and high cost, because of the repetitive sequencing to get the most accurate data by reducing error rates. Shorter reads make it difficult to be used for the *de novo* sequencing. Also, the technology does not generate the paired end sequences, which could help in the orientation and location determination of contigs for assembling of the data. In 2012, Helicos Biosciences announced the impoverishment. However, SeqLLCompany (Boston MA, USA) provides the services for DNA and RNA sequencing using this technology.

15.6.3 Ion Semiconductor Sequencing

Basically, this is an extension of the pyrosequencing technology as described by Ansorge in 2010. This technology uses a chip (semiconductor), which is fabricated with millions of micro wells. These wells capture the release of proton (H^+) during the sequencing followed by change in pH. Proton detected by the technology is the product other than PPi molecule released during polymerization. This is an amalgamation of semiconductor technique (digital) and chemistry, which allows the expression of chemical signals into digital data (to determine the base call/sequence). DNA Electronics in London licensed the principle of this technique, that is, the detection of proton (H^+).

This is a first commercial sequencing technology, which does not demand for costly optics, lasers and different fluorescently labeled nucleotides for complex sequencing chemistries (Ansorge 2016).

15.6.3.1 Ion Torrent

Ion Torrent released the Personal Genome Machine (PGM), a compact benchtop platform in late 2010 that was later acquired by Life Technologies, Carlsbad, USA. This machine uses the high-density arrays and generates the data usually 10–20,000 MB per run of up to 400 bp read length in 2–7 hours. It is according to the chip used and also based on application purpose. The basic methodology is almost similar to other NGS technology, which involves fragmentation and ligation of adapters. Hybridization to the complementary sequences (primers) bounded with the beads followed by emulsion PCR. After clonal amplification, beads flooded over the semiconductor chip. Each bead goes to the individual well present on the chip flowed by the floating of nucleotides sequentially. On incorporation of each nucleotide by DNA polymerase, a proton (H^+) is released, which results in change of the pH. This change in pH determines the base sequence by changing the chemical signal into digital signal. In homopolymers (more than one same nucleotide incorporated simultaneously), intensity of signal gets high, which is parallel to the pyrosequencing technology (Quail et al. 2012).

In the third quarter of 2012, Ion Torrent released its advanced and bigger platform named Ion Proton. This new platform play an important role in the sequencing of whole genomes, transcriptomes and exomes. The data output was up to 10,000 MB with 200 bases read length in a very short duration, that is, two to four hours. The platform has a number of applications including *de novo* sequencing, ChIP sequencing, sequencing analysis of the methylation in DNA, small RNA sequencing and gene expression analysis. Subsequently, other versions, namely, Ion S5 and Ion S5 XL, came with their broad range of applications having both low throughput and high throughput. In reference of throughput, these platforms can be compared to Illumina HiSeq platforms.

Advantage These platforms require very low input of (DNA or RNA) concentration (~10 ng) for the identification of mutations and expression profile. Also, the technology has simplified the analysis of the sequencing data because of the new

Ion Reporter Software. There are plug-ins and operating software available for data analysis from amplicon sequencing, microbial sequencing, etc. They are widely accepted because of the reasonable cost, though they generate shorter reads than some other platforms like PacBio. In simpler words, they are affordable, rapid (run completes in 2–4 h) and simple and so are suitable for laboratories.

Limitation Major limitation is lack of coverage in sequencing of genomes that contain very high-AT content using Ion Torrent Personal Genome Machine (Ballester et al. 2016). Another difficulty in homopolymer sequencing is stretches of more than six same nucleotides trigger deletion and insertion error rate (~1%) (Reuter et al. 2015).

15.7 Fourth-Generation Sequencing Technologies

This generation of the sequencing has the ability to in situ (perform sequencing directly in the cell) sequence the fixed tissue and cells by using second-generation methodology (Mignardi and Nilsson 2014). Targeted and untargeted methods developed for in situ RNA sequencing are based on the principle of ligation chemistry. Further, it has the quality to sequence the entire human genome rapidly and authentically at very low cost, that is, <$1000. Thus, fourth generation has come up with the use in numerous applications like validation of biomarkers and transcriptomic analysis. A group led by Church overcame the limitation present in the Ke's method where they gave an idea of partition sequencing to reduce read density. With this approach, determination of expression of large number of genes in the cell is possible in parallel fashion for several types of RNA. Example: mRNA, rRNA, antisense RNA and non-coding RNA (Mignardi and Nilsson 2014). Using in situ sequencing method, it is possible to screen the whole cell population with the resolution of single cell.

Fourth-generation platforms based on recent technique, namely, "Spatial Transcriptomics," are in their infantile stage (Stahl et al. 2016). Again, this technique is also based upon NGS chemistry for the sequencing. This technique offers the simultaneous visualization and quantitative analysis of the transcriptome (gene expression data) in the fixed tissues. Nanopore-based sequencing method is also available to sequence nucleic acid inexpensively in a short duration.

15.7.1 Nanopore Technology

Though this idea of nanopore sensors based sequencing is very old, it was first envisaged by David Deamer in 1989. This portable technology emerged from coulter counter and ion channels. On the supply of the voltage, particles of smaller size than the pore size circulated across the pore. The read lengths of > 150 Kb can be attained. Now, many companies have offered the strategies for nanopore-based sequencing. One is NanoTag sequencing by Genia where DNA strand gets excised in monomers followed by their channeling one after another, across a nanopore. Another is strand sequencing by Oxford Nanopore; here, whole single strand of DNA passes through a nanopore, which allows the pulling base by base in only one direction (ratcheted) via nanopore. Till now, Oxford Nanopore Technology is the most successful technology for strand sequencing by nanopore.

15.7.1.1 Oxford Nanopore (MinION)

MinION is the first nanopore device for sequencing. Oxford Nanopore Technologies, UK, licensed it in 2007 and commercialized in May 2014. A flow cell is present at the core of this device in which 2048 individual nanopores are present. They are divided into four groups of 512 nanopores in each group and controlled by application-specific integrated circuit (ASIC). Brief methodology for sequencing involves ligation of adapters to the fragments at each end. Adapters enable capture of the fragment and polymerase binding at the 5' ends of the fragments. Additionally, these adapters concentrate the DNA fragments closer to the nanopore, which enhances the rate of fragment capture thousandfold. Also, these hairpin-like adapters allow adjoined sequencing of two complementary strands by covalently attaching the strands to each other. On translocation of a fragment through nanopore, the polymerase processes along the template strand and the process repeats for complementary strand. The sensor identifies the change in ionic charge when fragments move through the nanopore. The change in the ionic charge or characteristic disruption in current is divided into separate events, which ensure associated duration, mean amplitude and variance. The series of events is finally interpreted using computer software/graphical models (e.g., MinKNOW) to identify the nucleotide sequence. Finally, the information collected from the template and complementary strands is merged to generate the "2D read." Another available method for the library preparation does not involve the hairpin adapters to covalently connect two strands of the fragments. This method generates the "1D reads"; so in this, nanopore reads only template strand. However, this allows high throughput but slightly less accuracy in data in comparison to 2D data (Jain et al. 2018).

Advantage The key advantage of this technology is that the device is able to produce long read length > 882 Kb (Quick et al. 2017; Jain et al. 2018). The ultra long read length provides the comfort in data alignment and assembly, which lowers the computational burden. Another advantage is that it is a portable device, which also provides us chance to see the data in real time. MinION is an economic and high throughput device for sequencing of nucleic acid. The biggest advantage of its portability and mini size is that it has given a chance to look for the opportunity of life in outer space (Castro-Wallace et al. 2017). This is the first DNA sequencing platform used in the cosmos. Major properties of MinION assists in rapid surveillance of epidemics like Ebola virus and Zika virus. Nanopore sequencers can also detect cytosine modifications in the native DNA (Rand et al. 2017).

Limitation One major drawback of the nanopore technology is the higher error rate. Recently in the mid of 2016, Oxford Nanopore Technologies launched a newer version of MinION. This platform is based on the newer chemistry known as R9 (R stands for reader) and providing the lesser error rates (https://nanoporetech.com/about-us/news/update-new-r9-nanoporefaster-more-accurate-sequencing-and-new-ten-minutepreparation). However, it is also not up to the mark for frontline applications. Currently, the latest version of the technology i.e. R9.4 is getting used in the flow cells of the MinION platform. Since the improvements in next-generation technologies are taking place very quickly, the limitations of this generation will also be taken care very soon.

Following are the features why Nanopore sequencers are more suitable for sequencing:

The biggest quality of them is no requirement of fluorescent labeling of the nucleotides in sequencing and also provides longer read length. Identification of nucleotide is based upon the chemical or electronic structure. Compact size (four-inch-long device) of the machine offers the in-field/natural environment experiments possible and it has lower cost of sequencing per run including higher throughput. This MinION platform has the capability to break the set market of $1000 target fixed by the NHGRI, USA.

15.7.1.2 Oxford Nanopore (ProMethION)

In the beginning of year 2017, Oxford Nanopore Technologies delivered its new highest throughput sequencing benchtop platform ProMethION to the laboratories, which has been commercially available in May 2018. Here, up to 48 flow cells can run independently, each consisting of 3000 channels (nanopores). Oxford Nanopore Technologies promised that ProMethION would perform even better than the best platform by Illumina. They have also assumed the data it will generate per run will be approximately 11 TB when the manufacturing was underway. Currently in 2018, it generates approximately 2 TB in 48 h.

Now this platform has been placed in many sites in many countries proving that results continue to rise. In June 2018, more than 100 GB from the individual flow cell of ProMethION at University of Aalborg was first achieved. In the University of Birmingham, at the time of writing of this chapter (first week of August 2018), ProMethION benchtop platform was able to break the record by producing >130 GB data/flow cell (https://nanoporetech.com/about-us/news/promethion-wild-2-data-yield-continues-climb).

Advantage The platform offers on-demand sequencing. It means the researcher can start and stop the run as and when required or utilize more than one flow cell for single experiment for high throughput and faster speed. Individual flow cells of 3000 nanopores can be used and it provides almost six times more data compared to MinION and GridION. Currently, ProMethION beta system module is in use that allows 192 different libraries within the whole device.

15.7.1.3 Oxford Nanopore (GridION X5)

In early 2017, Oxford Nanopore Technologies released another platform, namely, GridION X5. This is a grid collection of five units of MinION with built-in computing software for the base calling. This allows five queries simultaneously or individually at a time depending upon the requirement of the researcher.

This platform also uses the same core technology and is useful in generating huge data (~35 GB in 2017) of long read length along with immediate access to the data in real time like MinION and ProMethION (https://nanoporetech.com/products/gridion). Library preparation is very easy and fast and almost the same that of MinION, ProMethION and GridION. It requires very less concentration of sample (femtogramfor >40 Kb DNA) and also has versatile and complete range of cDNA and gDNA library preparation kits. There are two methods for library preparation based on amplification of library. PCR amplification-based preparation requires when the starting amount of DNA is low. If we use 20 GridION platforms at a time, then it can sequence the whole human genome in only 15 min with relatively lower cost. A compact microfluidic device, VolTrax (programmable Hand Off preparation of Sample), is made available by Oxford Nanopore Technologies for the automation of library preparation (Leggett and Clark 2017) that is made up of USB-powered base. A consumable cartridge can be placed onto this base with an array of fluid comprising pixels on the surface. The software controls movement of drops of fluid on the surface of the cartridge. In addition, Oxford Nanopore Technologies is in the making of other protocols and customized user protocols. Recently, a new protocol has been developed for direct RNA sequencing, which is the most awaited protocol. The initial versions are not as precise as those of DNA but there is high hope to see the potential method shortly for direct RNA sequencing.

15.7.1.4 Oxford Nanopore (SmidgION X5)

In late 2017, Oxford Nanopore Technologies released an even smaller platform than MinION that can be attached with the mobile phone, SmidgION. This device is also based on the same technology like MinION and ProMethION. This is a very small 128-nanopore channel flow cell platform for sequencing of clinical, environmental and ecological samples. It is useful in the monitoring of the outbreaks (pathogens) remotely (https://nanoporetech.com/products/smidgion).

Other than Oxford Nanopore Technologies, Hitachi (Goto et al. 2016) and Genia (maintained by Roche) are among the companies who are working on the biological nanopore technologies. But, there is no company till now that has launched its platform in the market. This Oxford Nanopore Technologies is competing with both long-standing PacBio longer read and Illumina's shorter read technologies. On the other hand, Oxford Nanopore Technologies is providing platform at almost no cost or very low cost. Laboratories have to just pay for the consumables.

Table 15.1 provides the overview and characteristics of commonly used next-generation sequencing platforms.

Table 15.1 Overview and characteristics of new and commonly used next generation sequencing platforms

Company	Platform	Maximum Read Length	Throughput/ run (Gb)	Run time	Error rate (%)	Error type
Illumina	MiSeq (Benchtop)	2 × 300	0.3–15	5–55 h	0.1	Mismatch
	HiSeq (Capital)	2 × 150	125–700	<1–4 days	0.1	Mismatch
	MiniSeq (Benchtop)	2 × 150	0.5–7.5	4–24 days	1	Mismatch
	NextSeq (Benchtop)	2 × 150	20–120	12–32 hr	1	Mismatch
	iSeq100 (Benchtop)	2 × 150	0.2–1.2	9–17.5 hr	<1	NA
Thermofisher (Ion Torrent)	PGM (Benchtop)	200–400	0.03–2	2–8 hr	1	Insertion/ deletion
	Ion S5 (Benchtop)	200–400	0.5–15	2–4 hr	1	Insertion/ deletion
	Ion S5 XL (Benchtop)	200–400	0.5–15	<24 hr	1	Insertion/ deletion
Pacific BioSciences	RS II (Capital)	>20,000	0.5–1	30 min– 24 hr	15	Insertion/ deletion
	Sequel (Capital)	>20,000	0.6–1.25	30 min–6 hr	NA	NA
Oxford Nanopore	MinION (Portable)	230,000– 300,000	21–42	1 min–48 hr	12	Insertion/ deletion
	ProMethION (Benchtop)	>200,000	3–6 Tb (>130 Gb/ flowcell)	1 min–64 hr	NA	NA

15.8 Applications of NGS to Address Public Health

To advance public health by unlocking the power of genome, we have just started utilizing high throughput sequencing technology, that is, next-generation sequencing. Schematic representation in Fig. 15.4 is illustrating that all the applications that address the public health utilize or take advantage of the different methods of high throughput sequencing technology like whole genome sequencing, targeted sequencing, ChIP sequencing, RNA sequencing, whole exome sequencing, transcriptome sequencing and amplicon sequencing. NGS is exhibiting a broad impact in public health welfare and clinical laboratories. This new high throughput technology holds remarkable promises with a wide range of applications in exploring biological questions, which includes management and surveillance of outbreaks, study of human microbiome to investigate the infectious organism/polymicrobial infections and taxonomic identification of microbiomes, diagnosis of infectious disease and investigation of zoonotic microbes transmission to humans from animals and so on. In this chapter, some key areas of NGS applications related to public health are summarized:

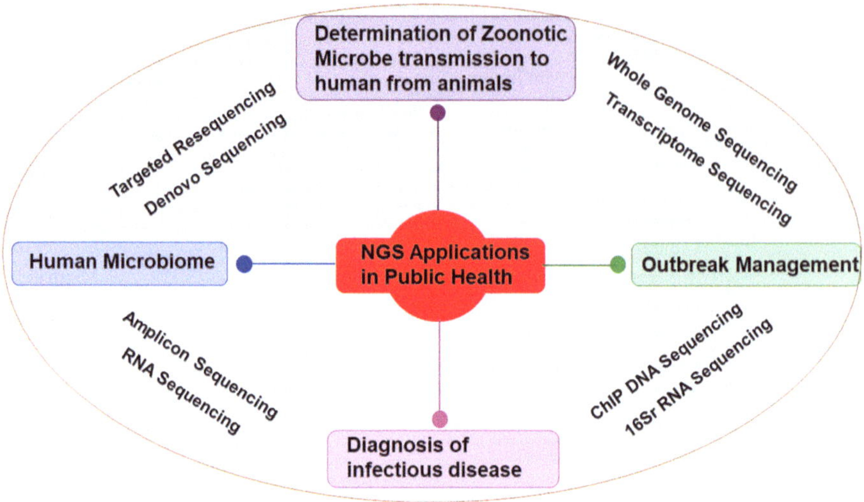

Fig. 15.4 Applications of Next-Generation Sequencing (NGS) in Public Health

15.8.1 Outbreak Management

"An outbreak anywhere is a risk everywhere"—Dr. Frieden. Outbreaks can be stressful for individuals and public. However, traditional epidemiology generally catches the source of an outbreak, for example, by case control studies (King et al. 2012). For the past so many decades, laboratory investigation played a significant role in investigation and management of outbreak (Sabat et al. 2013). Now, Whole Genome Sequencing (WGS)-based typing is encouraging the employment of next-generation sequencing for investigations of public health. WGS is very useful in the detection of outbreak and its management locally and globally and also in the monitoring of evolution of multidrug resistance pathogens (Albiger et al. 2016). First application of WGS in public health was to dissect the epidemiological connections in hospital-acquired infections, for example, bacterial (*Acinetobacterbaumannii*) outbreak in 2010 in a hospital in Birmingham, UK (Lewis et al. 2010). In a very short time, several studies have shown that WGS has taken charge for interpretation and stopping of the transmission pathways of pathogens in hospital outbreaks. Few examples of characterization of newly emerging pathogens, which helped to stop the transmission/spread between patients at the same center and inter health care centers transmission, are Methicillin Resistance *Staphylococcus aureus* (MRSA), carbapenem resistant *Klebsiellapneumonia*e (Harris et al. 2010; Snitkin et al. 2012) and early detection of *K. pneumoniae* high-risk clone (HiRiC) (Zhou et al. 2016).

A large outbreak of highly virulent Shiga toxin producing *Escherichia coli* (STEC) was also characterized by WGS. Upon characterization, WGS has the ability to reveal about isolates like species, strain, virulence, antibiotic resistance and much more information from the genome other than phylogenetic information to manage the case and outbreak. Investigation of foodborne disease and its outbreak management is the most important area for public health welfare, and WGS promises to identify those bacteria. According to a WHO survey, approximately 1900 million people get infected with foodborne pathogens every year and out of them, a big proportion (7,15,000) are not able to survive (2007–2015, Food Borne Disease Burden Epidemiology Reference Group). Great improvement has been found in the outbreak detected and outbreak solved (management) of listeriosis by traditional gold standard method in countries like France and US after implementation of WGS. Information about drug resistance and virulence characters can be taken into account for the clinical practice but more correlation among genotype and phenotype is required. Further, NGS data mining may disclose new targets, which may help in the investigation of outbreak by highly clonal pathogens. Remarkably, a major downside of NGS is that there is no standard guideline from controlling agencies for the sharing of data.

15.8.2 Human Microbiome

Human microbiome is one of the important players to affect the immunity of the host and metabolic functions that are not determined by human genome. Microbes, mainly bacteria, are both closest relative and enemy of our body. Yeast, single-celled eukaryotes, helminths and some viruses are also associated to our body. There are organisms in our microbiota, which cannot be cultured and identified. Some very well-known spots are found for the colonization in the human body like stomach, vagina, colon, skin, esophagus, oral cavity, nose and hair. In 2008, National Institute of Health (NIH), USA has started two high-profile human microbiome projects of international level, namely, Human Microbiome Project (HMP) and Metagenomics of the Human Intestinal Tract (MetaHit) on the foundation of NGS as a tool. These projects were initiated to isolate and characterize microbes present in the healthy individuals and diseased and to develop the new methods for computational analysis of sequenced genomes. In the MetaHit program, from the sequencing of gut microbiome of 124 healthy adults having obesity, it was found that there are >1000 bacterial species present in the human gut microbiota. This revolutionary project was based on the *de novo* assembly of short reads from human microbiome datasets (Qin et al. 2010; Arumugam et al. 2011).

The new sequencing technologies like NGS are facilitating researcher community to analyze the world of different microbial populations in varied environments and in human body from wider and deeper viewpoints. The ongoing advances in sequencing platforms are not only supporting the characterization of whole genome of microbes but also are a valuable tool for the taxonomic identification of microbiomes, which are an inhabitant of particular niche. This enables us to detect polymicrobial infections and colonization at better perspective. Also, identification of new species using sequencing of metagenomes by implementing whole genome sequencing method or 16S rRNA gene amplicon sequencing from a mixed population. A new approach, that is, metatranscriptomic method of sequencing, has made the contribution in functional analysis of the interactions among different microbes of a single microbiome. In case of whole genome sequencing, we get longer read length, which enables us in better assembling of genome from diverse organisms. By using reference sequences or denovo clustering, taxonomy profiling can also be performed. Nowadays, it is very much noticeable that imbalance of gut microbiome (also known as dysbiosis) is intensely connected with the immune disorder development and/or improper metabolic functioning (Kim et al. 2015). Here, we can say that microbiomics has its use in a very controlled manner at the point of care. For instance, if a person is suffering from meningitis and doctors/clinicians are not able to identify (pathogen eluding the clinicians like criminal eluding the police), then microbiomics plays pivotal role for the identification of the responsible pathogen, be it bacteria or amoeba. This identification helps the clinicians to precisely medicate (precision medicine) the patient with choice of antibiotics. In other words, genomics in these cases is life saving.

15.8.3 Diagnosis of Infectious Disease

Recently, the application of next-generation sequencing has also approached the field of infectious disease diagnosis like any other medical field to empower the public health globally. Earlier, the origin of the disease causing microbe and its diagnosis was only based upon the evidence of the existence of a given pathogen in a particular given sample. This standard time-taking method of culture-based identification/detection is still being used. However, this classical approach has many limitations, which include problem in cultivation of certain species of microbes like viruses and some other pathogens, which are hard to grow in culture. Further, culturing of microbes is time consuming and too expensive. Nucleic acid based diagnostic methods, for example, Polymerase Chain Reaction (PCR), have progressively replaced the culturing method. PCR-based method is cost effective, sensitive and specific but the main limitation is that it demands prior hypothetical knowledge. It can only identify the conserved targets of the pathogen and cannot distinguish between genotypes; this again restricts the detection of new emerging pathogens (Lecuit and Eloit 2014). Other methods for diagnosis were also developed, which include multiplex PCR assays, enzyme-linked immunoassays and pulse field gel electrophoresis (PFGE) to widen the pathogen detection sensitivity and specificity. But, these conventional methods also do not prove to be much useful because of insensitivity for clinical diagnosis.

On the other hand, NGS, which has revolutionized the diagnostic field, is different from other diagnostic assays because it does not require prior hypothetical knowledge. And now it has been commonly used to diagnose and discover the novel pathogens, for example. bacteria, fungi, virus or parasite (Frey et al. 2014). The field of "diagnosis genomics" or "pathogenomics" has unveiled emerging and re-emerging pathogens that translate the genomic technologies into methods for diagnosis. This new method has facilitated high-resolution mapping of the genetic determinants in microorganisms that uphold the pathogenicity. Whole genome sequencing and targeted amplicon sequencing of rRNA genes have come up as favored technologies for microbial identification from primary human specimens and to analyze the dynamic genomes with a high moldability that is a must for pathogens to cling to life in arduous environment (Edwards and Rohwer 2005; Weinstock 2012). Analysis of genomes using whole genome sequencing delivers high-resolution information to differentiate microbial strains that possess difference of as low as one Single Nucleotide Polymorphisms (SNP), thereby it can replace other multiple tests. For the detection of pathogenic microorganism from patients having suspected infections from uncultivable microbe and or not possible to diagnose by standard diagnostic method, whole genome sequencing method is an ideal option. In such cases, specimen directly from the patient can be sequenced. Here is an example of the strong ability of the whole genome sequencing method in diagnosis of a pediatric instance of alymphocytosis or severe combined immunodeficiency (SCID) and recurrent meningoencephalitis, where this whole genome sequencing together with a dedicated bioinformatics software (Naccache et al. 2014) diagnosed the *Leptospirasantarosai* in cerebrospinal fluid (CSF) (total DNA from CSF) using Illumina platform within 48 hours' time from the sample collection, where 475 sequence reads were aligned with the pathogen out of 3,063,784 total reads (Wilson et al. 2014). Similarly, there are a number of cases where whole genome sequencing diagnosed the pathogens from the uncultivable sample in patients having infectious syndrome of uncertain cause.

For the analysis of heterogeneity or identification of microbial species in a given medical sample, targeted amplicon sequencing of 16S rRNA genes by NGS is also a method of choice. Several studies have been performed in relation with the diagnosis of patient sample as well as healthy sample to identify the heterogeneity among the patient. Because from the studies it is now known that during progression of disease, bacterial diversity decreases, which may be due to the increased antibiotic exposure to the pathogen in the patients (Morgan et al. 2012).

Further, RNA sequencing technology, which is in its infancy, also holds the promise for its applicability in diagnosis of infectious disease causing organism. Based on the performed studies as of now, this is clear that next-generation sequencing is progressively working on its way for routine diagnostic purpose in the foundation for public health and clinical laboratories. The approach is applicable for diagnosis of all kinds of microorganisms (virus, fungi, bacteria and or parasite/eukaryotic organism), which participate in the infection process. But there are a number of challenges that still obstruct the widespread use of next-generation sequencing in the diagnosis of infectious disease. First and foremost, requirement is the development and improvement in the software that is required for the analysis of the sequencing data. However, now there are several open source pipelines for the diagnosis of pathogen by NGS available. But the limitation is again that they require a substantial knowledge of bioinformatics/bioinformatics expertise that is generally not available in clinical health laboratories.

15.8.4 Determination/Investigation of Zoonotic Microbes Transmission to Humans from Animals

Zoonotic diseases are a threat to public health, which is only recently acknowledged though the transmission of zoonotic microbial (infections that animals spread to humans) agents is on rise from last so many decades. From the studies it has been shown that since the Second World War, annually one pathogenic disease comes to light globally and developing countries like India significantly carry the ball.

A research group has reported that three out of four emerging pathogens to human are of zoonotic origin (Taylor et al. 2001). It has recently become apparent that zoonotic diseases involve worldwide devastating diseases like Ebola virus infection, bird flu (highly pathogenic avian influenza), severe acute respiratory syndrome (SARS) and so many other (Heymann and Dar 2014). These newly apparent threats for public health are linked with considerable economic cost, which includes direct or indirect impact on our healthcare system.

Next-generation sequencing (NGS) is revealing more understanding on the transmission of zoonotic microbes and the method of choice nowadays (Chatterjee et al. 2017). Earlier studies in this area were done by serotyping (Tenover et al. 1997). Recently, studies are based on newer methods like pulse field gel electrophoresis (PFGE) or multilocus variable number tandem repeat (MLVNTR) analysis to identify the specific species of microbe in animals and humans (Sabat et al. 2013). Nonetheless, still so much is left to understood, mainly frequency of transmission (number of contacts needed for the transmission from animals), risk factor linked with the acquiring of zoonotic agent and how the transfer of pathogen from animal to human is affected by antibiotic use in animal. For all these topics, NGS shows a newer perspective. This is also unraveling the difference in previously indistinguishable strains of human and animal (Harrison et al. 2013). Besides, NGS also permits a wide range analysis of how the use of antibiotics changes the specific microbiome and effects on interspecies transmission. Current projects are based on whole genome sequencing methods that will help to understand the dynamics and mechanism of transmission of pathogen between animals, human and environment. A recent study by a veterinary research group has revealed the presence of *mcr-1*gene (using whole genome sequencing method) in three *E. coli* strains isolated from poultry meat. Although this gene is not carried by any human strain, two of the three strains were related to ST117 (avian pathogenic *E. coli*), which is a common strain between human and poultry animal. This represents a potential concern in public health (Kluytmans-van den Bergh et al. 2016).

According to a latest published study in 2018, the next-generation sequencing technique has been used to diagnosed the neurobrucellosis from cerebrospinal fluid (Fan et al. 2018). Neurobrucellosis is the condition of brucellosis where central nervous system gets involved and it is very common zoonotic disease globally. Its diagnosis is challenging with clinical indications it shows because of their non-specificity and low sensitivity of routine culture test. This group has shown the command of NGS for diagnosis from CSF together with bioinformatics analysis.

Using whole genome sequencing on Illumina HiSeq platform, molecular epidemiology of the related isolates of methicillin-resistant *Staphylococcus aureus* (MRSA) has been investigated. Also, it was revealed that human and animal isolates of the same farm were only varied by few number of Single Nucleotide Polymorphisms (SNP). This analysis supports the possibility of zoonotic transmission of MRSA isolates. The study further shows that mecC-MRSA ST130 isolates can transmit between animal and human (Harrison et al. 2013). And this *mecC* gene is liable for the resistance to the penicillin like antibiotic methicillin. This study emphasizes the role of farm animal (livestock) as a likely reservoir of antibiotic resistant pathogen.

A bacterium, *Corynebacteriumulcerans,* that causes diphtheria-like infection in humans and present in pets was thought to work as reservoir for zoonotic transmission. Furthermore, the reports reveal that this new bacteria is now playing the lead role to cause diphtheria in past years in several economically developed countries. Here also, next-generation sequencing approach enables us to identify the novel virulence genes rapidly acquired by *C. ulcerans* and a putative pathogenicity island, which possesses the diphtheria toxin gene. This rapid acquisition of genes changes the virulence of the strain even in single round of zoonotic transmission. During genomewide sequencing/SNP profiling of pair of patient and domestic animal companion, it was revealed that there is very less or almost no difference between their profiles. This supports the idea that *C. ulcerans* encounter zoonotic transmission between human and animal. In addition to this, these results demonstrate that NGS helps in improving the phylogenetic and epidemiological studies by giving insights between closely relative isolates.

15.9 Future Perspectives

Next-generation sequencing is making possible the term "One test fits for all" as clinical laboratories, public health laboratories and researchers are progressively embracing it. Further, this technology has revolutionized each and every field of medical science and life sciences imparting numerous benefits in terms of massive parallel sequencing. Earlier, high cost of sequencing was also the barrier and now the reduced cost by several folds is enormously attracting the researchers and making it feasible to plan their research based on sequencing.

In time to come, sequencing of individual genomes of importance living in different conditions, having different nutritional intake and/or under different treatment conditions will pave the way for disease control and its prevention thereby facilitates the social security. It is thought that genome sequencing of livestock will enable scientific community for more precise identification of genetic markers of important traits. However, the information obtained from NGS data will also bring advantage to the medical field with respect to better diagnostics and therapeutics. Also, the data generated from NGS will help agricultural society to improve the breed of dairy cattle and beef cattle. Earlier, the sequencing of animal was used to be done for its use as model system to study the human/public health issues.

Further, sequencing of human microbiomes and parasites of agricultural animals can benefit in the development of therapeutics and new vaccines for social welfare. Cell-free fetus DNA sequencing (which simply requires only blood sample of six-week pregnant mother) is showing the new opportunities for prenatal diagnosis, which ends the risk for the fetus.

In the end, these rapidly growing new technologies are accelerating the process of drug discovery and personalized medicine for public health welfare. Now, it is not surprising to say that only our imagination can put the limits of what is possible to be done by next-generation sequencing technologies!

References

Adessi C, Matton G, Ayala G et al (2000) Solid phase DNA amplification: characterisation of primer attachment and amplification mechanisms. Nucleic Acids Res 28:E87

Albiger B, Leitmeyer K, Struelens M et al (2016) European Centre for Disease Prevention and Control. ECDC roadmap for integration of molecular and genomic typing into European-level surveillance and epidemic preparedness, Version 2.1, 2016–19. ECDC, Stockholm

Anderson S (1981) Shotgun DNA sequencing using cloned DNase I-generated fragments. Nucleic Acids Res 9:3015–3027

Ansorge W (2016) Next generation DNA sequencing (II): techniques, applications. Next Gene Seq App. https://doi.org/10.4172/2469-9853.1000S1-005

Arumugam M, Raes J, Pelletier E et al (2011) Enterotypes of the human gut microbiome. Nature 473:174–180

Bahassi el M, Stambrook PJ (2014) Next-generation sequencing technologies: breaking the sound barrier of human genetics. Mutagenesis 29:303–310

Ballester LY, Luthra R, Kanagal-Shamanna R, Singh RR (2016) Advances in clinical next-generation sequencing: target enrichment and sequencing technologies. Expert Rev Mol Diagn 16:357–372

Bentley DR, Balasubramanian S, Swerdlow HP et al (2008) Accurate whole human genome sequencing using reversible terminator chemistry. Nature 456:53–59

Bowers J, Mitchell J, Beer E et al (2009) Virtual terminator nucleotides for next-generation DNA sequencing. Nat Methods 6:593–595

Braslavsky I, Hebert B, Kartalov E, Quake SR (2003) Sequence information can be obtained from single DNA molecules. Proc Natl Acad Sci U S A 100:3960–3964

Buermans HP, den Dunnen JT (2014) Next generation sequencing technology: advances and applications. Biochim Biophys Acta 1842:1932–1941

Castro-Wallace SL, Chiu CY, John KK et al (2017) Nanopore DNA sequencing and genome assembly on the International Space Station. Sci Rep 7:18022

Chatterjee P, Sarma N, Hansda S (2017) Tropical diseases on insurgence: clinician's perspective. Indian J Dermatol 62:468–477

Edwards RA, Rohwer F (2005) Viral metagenomics. Nat Rev Microbiol 3:504–510

Eid J, Fehr A, Gray J et al (2009) Real-time DNA sequencing from single polymerase molecules. Science 323:133–138

Fan S, Qiao X, Liu L et al (2018) Next-generation sequencing of cerebrospinal fluid for the diagnosis of neurocysticercosis. Front Neurol 9:471

Fang G, Munera D, Friedman DI et al (2012) Genome-wide mapping of methylated adenine residues in pathogenic Escherichia coli using single-molecule real-time sequencing. Nat Biotechnol 30:1232–1239

Fleischmann RD, Adams MD, White O et al (1995) Whole-genome random sequencing and assembly of Haemophilus influenzae Rd. Science 269:496–512

Flusberg BA, Webster DR, Lee JH et al (2010) Direct detection of DNA methylation during single-molecule, real-time sequencing. Nat Methods 7:461–465

Frey KG, Herrera-Galeano JE, Redden CL et al (2014) Comparison of three next-generation sequencing platforms for metagenomic sequencing and identification of pathogens in blood. BMC Genomics 15:96

Goffeau A, Barrell BG, Bussey H et al (1996) Life with 6000 genes. Science 274(546):563–567

Goto Y, Yanagi I, Matsui K, Yokoi T, Takeda K (2016) Integrated solid-state nanopore platform for nanopore fabrication via dielectric breakdown, DNA-speed deceleration and noise reduction. Sci Rep 6:31324

Gullapalli RR, Desai KV, Santana-Santos L, Kant JA, Becich MJ (2012) Next generation sequencing in clinical medicine: challenges and lessons for pathology and biomedical informatics. J Pathol Inform 3:40

Guo J, Xu N, Li Z et al (2008) Four-color DNA sequencing with 3'-O-modified nucleotide reversible terminators and chemically cleavable fluorescent dideoxynucleotides. Proc Natl Acad Sci U S A 105:9145–9150

Guo J, Yu L, Turro NJ, Ju J (2010) An integrated system for DNA sequencing by synthesis using novel nucleotide analogues. Acc Chem Res 43:551–563

Harris TD, Buzby PR, Babcock H et al (2008) Single-molecule DNA sequencing of a viral genome. Science 320:106–109

Harris SR, Feil EJ, Holden MT et al (2010) Evolution of MRSA during hospital transmission and intercontinental spread. Science 327:469–474

Harrison EM, Paterson GK, Holden MT et al (2013) Whole genome sequencing identifies zoonotic transmission of MRSA isolates with the novel mecA homologue mecC. EMBO Mol Med 5:509–515

Heymann DL, Dar OA (2014) Prevention is better than cure for emerging infectious diseases. BMJ 348:g1499

Ho A, Murphy M, Wilson S, Atlas SR, Edwards JS (2011) Sequencing by ligation variation with endonuclease V digestion and deoxyinosine-containing query oligonucleotides. BMC Genomics 12:598

Jain M, Koren S, Miga KH et al (2018) Nanopore sequencing and assembly of a human genome with ultra-long reads. Nat Biotechnol 36:338–345

Karamohamed S, Nilsson J, Nourizad K, Ronaghi M, Pettersson B, Nyren P (1999) Production, purification, and luminometric analysis of recombinant Saccharomyces cerevisiae MET3 adenosine triphosphate sulfurylase expressed in Escherichia coli. Protein Expr Purif 15:381–388

Kim Y, Koh I, Rho M (2015) Deciphering the human microbiome using next-generation sequencing data and bioinformatics approaches. Methods 79-80:52–59

King LA, Nogareda F, Weill FX et al (2012) Outbreak of Shiga toxin-producing Escherichia coli O104:H4 associated with organic fenugreek sprouts, France, June 2011. Clin Infect Dis 54:1588–1594

Kluytmans-van den Bergh MF, Huizinga P, Bonten MJ et al (2016) Presence of mcr-1-positive Enterobacteriaceae in retail chicken meat but not in humans in the Netherlands since 2009. Euro Surveill 21:30149

Korlach J, Bjornson KP, Chaudhuri BP et al (2010) Real-time DNA sequencing from single polymerase molecules. Methods Enzymol 472:431–455

Krause J, Fu Q, Good JM et al (2010) The complete mitochondrial DNA genome of an unknown hominin from southern Siberia. Nature 464:894–897

Lecuit M, Eloit M (2014) The diagnosis of infectious diseases by whole genome next generation sequencing: a new era is opening. Front Cell Infect Microbiol 4:25

Leggett RM, Clark MD (2017) A world of opportunities with nanopore sequencing. J Exp Bot 68:5419–5429

Levene MJ, Korlach J, Turner SW, Foquet M, Craighead HG, Webb WW (2003) Zero-mode waveguides for single-molecule analysis at high concentrations. Science 299:682–686

Lewis T, Loman NJ, Bingle L et al (2010) High-throughput whole-genome sequencing to dissect the epidemiology of Acinetobacter baumannii isolates from a hospital outbreak. J Hosp Infect 75:37–41
Li Z, Bai X, Ruparel H, Kim S, Turro NJ, Ju J (2003) A photocleavable fluorescent nucleotide for DNA sequencing and analysis. Proc Natl Acad Sci U S A 100:414–419
Mardis ER (2008) Next-generation DNA sequencing methods. Annu Rev Genomics Hum Genet 9:387–402
Mardis ER (2011) A decade's perspective on DNA sequencing technology. Nature 470:198–203
Mardis ER (2013) Next-generation sequencing platforms. Annu Rev Anal Chem (Palo Alto, Calif) 6:287–303
Margulies M, Egholm M, Altman WE et al (2005) Genome sequencing in microfabricated high-density picolitre reactors. Nature 437:376–380
Maxam AM, Gilbert W (1977) A new method for sequencing DNA. Proc Natl Acad Sci U S A 74:560–564
Metzker ML (2010) Sequencing technologies - the next generation. Nat Rev Genet 11:31–46
Mignardi M, Nilsson M (2014) Fourth-generation sequencing in the cell and the clinic. Genome Med 6:31
Morey M, Fernandez-Marmiesse A, Castineiras D, Fraga JM, Couce ML, Cocho JA (2013) A glimpse into past, present, and future DNA sequencing. Mol Genet Metab 110:3–24
Morgan XC, Tickle TL, Sokol H et al (2012) Dysfunction of the intestinal microbiome in inflammatory bowel disease and treatment. Genome Biol 13:R79
Naccache SN, Federman S, Veeraraghavan N et al (2014) A cloud-compatible bioinformatics pipeline for ultrarapid pathogen identification from next-generation sequencing of clinical samples. Genome Res 24:1180–1192
Nakamura K, Oshima T, Morimoto T et al (2011) Sequence-specific error profile of Illumina sequencers. Nucleic Acids Res 39:e90
Obenrader S (2003) The sanger method, Dept. of biology, Davidson college. Retrieved from http://www.bio.davidson.edu/Courses/Molbio/MolStudents/spring2003/Obenrader/sanger_method_page.htm
Qin J, Li R, Raes J et al (2010) A human gut microbial gene catalogue established by metagenomic sequencing. Nature 464:59–65
Quail MA, Smith M, Coupland P et al (2012) A tale of three next generation sequencing platforms: comparison of Ion Torrent, Pacific Biosciences and Illumina MiSeq sequencers. BMC Genomics 13:341
Quick J, Grubaugh ND, Pullan ST et al (2017) Multiplex PCR method for MinION and Illumina sequencing of Zika and other virus genomes directly from clinical samples. Nat Protoc 12:1261–1276
Rand AC, Jain M, Eizenga JM et al (2017) Mapping DNA methylation with high-throughput nanopore sequencing. Nat Methods 14:411–413
Reuter JA, Spacek DV, Snyder MP (2015) High-throughput sequencing technologies. Mol Cell 58:586–597
Roberts RJ, Carneiro MO, Schatz MC (2013) The advantages of SMRT sequencing. Genome Biol 14:405
Ronaghi M (2001) Pyrosequencing sheds light on DNA sequencing. Genome Res 11:3–11
Ronaghi M, Karamohamed S, Pettersson B, Uhlen M, Nyren P (1996) Real-time DNA sequencing using detection of pyrophosphate release. Anal Biochem 242:84–89
Ronaghi M, Uhlen M, Nyren P (1998) A sequencing method based on real-time pyrophosphate. Science 281(363):365
Sabat AJ, Budimir A, Nashev D et al (2013) Overview of molecular typing methods for outbreak detection and epidemiological surveillance. Euro Surveill 18:20380
Sanger F, Air GM, Barrell BG et al (1977a) Nucleotide sequence of bacteriophage phi X174 DNA. Nature 265:687–695
Sanger F, Nicklen S, Coulson AR (1977b) DNA sequencing with chain-terminating inhibitors. Proc Natl Acad Sci U S A 74:5463–5467

Smith LM, Sanders JZ, Kaiser RJ et al (1986) Fluorescence detection in automated DNA sequence analysis. Nature 321:674–679

Snitkin ES, Zelazny AM, Thomas PJ et al (2012) Tracking a hospital outbreak of carbapenem-resistant Klebsiella pneumoniae with whole-genome sequencing. Sci Transl Med 4:148ra116

Stahl PL, Salmen F, Vickovic S et al (2016) Visualization and analysis of gene expression in tissue sections by spatial transcriptomics. Science 353:78–82

Taylor LH, Latham SM, Woolhouse ME (2001) Risk factors for human disease emergence. Philos Trans R Soc Lond Ser B Biol Sci 356:983–989

Tenover FC, Arbeit RD, Goering RV (1997) How to select and interpret molecular strain typing methods for epidemiological studies of bacterial infections: a review for healthcare epidemiologists. Molecular Typing Working Group of the Society for Healthcare Epidemiology of America. Infect Control Hosp Epidemiol 18:426–439

Thompson JF, Steinmann KE (2010) Single molecule sequencing with a HeliScope genetic analysis system. Curr Protoc Mol Biol 7:Unit7 10

Travers KJ, Chin CS, Turner SW et al (2010) A flexible and efficient template format for circular consensus sequencing and SNP detection. Nucleic Acids Res 38(15):e159. https://doi.org/10.1093/nar/gkq543

Uemura S, Aitken CE, Korlach J, Flusberg BA, Turner SW, Puglisi JD (2010) Real-time tRNA transit on single translating ribosomes at codon resolution. Nature 464:1012–1017

Venter JC, Adams MD, Myers EW et al (2001) The sequence of the human genome. Science 291:1304–1351

Voelkerding KV, Dames SA, Durtschi JD (2009) Next-generation sequencing: from basic research to diagnostics. Clin Chem 55:641–658

Weinstock GM (2012) Genomic approaches to studying the human microbiota. Nature 489:250–256

Wilson MR, Naccache SN, Samayoa E et al (2014) Actionable diagnosis of neuroleptospirosis by next-generation sequencing. N Engl J Med 370:2408–2417

Zhou K, Lokate M, Deurenberg RH et al (2016) Use of whole-genome sequencing to trace, control and characterize the regional expansion of extended-spectrum beta-lactamase producing ST15 Klebsiella pneumoniae. Sci Rep 6:20840

Correction to: Molecular Diagnosis of Acute and Chronic Brucellosis in Humans

Maryam Dadar, Youcef Shahali, and Gamal Wareth

Correction to:
Chapter 10 in: P. K. Arora (ed.), *Microbial Technology for the Welfare of Society*, Microorganisms for Sustainability 17, https://doi.org/10.1007/978-981-13-8844-6_10

The chapter was originally published with an additional affiliation for the authors "Maryam Dadar and Youcef Shahali" which has been removed now. The correct affiliation of these authors are "Razi Vaccine and Serum Research Institute (RVSRI), Agricultural Research, Education and Extension Organization (AREEO), Karaj, Iran".

The updated version of this chapter can be found at
https://doi.org/10.1007/978-981-13-8844-6_10

P. K. Arora (ed.), *Microbial Technology for the Welfare of Society*,
Microorganisms for Sustainability 17,
https://doi.org/10.1007/978-981-13-8844-6_16

MIX
Papier aus verantwortungsvollen Quellen
Paper from responsible sources
FSC® C105338

If you have any concerns about our products,
you can contact us on
ProductSafety@springernature.com

In case Publisher is established outside the EU,
the EU authorized representative is:
Springer Nature Customer Service Center GmbH
Europaplatz 3, 69115 Heidelberg, Germany

Printed by Libri Plureos GmbH
in Hamburg, Germany